T0093940

Springer Textbooks in Earth Sciences, Geography and Environment

The Springer Textbooks series publishes a broad portfolio of textbooks on Earth Sciences, Geography and Environmental Science. Springer textbooks provide comprehensive introductions as well as in-depth knowledge for advanced studies. A clear, reader-friendly layout and features such as end-of-chapter summaries, work examples, exercises, and glossaries help the reader to access the subject. Springer textbooks are essential for students, researchers and applied scientists.

More information about this series at http://www.springer.com/series/15201

Emin Özsoy

Geophysical Fluid Dynamics I

An Introduction to Atmosphere—Ocean Dynamics: Homogeneous Fluids

Emin Özsoy
Eurasia Institute of Earth Sciences
Istanbul Technical University
Istanbul, Turkey

ISSN 2510-1307 ISSN 2510-1315 (electronic)
Springer Textbooks in Earth Sciences, Geography and Environment
ISBN 978-3-030-16972-5 ISBN 978-3-030-16973-2 (eBook)
https://doi.org/10.1007/978-3-030-16973-2

© Springer Nature Switzerland AG 2020
This work is subject to copyright. All rights are reserved by the Publisher, whether the whole or part of the material is concerned, specifically the rights of translation, reprinting, reuse of illustrations, recitation, broadcasting, reproduction on microfilms or in any other physical way, and transmission or information storage and retrieval, electronic adaptation, computer software, or by similar or dissimilar methodology now known or hereafter developed.
The use of general descriptive names, registered names, trademarks, service marks, etc. in this publication does not imply, even in the absence of a specific statement, that such names are exempt from the relevant protective laws and regulations and therefore free for general use.
The publisher, the authors and the editors are safe to assume that the advice and information in this book are believed to be true and accurate at the date of publication. Neither the publisher nor the authors or the editors give a warranty, expressed or implied, with respect to the material contained herein or for any errors or omissions that may have been made. The publisher remains neutral with regard to jurisdictional claims in published maps and institutional affiliations.

This Springer imprint is published by the registered company Springer Nature Switzerland AG
The registered company address is: Gewerbestrasse 11, 6330 Cham, Switzerland

Bernouilli equation displayed at the exterior wall of İstanbul Museum of Modern Art, overlooking the Bosphorus. "Hydrodynamica (Applied)", size 4.2 m x 62 m, presented by Liam Gillick and Esther Shipper Gallery, during the 14th İstanbul Biennial "TUZLU SU / SALT WATER", catalogue, pages 408–409, 5.9.2015–1.11.2015. Photo by Emin Özsoy, taken during a sailboat trip to the site organized by Tülay Çokacar

A Miniature Ocean: spiral eddies, fronts, segments of the Bosphorus Jet, hydraulic controls at straits, inter-basin transports, complex hydrodynamic and turbulent structures coinciding with a dinoflagellate bloom (red tide) in the Marmara Sea, presented by Emin Özsoy during the 14th Istanbul Biennial, "TU- ZLU SU / SALTWATER", catalogue, pages 308–309, 5.9.2015–1.11.2015, by courtesy of NASA Earth Observatory. This is a color enhanced version of the image on May 17, 2015 (https://earthobservatory.nasa.gov/images/85947/blooms-in-the-sea-of-marmara). Credits: NASA Earth Observatory

... to first captains, seafarers, geographers, and oceanographers of the Seas of the Old World, *where all has begun ...*

Preface

The present book is the first of two volumes aiming basic review of Geophysical Fluid Dynamics (GFD) for the student and researcher interested in basic, step-wise development from first principles and simple models that mimic the complex behavior of our ocean and atmosphere.

GFD aims a theoretical description of fluid dynamics at earthly scales. New age of super-computing, modeling, satellite and communication technologies present a "big science" complement to GFD, originally built upon empirical observations by travelers, navigators, and geographers across the globe, interpreted in recent times through mathematical analyses. An articulate review of the history, rationale for study and status of GFD by Vallis[1] contends that some of the main goals and past triumphs of GFD lie in explaining "fluid-dynamical emergent phenomena" (system behavior very different from that of individual components, for instance, in the outstanding cases of the Gulf Stream, hurricanes, etc.). The needed explanation has often called for theory leading to advances in GFD. Perhaps it is part of our fortune that chances and means to observe emergent phenomena at geophysical scales have only become available in recent history, with increased human activity, technical capacity, and skill.

Historically, any significant advance in GFD had to wait for advances in fluid mechanics in the later part of eighteenth century. Its major development was accomplished in the second half of twentieth century. While soundly moving into the twenty-first century, GFD has yet to tackle humanity's greatest problems involving climate dynamics.

The recognition of earth's rotation as the essential element seems to be due to Hadley in 1735, and Laplace in 1776. The scientific roots of oceanography, on the other hand, go back to 1679 when Marsili[2] used the "scientific method" of the Galilean revolution, to perform actual measurements at sea to explain "emergent

---

[1]Vallis G. K., 2016. *Geophysical fluid dynamics: whence, whither and why?*, Proc. R. Soc. A 472: 20160140.

[2]Marsili, L. F., 1681: *Osservazioni intorno al Bosforo Tracio overo Canale di Costantinopoli*, Rappresentate in Lettera alla Sacra Real Maesta Cristina Regina di Svezia da Luigi Ferdinando Marsili. Nicolo Angelo Tinassi, 108 pp.

phenomena" of the Bosphorus exchange currents,[3] to answer scientific questions that were left uncertain at Gibraltar Strait and elsewhere till few centuries later.

The puzzle on the theory of ocean currents has survived in the "Seas of the Old World" since the Galilean age and perhaps not fully solved until the last century. The book cover and the dedicated images in the first pages contrast the simplicity of Bernouilli principle with the defying complexity of "a miniature ocean", spanning multiple time scales.

Such is the basic nature of fluid dynamics, full of surprises and always capable of creating and resolving new problems at all scales, despite the simple physics involved in the Navier–Stokes equations. Yet, the expansive nature of fluid dynamics is perhaps best illustrated by the standing challenge by the Clay Mathematics Institute to prove uniqueness of solutions to Navier–Stokes equations.

Now we appreciate that instability is an essential property of fluid motions, leading to increasing awareness of the need to combine observations and models via data assimilation, for increased predictability of forecast states.

Surprises are even more abundant in living earth compared to fluid earth, as well for the solid earth that lies beneath. Complexity of these systems often defies deep understanding because surprise is a rule rather than exception in nature, with imminent new findings often overruling what is learned by incomplete observations. Human activity since the beginning of anthropocene has altered global ecosystems, making those interpretations even harder. Understanding earth systems has often been influenced by social history,[4] leaving future generations the task to respond to environmental burdens of all past human activities.

The book provides essential mathematical development that walks through elementary description of fluid dynamics toward a review of complex geophysical flows, but restricting attention to homogeneous fluids in this volume. While doing so, care is taken to provide essential steps in derivations often briefly stated elsewhere in expert literature. Yet, the area of interest is quite wide and perhaps our analyses can only provide an elementary view for further inquiry. In this regard, the reader is directed to expert literature of greater coverage on GFD, partially listed in the bibliography section.

This volume is devoted to homogeneous fluids, both at laboratory and geophysical scales, providing essentials for a reader who may either wish to settle for fundamentals or access a road map to safely proceed into the labyrinth of more advanced topics. The following volume will then build on this basis to add thermodynamics and stratification effects.

The presented material has been selectively developed from past lectures on dynamical oceanography and meteorology, offered as part of graduate programs at the Institute of Marine Sciences and Eurasia Institute of Earth Sciences

[3]Pinardi, N., Özsoy, E., Latif, M. A., Moroni, F., Grandi, A., Manzella, G., De Strobel, F. and V. Lyubartsev (2018). Measuring the sea: the first oceanographic cruise (1679–1680) and the roots of oceanography, Phys. Oceanogr., 48 (4), 845-860.

[4]Attali, J., 2017. *Histoires de la mer*, Librairie Artheme Fayard, 327 p.

(respectively of the Middle East Technical University, Mersin and İstanbul Technical University, İstanbul), and also as part of a school at the International Center for Theoretical Physics, Trieste.

Many thanks are due to various individuals who either generously provided motivation, encouragement, or positively contributed to the knowledge base of the book. Ümit Ünlüata, former advisor and mentor of the author, was foremost among those gifted people whose leadership, scientific collaboration, and sharing GFD knowledge around the "Seas of the Old World" have had fundamental impact. Participation of former graduate students and researchers in the various courses that helped development of the material is gratefully acknowledged.

Erdemli, Mersin, İstanbul, Ankara, Didyma, Turkey Emin Özsoy
February 2019

Contents

1	**A Brief Review of Algebra and Calculus**	1
1.1	Vectors and Cartesian Tensors	1
1.2	Algebraic Operations	3
1.2.1	Scalars	3
1.2.2	Vectors	3
1.2.3	Second-Order Tensors (dyads)	5
1.3	Differentiation and Integration	6
1.3.1	Differentiation and Integration with Respect to a Scalar	6
1.3.2	The 'del' Operator	6
1.3.3	Identities Associated with the 'del' Operator	8
1.3.4	Integral Theorems	9
1.4	The Material Derivative	10
1.5	Leibnitz' Rules	13
2	**Fluid Properties and Kinematics**	19
2.1	Definition of a Fluid	19
2.2	Continuum Hypothesis	19
2.3	Properties of Fluids	20
2.4	Volume and Surface Forces	20
2.4.1	Stress in a Fluid that Is at Rest	24
2.4.2	Stress in a Moving Fluid	25
2.5	Specification of the Flow Field	25
2.6	Conservation of Mass (Continuity Equation)	26
2.6.1	Continuity Equation	26
2.6.2	Stream Function–2-D, Incompressible, Steady Flow	28
2.7	Analysis of Relative Motion Near a Point	31
2.7.1	Pure-Straining Motion	32
2.7.2	Rigid Body Rotation	34
2.7.3	Modes of Motion in a Fluid	35

3 Equations of Motion of a Fluid . . . 39
3.1 Continuity Equation–Mass Conservation . . . 39
3.2 Equation of Motion–Momentum Equation . . . 39
3.2.1 Integral Form of the Momentum Equation . . . 41
3.3 Stress Tensor in a Newtonian Fluid . . . 42
3.4 Navier–Stokes Equation . . . 45
3.5 Pressure in a Fluid . . . 47
3.5.1 Pressure in a Fluid at Rest . . . 47
3.5.2 Pressure in a Moving Homogeneous Fluid . . . 48
3.6 Equation of Motion Relative to Moving Frame of Reference . . . 48
3.6.1 Modification of Gravity and Pressure Due to Centrifugal Acceleration . . . 51
3.7 Complete Set of Governing Equations . . . 53
3.8 Vorticity Dynamics . . . 54
3.8.1 The Vorticity Diffusion Term $\nu\nabla^2\omega_A$ . . . 56
3.8.2 The Tipping and Stretching Term $\omega_A \cdot \nabla \mathrm{u}$ . . . 56
3.8.3 The Divergence Term $-\omega_A(\nabla \cdot \mathrm{u})$ . . . 57
3.8.4 The Solenoidal Term $\frac{1}{\rho^2}\nabla\rho \times \nabla p$ . . . 57
3.9 Kelvin's Circulation Theorem . . . 58
3.10 Bernoulli's Theorem . . . 60

4 Flow of a Homogeneous Incompressible Fluid . . . 63
4.1 Governing Equations . . . 63
4.2 Steady Unidirectional Flow . . . 64
4.2.1 Poiseuille Flow . . . 65
4.2.2 Flow Between Plates . . . 67
4.3 Unsteady Unidirectional Flow . . . 68
4.3.1 Example—Smoothing of a Velocity Discontinuity . . . 72
4.3.2 Flow Due to an Oscillating Plane Boundary . . . 74
4.4 Unidirectional Flow Including Coriolis Effects . . . 76
4.4.1 The Steady Ekman Layer at the Surface . . . 76
4.4.2 The Steady Ekman Layer at the Bottom . . . 79
4.4.3 The Oscillatory Ekman Boundary Layer at the Surface . . . 82

5 Rotating, Homogeneous, Incompressible Fluids . . . 89
5.1 Equations of Motion for an Incompressible Homogeneous Fluid . . . 89
5.2 Inviscid Rotating Flows . . . 90
5.2.1 Inertial Motion—Unsteady (Periodic) Uniform Flow . . . 92

5.2.2 "Elasticity" in a Rotating Fluid—Restoring Effects of Coriolis . . . 95
5.2.3 Geostrophic Motion: Steady Flow with a Pressure Gradient . . . 96

6 Shallow Water Theory . . . 105
6.1 Tangent Plane Approximation . . . 105
6.2 Shallow Water Approximations . . . 107
6.2.1 Scaling of the Equations . . . 107
6.2.2 Continuity of Surface Forces (Dynamic Boundary Conditions) . . . 112
6.2.3 Hydrostatic Pressure . . . 116
6.2.4 Kinematic Boundary Conditions . . . 117
6.2.5 Shallow Water Equations . . . 117
6.2.6 Conservation Properties . . . 123
6.2.7 Vorticity Conservation . . . 125
6.2.8 Energy Conservation . . . 127
6.3 The f-Plane and the β-Plane Approximations . . . 128
6.4 Simple Applications of the Potential Vorticity Conservation . . . 130
6.4.1 Geostrophic Flow . . . 130
6.4.2 Flow over a Topographic Ridge (f-Plane) . . . 133
6.5 Topographic Effects . . . 135
6.5.1 Solutions for Uniform Shelf . . . 139
6.5.2 Nonlinear Inviscid Solutions . . . 141
6.5.3 Flow over a Depth Discontinuity (β-Plane) . . . 142
6.6 Oscillatory Motions . . . 147
6.6.1 Planetary (Rossby) Waves . . . 147
6.6.2 Planetary Waves with a Mean Current . . . 153
6.6.3 Small Amplitude Motions with a Free Surface . . . 154
6.6.4 Plane Waves for Constant Depth . . . 156
6.6.5 Poincaré and Kelvin Waves . . . 158
6.6.6 Topographic Rossby Waves . . . 165

7 Quasigeostrophic Theory . . . 183
7.1 An Overview and Derivation of Quasi-geostrophic Equations . . . 183

8 Elements of Ocean General Circulation . . . 193
8.1 The Ocean Circulation at Global Scale . . . 193
8.2 Elements of the Steady, Rigid Lid Ocean Circulation (f-Plane) . . . 194
8.2.1 Governing Equations in the f-Plane . . . 194
8.2.2 The Ocean Interior (f-Plane) . . . 196
8.2.3 Ekman Boundary Layers (f-Plane) . . . 198

8.2.4 Surface Ekman Layer (f-Plane)–Wind-Driven Flow 200
8.2.5 Bottom Ekman Layer (f-Plane)–Interior-Driven Flow ... 203
8.2.6 Ocean Circulation on the f-Plane ... 204
8.3 Ocean Circulation on the β-Plane ... 207
8.3.1 Development of the Model ... 207
8.3.2 The Sverdrup Interior and an Approximate Solution for the Circulation ... 211
8.3.3 Approximate Solution for the Circulation with a Western Boundary Layer ... 214

9 Two-Layer Models ... 221
9.1 Introduction ... 221
9.2 Two-Layer Shallow Water Equations ... 221
9.3 Free Periodic Motions ... 225
9.4 Normal Modes of Oscillation ... 225
9.4.1 Case I: Constant Depth ... 227
9.4.2 Case II: Variable Bottom Topography ... 228
9.5 Equations for Normal Modes ... 230
9.5.1 Approximate Modes ... 230
9.5.2 Barotropic Normal Mode—Leading Order ... 231
9.5.3 Baroclinic Normal Mode—Leading Order ... 232
9.5.4 Basin Lateral Boundary Conditions ... 233
9.5.5 Periodic Motions in a Basin ... 234

10 Exercises ... 235

Recommended Text Books ... 285

A Brief Review of Algebra and Calculus 1

1.1 Vectors and Cartesian Tensors

Scalars, vectors and tensors will be used in this text to define physical quantities.

Indicial notation:

In general an nth order tensor $\mathbf{A}$ can be denoted by n indices (subscripts) as

$$\mathbf{A} = A_{ijk...}$$

where each index $(i, j, k, \ldots)$ has a range of $i = 1, 2, 3,\ \ j = 1, 2, 3.$ etc. Therefore an nth order tensor has 3^n components. For example a second order tensor (dyad) has 9 components

$$\mathbf{A} = A_{ij} = \begin{pmatrix} A_{11} & A_{12} & A_{13} \\ A_{21} & A_{22} & A_{23} \\ A_{31} & A_{32} & A_{33} \end{pmatrix}$$

A first order tensor $A_i = (A_1\ A_2\ A_3)$ is equivalent to a vector, and a zeroth order tensor is a scalar. When referred to a cartesian (rectangular) system of coordinates a tensor is cartesian. Only cartesian tensors are considered in this text and tensors of up to second order will be our main interest.

Range Convention:

Whenever a subscript appears unrepeated in a term, it is understood to take on the values $1, 2, 3$. For example, the vector equation

$$F_i = ma_i$$

implies that each all component for i = 1,2,3 satisfies the equation separately. We note that this is equivalent to the vector equation

$$\mathbf{F} = m\mathbf{a}$$

where $\mathbf{F}$ and $\mathbf{a}$ are vectors with three components.

© Springer Nature Switzerland AG 2020

E. Özsoy, *Geophysical Fluid Dynamics I*, Springer Textbooks in Earth Sciences, Geography and Environment, https://doi.org/10.1007/978-3-030-16973-2_1

Summation convention:

Whenever a subscript occurs repeatedly in a term or an expression it is understood to represent a summation over the range 1,2,3. For example the expression

$$p = cA_{ii}$$

is equivalent to

$$p = c\sum_{i=1}^{3} A_{ii} = c(A_{11} + A_{22} + A_{33})$$

On the order hand, regular matrix multiplication (**A**, **B**, **C** are second order tensors, i.e. 3×3 matrices)

$$\mathbf{C} = \mathbf{AB}$$

can be written as

$$C_{ij} = A_{ik}B_{kj}$$

and implies

$$\begin{bmatrix} (A_{11}B_{11}+A_{12}B_{21}+A_{13}B_{31}) & (A_{11}B_{12}+A_{12}B_{22}+A_{13}B_{32}) & (A_{11}B_{13}+A_{12}B_{23}+A_{13}B_{33}) \\ (A_{21}B_{11}+A_{22}B_{21}+A_{23}B_{31}) & (A_{21}B_{12}+A_{22}B_{22}+A_{23}B_{32}) & \dots \\ (A_{31}B_{11}+A_{32}B_{21}+A_{33}B_{31}) & \dots & \dots \end{bmatrix}$$

Definitions—special tensors:

The *Kronecker delta tensor* δ_{ij} is a second order tensor (9 components) defined as,

$$\delta_{ij} = \begin{cases} 1 & \text{if } i = j \\ 0 & \text{if } i \neq j \end{cases} = \begin{bmatrix} 1\,0\,0 \\ 0\,1\,0 \\ 0\,0\,1 \end{bmatrix} = I \tag{1.1}$$

where I is called the identity matrix.

The alternating unit tensor ϵ_{ijk} is a third order tensor (27 components) defined as

$$\epsilon_{ijk} = \begin{cases} +1 & \text{if } i, j, k = 1, 2, 3 \text{ or } 2, 3, 1 \text{ or } 3, 1, 2 \ (\textit{even permutation}) \\ -1 & \text{if } i, j, k = 3, 2, 1 \text{ or } 1, 3, 2 \text{ or } 2, 1, 3 \ (\textit{odd permutation}) \\ 0 & \text{if } i = j \text{ or } i = k \text{ or } j = k, \ (\textit{if any two indices coincide}) \end{cases} \tag{1.2}$$

Definitions—transpose, symmetric and antisymmetric tensors:

(a) The transpose of a second order tensor (matrix) **A** is defined as

$$\mathbf{A}^T = (A_{ij})^T = A_{ji},$$

such that the rows and columns are interchanged. It is easily proved that

$$(\mathbf{A}^T)^T = \mathbf{A} \textit{ and } (\mathbf{AB})^T = \mathbf{B}^T\mathbf{A}^T.$$

(b) If $\mathbf{A}^{\mathrm{T}} = \mathbf{A}$ then $\mathbf{A}$ is a *symmetric* tensor.

(c) If $\mathbf{A}^{\mathrm{T}} = -\mathbf{A}$ then $\mathbf{A}$ is an *anti-symmetric* (or skew-symmetric) tensor.

(d) Any second order tensor can be represented as a sum of symmetric and anti-symmetric tensors:

$$\mathbf{A} = \frac{1}{2}(\mathbf{A} + \mathbf{A}^T) + \frac{1}{2}(\mathbf{A} - \mathbf{A}^T) = \mathbf{A}_s + \mathbf{A}_u. \tag{1.3}$$

where $\mathbf{A}_s = \frac{1}{2}(\mathbf{A} + \mathbf{A}^T)$ is the symmetric part satisfying $\mathbf{A}_s = \mathbf{A}_s^T$, and $\mathbf{A}_u = \frac{1}{2}(\mathbf{A} - \mathbf{A}^T)$ is the antisymmetric part satisfying $\mathbf{A}_u = -\mathbf{A}_u^T$.

Definition—Principal Coordinates:

For a symmetric cartesian second order tensor, it is always possible to make a coordinate transformation to another cartesian system, to make the off-diagonal components vanish. The new coordinates in which this becomes possible are called *principal coordinates*. The trace $A_{ii} = A_{11} + A_{22} + A_{33}$ of a symmetric tensor is invariant under any cartesian coordinate transformation.

Definition—Isotropic Tensor:

An *isotropic tensor* is one in which there is no directional preference. Under all cartesian coordinate transformations, the components of an isotropic tensor remain unchanged. The most basic isotropic tensor is the Kronecker delta (identity matrix). All even order isotropic tensors can be expressed in terms of Kronecker delta tensors. A second example of an isotropic tensor is a scalar (i.e., a zeroth order tensor) since it remains the same under all coordinate transformations.

1.2 Algebraic Operations

1.2.1 Scalars

Algebraic operations are addition, subtraction, multiplication and division.

1.2.2 Vectors

A vector can be denoted as $(\mathbf{q})$ or in the form of a row $((q_i), i = 1, 2, 3)$ or column $((q_j), j = 1, 2, 3)$ matrix. A vector $\mathbf{q}$ has three components, each of which is parallel to the coordinates referenced. For example

$$\mathbf{q} = q_1\hat{e}_1 + q_2\hat{e}_2 + q_3\hat{e}_3,$$

where $\hat{e}_i$ are unit vectors along the coordinate axes. In summation notation this can be written as

$$\mathbf{q} = q_i\hat{e}_i. \tag{1.4}$$

The absolute value of the vector (a scalar) is denoted as,

$$|\mathbf{q}| = \sqrt{q_1^2 + q_2^2 + q_3^2}.$$

Addition, subtraction and multiplication are defined for vectors but division is not defined.

Addition and subtraction is made by operating individually on the components

$$\mathbf{a} \pm \mathbf{b} = a_i \hat{e}_i \pm b_i \hat{e}_i = (a_i \pm b_i)\hat{e}_i. \tag{1.5}$$

In vector multiplication, more than one kind of product is possible.

(a) *The scalar product* (dot product):

$$\mathbf{a} \cdot \mathbf{b} = a_i b_i = a_1 b_1 + a_2 b_2 + a_3 b_3 = |\mathbf{a}||\mathbf{b}| \cos \theta, \tag{1.6}$$

where θ is the smaller angle between the two vectors. If $\mathbf{a} \cdot \mathbf{b} = 0$, the two vectors are said to be *orthogonal* ($\theta = \pi/2$). The scalar multiplication is *commutative*

$$\mathbf{a} \cdot \mathbf{b} = \mathbf{b} \cdot \mathbf{a},$$

and *distributive*

$$\mathbf{a} \cdot (\mathbf{b} + \mathbf{c}) = \mathbf{a} \cdot \mathbf{b} + \mathbf{a} \cdot \mathbf{c},$$

but *not associative*

$$(\mathbf{a} \cdot \mathbf{b})\mathbf{c} \neq \mathbf{a}(\mathbf{b} \cdot \mathbf{c}).$$

(b) *The vector product* (cross product):

$$\mathbf{a} \times \mathbf{b} = (a_2 b_3 - b_2 a_3)\hat{e}_1 + (a_1 b_3 - b_1 a_3)\hat{e}_2 + (a_1 b_2 - b_1 a_2)\hat{e}_3 = (|\mathbf{a}||\mathbf{b}| \sin \theta)\hat{n}, \tag{1.7}$$

where $\hat{e}_i$ are unit vectors $\hat{n}$ is the unit vector normal to the plane of the vectors **a** and **b**. The cross product can also be expressed in the form of a determinant

$$\mathbf{a} \times \mathbf{b} = \begin{vmatrix} \hat{e}_1 & \hat{e}_2 & \hat{e}_3 \\ a_1 & a_2 & a_3 \\ b_1 & b_2 & b_3 \end{vmatrix}. \tag{1.8}$$

By making use of the alternating unit tensor (1.2), the cross product can be expressed in indicial notation as

$$\mathbf{c} = c_i \hat{e}_i = \mathbf{a} \times \mathbf{b} = \epsilon_{ijk} a_j b_k \hat{e}_i. \tag{1.9}$$

The cross product is *not commutative* (a sign change occurs if the order of multiplication is reversed)

$$\mathbf{a} \times \mathbf{b} = -\mathbf{b} \times \mathbf{a},$$

not associative

$$\mathbf{a} \times (\mathbf{b} \times \mathbf{c}) \neq (\mathbf{a} \times \mathbf{b}) \times \mathbf{c},$$

but distributive

$$(\mathbf{a} + \mathbf{b}) \times \mathbf{c} = \mathbf{a} \times \mathbf{c} + \mathbf{b} \times \mathbf{c}.$$

(c) *The dyadic product* (indefinite product):

$$\mathbf{a} \circ \mathbf{b} = \mathbf{a}\,\mathbf{b} = a_i b_j = C_{ij}. \tag{1.10}$$

The product is a second order tensor

$$C_{ij} = a_i b_j$$

Vector identities:
The following are common identities of vector algebra:

$$\begin{aligned}
(\mathbf{a} \times \mathbf{b}) \cdot \mathbf{c} &= (\mathbf{b} \times \mathbf{c}) \cdot \mathbf{a} = (\mathbf{c} \times \mathbf{a}) \cdot \mathbf{b} \\
(\mathbf{a} \times \mathbf{b}) \times \mathbf{c} &= (\mathbf{a} \cdot \mathbf{c})\mathbf{b} - (\mathbf{a} \cdot \mathbf{b})\mathbf{c} \\
(\mathbf{a} \times \mathbf{b}) \cdot (\mathbf{c} \times \mathbf{d}) &= (\mathbf{a} \cdot \mathbf{c})(\mathbf{b} \cdot \mathbf{d}) - (\mathbf{a} \cdot \mathbf{d})(\mathbf{b} \cdot \mathbf{c}) \\
(\mathbf{a} \circ \mathbf{b})\mathbf{c} &= \mathbf{a}(\mathbf{b} \cdot \mathbf{c})
\end{aligned} \tag{1.11a–d}$$

1.2.3 Second-Order Tensors (dyads)

The algebraic operations defined for tensors are addition, subtraction and multiplication. These operations will be demonstrated for second-order tensors, since we will only be using those in the present context.

Addition and subtraction is similar to vectors, where each component is added or subtracted individually

$$\mathbf{C_{ij}} = \mathbf{A} \pm \mathbf{B} = A_{ij} \pm B_{ij}$$

Several kinds of multiplication are possible, classified according to the end result:
(a) *Single dot product (matrix multiplication)*

$$\mathbf{C} = \mathbf{A} \cdot \mathbf{B} = \mathbf{AB} = C_{ij} = A_{ik} B_{kj} \tag{1.13}$$

the result is a second order tensor.
(b) *Double dot product (inner product)*

$$C = \mathbf{A} : \mathbf{B} = a_{ij} b_{ij} \tag{1.14}$$

the result is a scalar.

(c) *Dot product of a tensor and vector*
A tensor dotted with a vector (post multiplication) results in a vector:

$$\mathbf{P} = \mathbf{A} \cdot \mathbf{q} = p_i \hat{e}_i = A_{ij} q_j \hat{e}_i . \tag{1.15}$$

Similarly, a vector dotted with a tensor (premultiplication) yields a vector:

$$\mathbf{P} = \mathbf{q} \cdot \mathbf{A} = p_i \hat{e}_i = q_j A_{ji} \hat{e}_i , \tag{1.16}$$

Clearly,

$$\begin{gathered} \mathbf{A} \cdot \mathbf{q} \neq \mathbf{q} \cdot \mathbf{A}, \\ (A_{ij} q_i \neq q_j A_{ji}). \end{gathered} \tag{1.17}$$

unless the tensor **A** is symmetric, i.e. unless $A_{ij} = A_{ji}$ $(\mathbf{A}^{\mathrm{T}} = \mathbf{A})$.

1.3 Differentiation and Integration

1.3.1 Differentiation and Integration with Respect to a Scalar

Differentiation and integration with respect to a scalar is quite straightforward and applies to each component of the tensor, vector and scalar separately. For the second order tensor, these can be written as,

$$\frac{d\mathbf{A}}{dt} = \frac{d\mathbf{A_{ij}}}{dt} \quad \text{and} \quad \int \mathbf{A} dt = \int A_{ij} dt \tag{1.18}$$

Differentiation and integration of an added or subtracted quantity applies to each term separately. In the differentiation and integration of a product, the rules for scalars can be applied,

$$\frac{d\mathbf{A} \cdot \mathbf{B}}{dt} = \frac{dA_{ik} B_{kj}}{dt} = A_{ik} \frac{dB_{kj}}{dt} + \frac{dA_{ik}}{dt} B_{kj} = \mathbf{A} \cdot \frac{d\mathbf{B}}{dt} + \frac{d\mathbf{A}}{dt} \cdot \mathbf{B} , \tag{1.19a}$$

$$\int \mathbf{A} \cdot \frac{d\mathbf{B}}{dt} dt = \mathbf{A} \cdot \mathbf{B} - \int \mathbf{B} \cdot \frac{d\mathbf{A}}{dt} dt . \tag{1.19b}$$

1.3.2 The 'del' Operator

Differentiation with respect to a vector quantity is more complex. Specifically, the differentiation with respect to the position vector $\mathbf{x} = x_i \hat{e}_i$ is defined by the 'Del' operator.

$$\nabla = \frac{\partial}{\partial \mathbf{x}} = \hat{e}_i \frac{\partial}{\partial x_i} = \hat{e}_1 \frac{\partial}{\partial x_1} + \hat{e}_2 \frac{\partial}{\partial x_2} + \hat{e}_3 \frac{\partial}{\partial x_3} \tag{1.20}$$

where $Ox_1x_2x_3$ is a cartesian coordinate system.

As more than one kind of product is possible for vectors and tensors; similarly the 'del' operation yields different results according to the type of operation. The possibilities are as follows

(a) Gradient of a scalar:

Let ϕ be a scalar. Then gradient is defined as,

$$\nabla\phi = grad\ \phi = \hat{e}_i \frac{\partial \phi}{\partial x_i} = \hat{e}_1 \frac{\partial \phi}{\partial x_1} + \hat{e}_2 \frac{\partial \phi}{\partial x_2} + \hat{e}_3 \frac{\partial \phi}{\partial x_3} \tag{1.21}$$

the result being a vector.

(b) Gradient of a vector:

Let $\mathbf{q}$ be a vector. Then the gradient of this vector is a dyadic operation yielding second-order tensor,

$$\begin{aligned} \nabla \circ \mathbf{q} = \nabla\mathbf{q} = grad\ \mathbf{q} &= \frac{\partial q_i}{\partial x_j} \\ &= \begin{bmatrix} \frac{\partial q_1}{\partial x_1} & \frac{\partial q_1}{\partial x_2} & \frac{\partial q_1}{\partial x_3} \\ \frac{\partial q_2}{\partial x_1} & etc. & \\ \frac{\partial q_3}{\partial x_1} & & \end{bmatrix} \end{aligned} \tag{1.22}$$

(c) Divergence of a vector:

The divergence of a vector is defined as,

$$\nabla \cdot \mathbf{q} = div\ \mathbf{q} = \frac{\partial q_i}{\partial x_i} = \frac{\partial q_1}{\partial x_1} + \frac{\partial q_2}{\partial x_2} + \frac{\partial q_3}{\partial x_3} \tag{1.23}$$

yielding a scalar.

(d) Divergence of a tensor:

The divergence of the tensor $\mathbf{A}$ is defined as,

$$\begin{aligned} \nabla \cdot \mathbf{A} = div\ \mathbf{A} &= \frac{\partial A_{ij}}{\partial x_j} \hat{e}_i \\ &= \begin{bmatrix} \frac{\partial A_{11}}{\partial x_1} + \frac{\partial A_{12}}{\partial x_2} + \frac{\partial A_{13}}{\partial x_3} \\ \frac{\partial A_{21}}{\partial x_1} + \frac{\partial A_{22}}{\partial x_2} + \frac{\partial A_{23}}{\partial x_3} \\ \frac{\partial A_{31}}{\partial x_1} + \frac{\partial A_{32}}{\partial x_2} + \frac{\partial A_{33}}{\partial x_3} \end{bmatrix} \end{aligned} \tag{1.24}$$

yielding a vector.

(e) Curl of a vector:

The curl of a vector is defined as,

$$\nabla \times \mathbf{q} = curl\ \mathbf{q} = \epsilon_{ijk} \frac{\partial q_k}{\partial x_j} \hat{e}_i$$
$$= \begin{vmatrix} \hat{e}_1 & \hat{e}_2 & \hat{e}_3 \\ \frac{\partial}{\partial x_1} & \frac{\partial}{\partial x_2} & \frac{\partial}{\partial x_3} \\ q_1 & q_2 & q_3 \end{vmatrix} \tag{1.25}$$

the result is a vector.

(f) The Laplacian operator:

The Laplacian operator is defined as,

$$\nabla^2 = \nabla \cdot \nabla = \frac{\partial}{\partial x_i} \frac{\partial}{\partial x_i} = \frac{\partial^2}{\partial x_i^2} = \frac{\partial^2}{\partial {x_1}^2} + \frac{\partial^2}{\partial {x_2}^2} + \frac{\partial^2}{\partial {x_3}^2} \tag{1.26}$$

This operator can be applied to scalars, vectors and tensors alike, and the result is the time same type as the quantity that is operated upon.

1.3.3 Identities Associated with the 'del' Operator

Certain identities relating different forms of the 'del' operator will be used in the present context. These identities can be classified in to two groups.

(i) Those involving vectors and scalars only (s, p are scalars, $\mathbf{a}$, $\mathbf{b}$ are vectors)

$$\nabla \cdot s\mathbf{a} = s\nabla \cdot \mathbf{a} + \mathbf{a} \cdot \nabla s$$

$$\nabla \times s\mathbf{a} = s\nabla \times \mathbf{a} + \nabla s \times \mathbf{a}$$

$$\nabla \cdot \mathbf{a} \times \mathbf{b} = \mathbf{b} \cdot \nabla \times \mathbf{a} - \mathbf{a} \cdot \nabla \times \mathbf{b}$$

$$\nabla \times (\mathbf{a} \times \mathbf{b}) = (\mathbf{b} \cdot \nabla)\mathbf{a} - (\mathbf{a} \cdot \nabla)\mathbf{b} + \mathbf{a}(\nabla \cdot \mathbf{b}) - \mathbf{b}(\nabla \cdot \mathbf{a})$$

$$\nabla(\mathbf{a} \cdot \mathbf{b}) = (\mathbf{a} \cdot \nabla)\mathbf{b} + (\mathbf{b} \cdot \nabla)\mathbf{a} + \mathbf{a} \times (\nabla \times \mathbf{b}) + \mathbf{b} \times (\nabla \times \mathbf{a})$$

$$\nabla \times (\nabla \times \mathbf{a}) = \nabla(\nabla \cdot \mathbf{a}) - \nabla \cdot \nabla\mathbf{a} = \nabla(\nabla \cdot \mathbf{a}) - \nabla^2\mathbf{a}$$

$$\nabla \cdot \nabla^2\mathbf{a} = \nabla^2\nabla \cdot \mathbf{a}$$

$$\nabla \times \nabla^2\mathbf{a} = \nabla^2\nabla \times \mathbf{a}$$

$$\nabla \times \nabla s = \mathbf{O}$$

$$\nabla \cdot (\nabla \times \mathbf{a}) = 0$$
$$\nabla \cdot (\nabla s \times \nabla p) = 0 \tag{1.27a–k}$$

(ii) Those involving tensors and dyadic operations (s is a scalar, $\mathbf{a}$, $\mathbf{b}$ are vectors, $\mathbf{A}$ is a tensor, and $\mathbf{I}=\delta_{ij}$ is Kronecker delta)

$$\nabla \cdot (\mathbf{a} \circ \mathbf{b}) = \mathbf{a}(\nabla \cdot \mathbf{b}) + \mathbf{b} \cdot (\nabla \circ \mathbf{a}) = \mathbf{a}(\nabla \cdot \mathbf{b}) + (\mathbf{b} \cdot \nabla)\mathbf{a}$$

$$\nabla \cdot (\mathbf{A} \cdot \mathbf{a}) = (\nabla \cdot \mathbf{A}^T) \cdot \mathbf{a} + \mathbf{A}^T : (\nabla \circ \mathbf{a})$$

$$\nabla \cdot (s\mathbf{I}) = \nabla s$$

$$\nabla \cdot (\nabla \circ \mathbf{a})^T = \nabla \circ (\nabla \cdot \mathbf{a})$$

$$\nabla \times \nabla \times \mathbf{a} = \nabla \cdot [(\nabla \circ \mathbf{a})^T - (\nabla \circ \mathbf{a})]$$

$$\nabla \cdot (\nabla \circ \mathbf{a}) = \nabla \circ (\nabla \cdot \mathbf{a}) - \nabla \times \nabla \times \mathbf{a} = \nabla^2 \mathbf{a}$$

$$\nabla(\mathbf{a} \cdot \mathbf{b}) = (\nabla \circ \mathbf{a})^T \cdot \mathbf{b} + (\nabla \circ \mathbf{b})^T \cdot \mathbf{a}$$

$$(\nabla \circ \mathbf{a})^T \cdot \mathbf{a} = \nabla(\frac{1}{2}\mathbf{a} \cdot \mathbf{a})$$

$$\begin{aligned}(\nabla \circ \mathbf{a}) \cdot \mathbf{a} &= \nabla(\frac{1}{2}\mathbf{a} \cdot \mathbf{a}) + [(\nabla \circ \mathbf{a}) - (\nabla \circ \mathbf{a})^T] \cdot \mathbf{a} \\ &= \nabla(\frac{1}{2}\mathbf{a} \cdot \mathbf{a}) + (\nabla \times \mathbf{a}) \times \mathbf{a} \\ &= (\mathbf{a} \cdot \nabla)\mathbf{a}\end{aligned}$$

$$\begin{aligned}\nabla \cdot [(\nabla \circ \mathbf{a}) \cdot \mathbf{a}] &= [\nabla \cdot (\nabla \circ \mathbf{a})^T] \cdot \mathbf{a} + (\nabla \circ \mathbf{a})^T \cdot (\nabla \circ \mathbf{a}) \\ &= [\nabla(\nabla \cdot \mathbf{a})] \cdot \mathbf{a} + (\nabla \circ \mathbf{a})^T \cdot (\nabla \circ \mathbf{a})\end{aligned} \tag{1.28a–i}$$

1.3.4 Integral Theorems

Two integral theorems which are of particular interest are given below

Green's Theorem (Divergence Theorem):

If $\mathbf{a}$ and its partial derivatives are continuous in V and S (where V is a fixed volume in space and S is the bounding closed surface) and if S is piecewise smooth, then the theorem states that

$$\int_V \nabla \cdot \mathbf{a}\, dV = \int_S \mathbf{a} \cdot \mathbf{n}\, dS \tag{1.29}$$

where $\mathbf{n}$ is the unit normal vector pointing outward on surface S.

Extensions of this theorem can be made as follows:

(a) if $\mathbf{a} = \nabla\phi$, with ϕ defined as a scalar, then

$$\int_V \nabla \cdot \nabla\phi \, dV = \int_V \nabla^2\phi \, dV = \int_S \nabla\phi \cdot \mathbf{n} \, dS = \int \frac{\partial\phi}{\partial n} \, dS, \tag{1.30}$$

where $\frac{\partial\phi}{\partial n}$ denotes a derivative in the direction of the outward normal vector $\mathbf{n}$.

(b) if $\mathbf{a} = \psi\nabla\phi$, with ϕ, ψ defined as scalars, then

$$\int_V (\psi\nabla^2\phi + \nabla\psi \cdot \nabla\phi) \, dV = \int_S \left(\psi\frac{\partial\phi}{\partial n}\right) dS \tag{1.31}$$

(c) From (a) and (b), the following can also be obtained

$$\int_V (\psi\nabla^2\phi - \phi\nabla^2\psi) \, dV = \int_S \left(\psi\frac{\partial\phi}{\partial n} - \phi\frac{\partial\psi}{\partial n}\right) dS \tag{1.32}$$

Analogues of the divergence theorem can also be obtained for scalars and tensors:

$$\int_V \nabla s \, dV = \int_S s\mathbf{n} \, dS \tag{1.33}$$

and

$$\int_V \nabla \cdot \mathbf{A} \, dV = \int_S \mathbf{n} \cdot \mathbf{A} \, dS \tag{1.34}$$

where s is a scalar and $\mathbf{A}$ is a tensor.

Stokes' Theorem:

If $\mathbf{a}$ is continuously differentiable in a simply and two-sided surface S bounded by a closed curve C, then the theorem states that

$$\int_S \mathbf{n} \cdot (\nabla \times \mathbf{a}) \, dS = \oint_C \mathbf{a} \cdot d\mathbf{r} \tag{1.35}$$

1.4 The Material Derivative

The material derivative is of particular use in the description of fluid motion. Consider a Cartesian reference frame oxyz in which a fluid motion is described. Let $\mathbf{r} = (x, y, z)$ be the position of a point (not necessarily moving with the fluid). The motion of the fluid is completely described if a velocity field $\mathbf{u}$ is specified at every point in space and as a function of time:

$$\mathbf{u} = \mathbf{u}(\mathbf{r}, t)$$

Definition:

A *fluid particle* or a *material point* is defined as a point identified to be moving with the fluid. The trajectory $\mathbf{r}(t)$ of such a fluid particle would then be obtained by solving

$$\frac{d\mathbf{r}(t)}{dt} = \mathbf{u}(\mathbf{r}(t), t) \ ,$$

Next, consider a scalar function $T(\mathbf{r}, t)$ in the same space. The values of this function will be seen as $T(\mathbf{r}(t), t)$ by an observer moving on a path $\mathbf{r}(t)$. The rate of change seen by the observer is then (using the chain rule):

$$\frac{dT(\mathbf{r}(t), t)}{dt} = \frac{dT(x(t), y(t), z(t), t)}{dt} = \frac{\partial T}{\partial t} + \frac{\partial T}{\partial x}\frac{dx}{dt} + \frac{\partial T}{\partial y}\frac{dy}{dt} + \frac{\partial T}{\partial z}\frac{dz}{dt} \ ,$$

Denoting the observer's velocity $\mathbf{v}$ as

$$\mathbf{v} = \frac{d\mathbf{r}}{dt} = \left(\frac{dx}{dt}, \frac{dy}{dt}, \frac{dz}{dt}\right) \ ,$$

the rate of change of T becomes

$$\frac{dT}{dt} = \frac{\partial T}{\partial t} + \mathbf{v} \cdot \nabla T \ .$$

If the observer is moving with a fluid particle, his velocity is the same as the fluid velocity, $\mathbf{v} = \mathbf{u}$. The rate of change seen by an observer riding on a fluid particle is then given by the *total derivative* or *material derivative*

$$\frac{DT}{Dt} = \frac{\partial T}{\partial t} + \mathbf{u} \cdot \nabla T \ ,$$

where capital D has been used to denote the material derivative.

Although this derivative has been considered for a scalar field, it can be generalized for vectors since it can be applied for each component separately. We define

$$\frac{D}{Dt} = \frac{\partial}{\partial t} + \mathbf{u} \cdot \nabla \qquad (1.36)$$

as the *material derivative operator* (also called *comoving, substantial, convective or particle derivative*). This operator acting on any scalar or vector quantity represents the time rate of change of that quantity with respect to a material point.

Note: The convention with capital D's, (D/Dt), is generally applied to *field variables* (variables that can be assigned to individual points in the space). Examples are temperature, density, velocity, momentum etc. The convention d/dt is applied to variables that can only be assigned to a lump of fluid in an infinitesimal or finite *volume*. Examples are energy, entropy etc.

Further definitions:

Surface: A mathematical surface is defined by the equation,

$$\phi(\mathbf{r}(t), t) = c \tag{1.37}$$

where ϕ is a function, c a scalar constant, and $\mathbf{r}(t)$ the position vector of a mathematical point on the surface. Differentiation with respect to time yields

$$\frac{d\phi}{dt} = \frac{\partial \phi}{\partial t} + \mathbf{v} \cdot \nabla\phi = 0 \tag{1.38}$$

where $\mathbf{v} = d\mathbf{r}/dt$ is the point velocity assigned to the points on the surface moves. A *unit normal vector* on the surface is defined as

$$\mathbf{n}(\mathbf{r}, t) = \frac{\nabla\phi}{|\nabla\phi|} \tag{1.39}$$

The *normal velocity* v_n is defined as the component of $\mathbf{v}$ normal to the surface:

$$v_n = \mathbf{v} \cdot \mathbf{n} = \mathbf{v} \cdot \frac{\nabla\phi}{|\nabla\phi|} \tag{1.40}$$

A *material surface* is defined as a surface made up of material points, i.e. moving with the fluid. The fact that fluid particles on a material surface may never get off the surface implies $\mathbf{u} \cdot \mathbf{n} = \mathbf{v} \cdot \mathbf{n}$. Through use of (1.38) and (1.40) it can therefore be shown that a material surface has to satisfy

$$\frac{D\phi}{Dt} = \frac{\partial \phi}{\partial t} + \mathbf{u} \cdot \nabla\phi = 0 \tag{1.41}$$

A *fluid interface* is defined as a material surface across which one more of the fluid properties (e.g. density, velocity) is discontinuous. However, note that the normal component of the fluid velocity may not be discontinuous across the interface, consistent with the definition of material surface.

$$\mathbf{u}_1 \cdot \mathbf{n} = \mathbf{u}_2 \cdot \mathbf{n}, \tag{1.42}$$

where $\mathbf{u}_1$ and $\mathbf{u}_2$ refer to the velocities assigned to the fluid particles on the two sides of the interface.

A *free surface* is defined as an interface on which pressure is constant.

A *boundary* is a material surface that is fixed with respect to time, i.e. $\mathbf{v} = d\mathbf{r}/dt = 0$. Since $\mathbf{u} \cdot \mathbf{n} = \mathbf{v} \cdot \mathbf{n}$ on any material surface; on a boundary, the fluid velocity must satisfy

$$\mathbf{u} \cdot \mathbf{n} = 0 \tag{1.43}$$

A *material volume* is defined as a volume that is made up of material points, i.e. a volume that moves with the fluid.

1.5 Leibnitz' Rules

Leibnitz' Rules from calculus are given below

(i) One dimensional integrals:

Consider functions $P(x,t)$, $\partial P(x,t)/\partial t$ that are continuous in some interval $a(t) \leq x \leq b(t)$, where the end points $a(t)$ and $b(t)$ are variable with respect to time. Leibnitz' rule states that

$$\frac{d}{dt}\int_{a(t)}^{b(t)} P(x,t)\,dx = \int_{a(t)}^{b(t)} \frac{\partial P(x,t)}{\partial t}\,dx - P(a(t),t)\,\frac{da}{dt} + P(b(t),t)\,\frac{db}{dt} \tag{1.44a}$$

(ii) Two dimensional analogue

Consider Q(x,y,t), $\frac{\partial Q(x,y,t)}{\partial t}$ continuously defined in same 2-D surface $S(t)$ bounded by a closed, moving curve C(t). Then,

$$\frac{d}{dt}\int_{S(t)} Q(x,y,t)\,dS = \int_{S(t)} \frac{\partial Q}{\partial t}\,dS + \oint_{C(t)} Q v_n\,dl \tag{1.44b}$$

where $v_n = \mathbf{n}\cdot\mathbf{v}$ is the normal velocity of points on C.

(iii) Three dimensional analogue.

Consider functions $R(\mathbf{r},t)$, $\frac{\partial R(\mathbf{r},t)}{\partial t}$ that are continuous in same volume $V(t)$ bounded by a moving surface $S(t)$. Then,

$$\frac{d}{dt}\int_{V(t)} R(\mathbf{r},t)\,dV = \int_{V(t)} \frac{\partial R}{\partial t}\,dV + \int_{S(t)} R v_n\,dS \tag{1.44c}$$

where $v_n = \mathbf{n}\cdot\mathbf{v}$ is the normal velocity of points on S.

Exercises

Exercise 1

Prove the following vector identities, by using tensor notation, $\mathbf{a} = a_i\hat{e}_i$, $\mathbf{b} = b_i\hat{e}_i$, $\mathbf{c} = c_i\hat{e}_i$.

$$\mathbf{a}\cdot\mathbf{b} = \mathbf{b}\cdot\mathbf{a}$$
$$\mathbf{a}\times\mathbf{b} = -\mathbf{b}\times\mathbf{a}$$
$$(\mathbf{a}\times\mathbf{b})\cdot\mathbf{c} = (\mathbf{b}\times\mathbf{c})\cdot\mathbf{a}$$

Exercise 2

Prepare a table listing values of the alternating unit tensor ϵ_{ijk} for all combinations of i, j and k:

$$\epsilon_{ijk} = \begin{cases} +1 & \text{if } i,j,k = 1,2,3 \text{ or } 2,3,1 \text{ or } 3,1,2 \ (\textit{even permutation}) \\ -1 & \text{if } i,j,k = 3,2,1 \text{ or } 1,3,2 \text{ or } 2,1,3 \ (\textit{odd permutation}) \\ 0 & \text{if } i=j \text{ or } i=k \text{ or } j=k, \ (\textit{if any two indices coincide}) \end{cases}$$

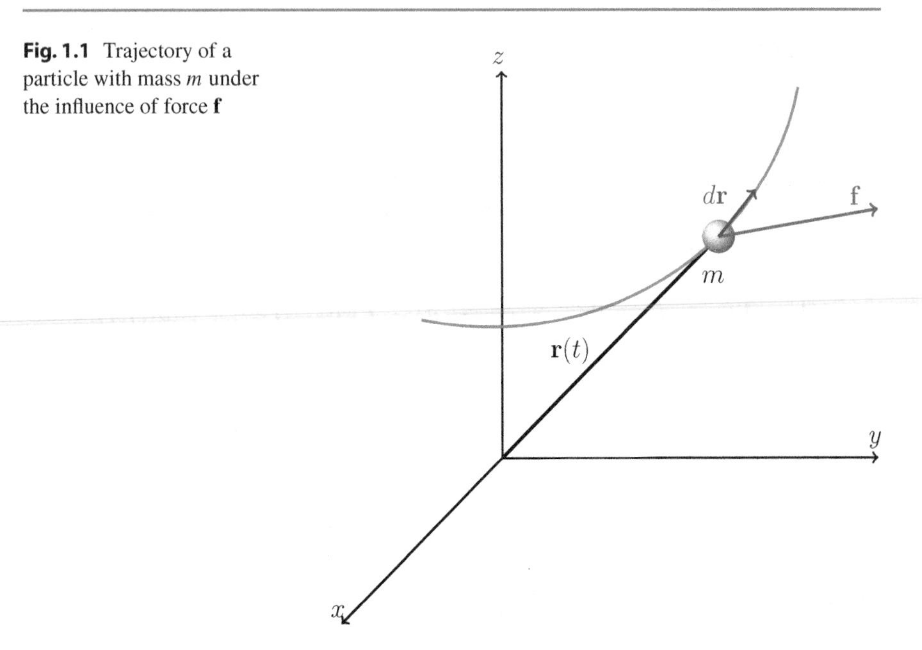

Fig. 1.1 Trajectory of a particle with mass m under the influence of force $\mathbf{f}$

and check the validity of the following:

$$\mathbf{c} = c_i \hat{e}_i = \mathbf{a} \times \mathbf{b} = \epsilon_{ijk} a_j b_k \hat{e}_i,$$

i.e.:

$$c_1 = a_2 b_3 - a_3 b_2, \quad c_2 = a_3 b_1 - a_1 b_3, \quad c_3 = a_1 b_2 - a_2 b_1.$$

`Exercise 3`

The motion of a particle with mass m in a fixed coordinate system $Oxyz$ under the influence of a force field $\mathbf{f}(\mathbf{x}, t)$ is determined by the *equation of motion* under (Newton's Second Law of classical mechanics):

$$m \frac{d^2 \mathbf{r}}{dt^2} = \mathbf{f},$$

where $\mathbf{r}(t)$ is the path followed by the particle in the coordinate system $Oxyz$ in Fig. 1.1.

The increment of work done on the particle by the force is

$$dW = \mathbf{f} \cdot d\mathbf{r}.$$

`(i)` Show that the work done can be expressed as

$$dW = d\left(\frac{1}{2} m \frac{d\mathbf{r}}{dt} \cdot \frac{d\mathbf{r}}{dt}\right) \equiv dT$$

where

$$\mathbf{u} \equiv \frac{d\mathbf{r}}{dt}$$

is the *particle velocity* and

$$T \equiv \frac{1}{2}m\frac{d\mathbf{r}}{dt} \cdot \frac{d\mathbf{r}}{dt} = \frac{1}{2}m\mathbf{u} \cdot \mathbf{u}$$

is defined as the *kinetic energy* of the particle.

Show that the work done by moving the particle from position $\mathbf{r}_1$ to $\mathbf{r}_2$ is

$$W_{12} = \int_1^2 dW = \int_1^2 dT = T_2 - T_1 = \Delta T$$

where $T_1 = T(\mathbf{r}_1)$ $T_2 = T(\mathbf{r}_2)$. Interpret this result.

(ii) If a particle starts from its initial position $\mathbf{r}(t_0)$ at time t_0 and moves around a closed orbit C in the time interval Δt, it would come back to its original position $\mathbf{r}(t_0 + \Delta t) = \mathbf{r}(t_0) \equiv \mathbf{r}_0$ at time $t_0 + \Delta t$. What would be the work done by moving the particle along the closed path C?

What is the relation between the initial and final velocities, respectively $\mathbf{u}(t_0)$ and $\mathbf{u}(t_0 + \Delta t)$ of the particle for the following cases: (a) $W_{00} = 0$, (b) $W_{00} > 0$, (c) $W_{00} < 0$? In what way can one interpret these cases in physical terms? In which case can $\mathbf{f}(\mathbf{r}(t), t)$ claimed to be a *conservative force field*?

(iii) Using the results developed in part ii for the case $W_{00} = 0$, show that a conservative force field $\mathbf{f}(\mathbf{r}(t), t)$ can be expressed as the gradient of a scalar function $V(\mathbf{r})$ that is solely a function of position (i.e., without any dependence on time t). Given any force field $\mathbf{f}$, can we check if this force field is conservative or not, based on the results of section (ii)?

(iv) Using the results from above sections for the conservative case, obtain an expression for $V(\mathbf{r})$ and show that

$$dW = dT = -dV.$$

What physical meaning do these quantities have? Does $V(\mathbf{r})$ depend on the path of integration?

(v) Again, if the force field $\mathbf{f}$ is conservative, show that

$$\frac{dE}{dt} == \frac{d}{dt}(T + V) = 0$$

where $E = T + V$. Interpret this result.

(vi) If the force field is a function of time $\mathbf{f} = \mathbf{f}(r(t), t)$, can it be a conservative field?

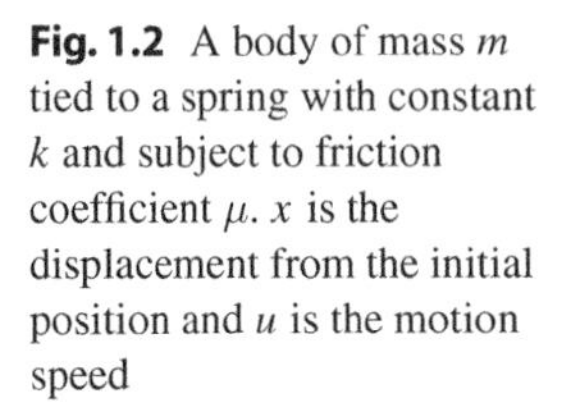
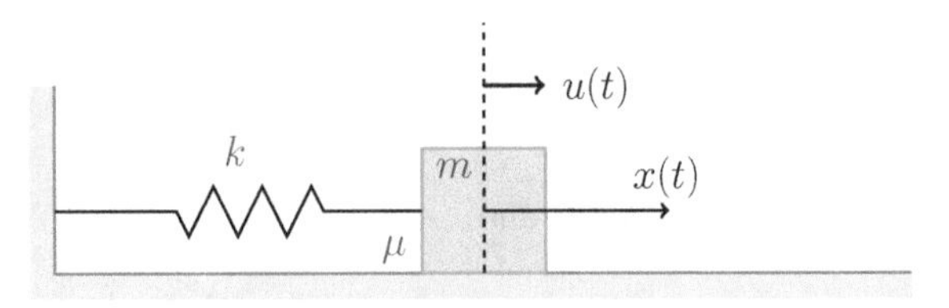

Fig. 1.2 A body of mass m tied to a spring with constant k and subject to friction coefficient μ. x is the displacement from the initial position and u is the motion speed

(vii) In general, a force field $\mathbf{f}(\mathbf{r})$ that is constant in time can be decomposed into its conservative ($\mathbf{f}_c$) and non-conservative ($\mathbf{f}_n$) parts $\mathbf{f} = \mathbf{f}_c + \mathbf{f}_n$. What would be the equation for total energy in this case? Is total energy conserved in the general case of non-conservative forces being present?

(a) If the non-conservative force component $\mathbf{f}_n$ is a *frictional force*, (assuming a linearly parameterized case) what would be its direction and magnitude? Is total energy conserved in this case? What happens to the total (kinetic + potential) energy?

(b) If the non-conservative force component $\mathbf{f}_n$ would be the *Coriolis force*, in what direction would this force act? Is total energy conserved in this case?

Exercise 4

Consider a spring-mass mechanical system comprised of a body with mass m, restrained by a spring with *spring constant* k as shown in Fig. 1.2.

The linear spring force acting on the body when it is displaced by a distance $\mathbf{r}$ from its equilibrium position is given by

$$\mathbf{f} = -k\mathbf{r}.$$

Note that, since the motion is only in one direction, vector variables can be replaced by scalar ones, setting u for velocity component, x for distance, and

$$f = -kx$$

for the force in x-direction.

(i) Is the spring force conservative?

(ii) How can we write the equation of motion (Newton's Second Law) for the system?

Obtain solutions to the equation of motion with initial conditions given as

$$x(t = 0) = 0; \quad u(t = 0) = u_0$$

What type of motion is created? Plot the solution $x(t)$ as a function of time. For how long is the motion sustained?

What is the maximum displacements $\pm x_m$ of the mass from equilibrium? What is the velocity at these maximum displacements?

(iii) What is the total energy of the system? What is the kinetic energy? What is the potential energy? Is total energy conserved? How do these energy components vary as a function of time t and as a function of position x?

`(iv)` If a friction force is added to the system in addition to the spring force, we can write the total force acting on the body as

$$f = -kx - \mu u = -kx - \mu \frac{dx}{dt}.$$

Is the force field in this case conservative? Interpret what happens when both force components are present. What is the equation of motion? What is the possible solution with the same initial conditions as in part `(ii)`? Plot and describe the motion. Is total energy conserved? Plot and describe variations of kinetic and potential energy components with respect to time, t.

2 Fluid Properties and Kinematics

2.1 Definition of a Fluid

A strict definition of a fluid (gas or liquid) is a difficult one, since the distinction between solids and fluids is not so sharp one. Often a solid (such as plastics) may act like a fluid after applying a strong force for a sufficient time, or a liquid (such as paint) may solidify if it is allowed to stand for a time. Some polymer solutions can simultaneously exhibit solid-like and fluid-like behavior. In general, a fluid is defined by the relative ease with which it may be deformed, and by the fact that it does not have a preferred shape.

A *simple fluid* is defined as a substance which cannot withstand any tendency by applied forces to deform it in any way which leaves the volume unchanged. Since most fluids such as water and air luckily behave in this manner, simple fluids will be considered in the present context.

More specifically, the distinction between solids and fluids arise from their mechanical behavior which is determined by the relation between stress (applied force per unit area) and strain (proportional deformation). For a solid, strain is a function of stress; and in the case of perfectly elastic solids, a linear relationship exists (e.g. Hooke's Law). As compared to solids, fluids can be deformed indefinitely under an applied force, no matter how small the force may be, as long as the force applied creates deformations other than a simple a change in volume. In other words, that part of stress which does not directly lead to a volumetric change is proportional to the time rate of strain.

2.2 Continuum Hypothesis

Since a fluid is made up of molecules, a full description would depend on the dynamics of each molecule. However a description at this level is quite difficult since this would mean a wildly non-uniform distribution of properties. The theory of fluid

© Springer Nature Switzerland AG 2020

E. Özsoy, *Geophysical Fluid Dynamics I*, Springer Textbooks in Earth Sciences, Geography and Environment, https://doi.org/10.1007/978-3-030-16973-2_2

dynamics is constructed on the *continuum hypothesis* which assumes that *locally* the fluid forms a continuum (i.e. are uniform mass distribution) with in an infinitesimal volume which is still large compared to the molecular scale. This allows a macroscopic description of the fluids and the development of equations that are independent of the molecular structure.

2.3 Properties of Fluids

A full description of fluid properties requires utilization of the theory of thermodynamics. Since we need to consider incompressible, homogeneous fluids in this introduction, we do not need to review thermodynamics at present. A review of thermodynamics will eventually become necessary in the subsequent stratified fluids volume. Mechanical properties such as stress and rate of deformation will be studied within the present context.

2.4 Volume and Surface Forces

It is possible to classify forces acting on a fluid parcel of Fig. 2.1 into two groups:

Volume forces (or *body forces*) penetrate into the interior and act on distributed elements of fluid. Examples are gravity and electromagnetic forces (in conducting fluids). The body force acting on a volume element δV surrounding a point $\mathbf{r}$ at time t is $\mathbf{F}(\mathbf{r}, t)\rho\delta V$ where $\mathbf{F}$ is a body force per unit mass and ρ the density. In the case of gravity $\mathbf{F} = \mathbf{g}$ is the gravity force per unit mass.

Surface forces are forces of direct molecular origin and therefore decrease extremely rapidly with distance between interacting elements. Therefore the penetration depth of surface forces even in an infinitesimal volume element of fluid is small, since this volume is still large compared to the molecular scale, (continuum hypothesis). Since the penetration depth is small, these forces can be considered to apply on the surfaces of a fluid element. The total force exerted across the element is proportional to its area δA and is given by $\mathbf{\Sigma}(\mathbf{n}, \mathbf{r}, t)\delta A$ where $\mathbf{\Sigma}$ is the local *stress* (surface force per unit area) and $\mathbf{n}$ the unit normal to the surface. The surface force on the other side of the area element is $\mathbf{\Sigma}(-\mathbf{n}, \mathbf{r}, t) = -\mathbf{\Sigma}(\mathbf{n}, \mathbf{r}, t)$ since the forces on the two sides must be balanced, therefore $\mathbf{\Sigma}$ is an odd function of $\mathbf{n}$.

Consider surface forces acting instantaneously on a material element δV in the shape of a tetrahedron in Fig. 2.2, where the three perpendicular surfaces with unit vectors $\mathbf{a}, \mathbf{b}, \mathbf{c}$ are defined in the planes of an orthogonal (cartesian) coordinate system.

The sum of the surface forces acting on the tetrahedral element of volume δV is,

$$\mathbf{S} = \mathbf{\Sigma}(\mathbf{n})\delta A + \mathbf{\Sigma}(-\mathbf{a})\delta A_1 + \mathbf{\Sigma}(-\mathbf{b})\delta A_2 + \mathbf{\Sigma}(-\mathbf{c})\delta A_3$$

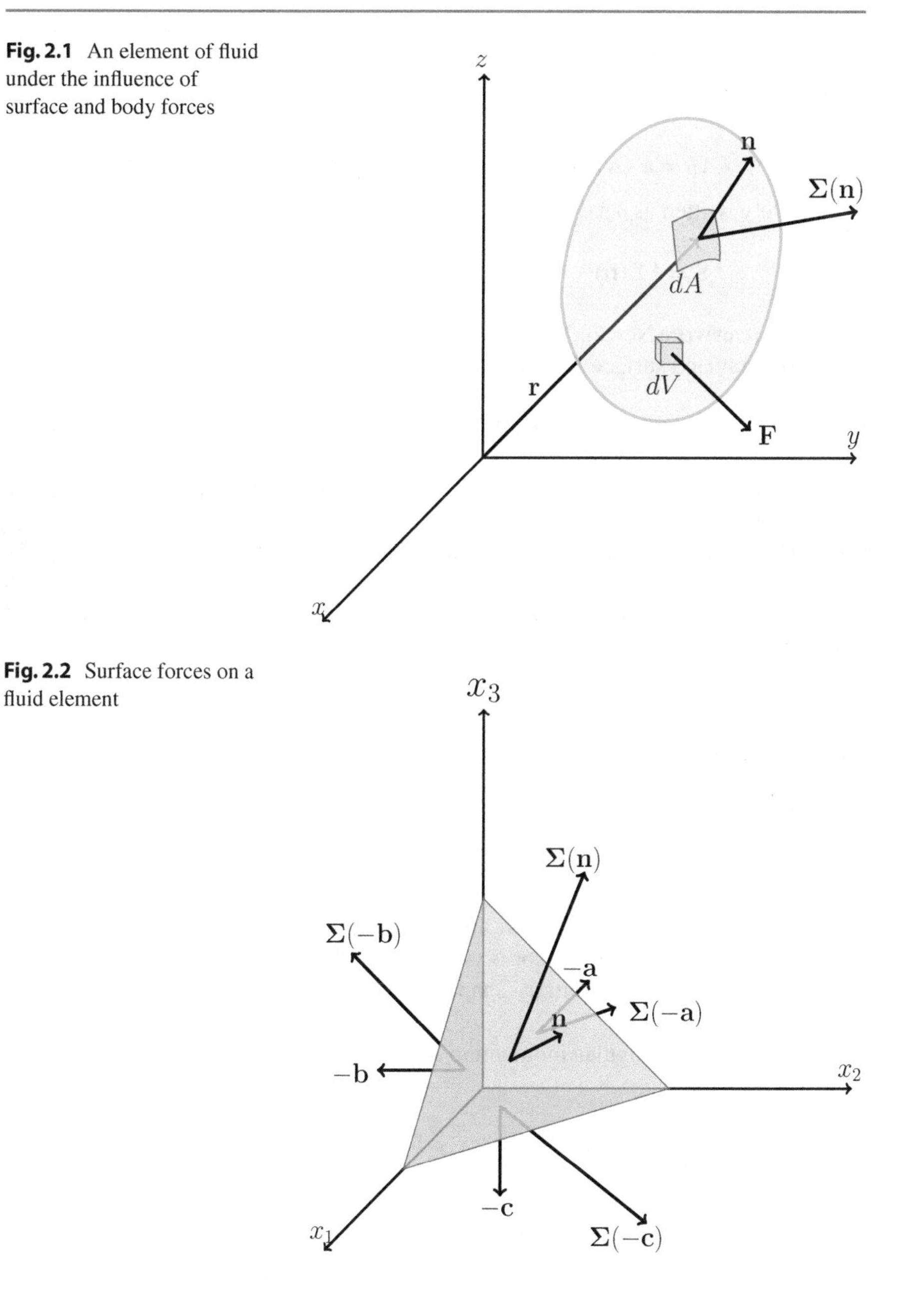

Fig. 2.1 An element of fluid under the influence of surface and body forces

Fig. 2.2 Surface forces on a fluid element

But since the surfaces $\delta A_1, \delta A_2, \delta A_3$ are projections of δA on the corresponding planes

$$\delta A_1 = \mathbf{a} \cdot \mathbf{n}\, \delta A, \quad \delta A_2 = \mathbf{b} \cdot \mathbf{n}\, \delta A \quad \text{and} \quad \delta A_3 = \mathbf{c} \cdot \mathbf{n}\, \delta A,$$

alternatively written as $\delta A_1 = a_j n_j \delta A$, etc. It follows that

$$\mathbf{S} = \left\{ \boldsymbol{\Sigma}(\mathbf{n}) - \left[a_j \boldsymbol{\Sigma}(\mathbf{a}) + b_j \boldsymbol{\Sigma}(\mathbf{b}) + c_j \boldsymbol{\Sigma}(\mathbf{c}) \right] n_j \right\} \delta A \tag{2.1}$$

Now we can write Newton's second law for this fluid element (mass × acceleration = total of body and surface forces) as

$$\rho \delta V \mathbf{a} = \rho \mathbf{F} \delta V + \mathbf{S} \tag{2.2}$$

where $\rho \delta V$ is the elemental mass, $\mathbf{a}$ the acceleration of the element. If the linear dimensions of the tetrahedron are made to shrink to zero without change of shape, the first two terms of (2.2) approach zero as δV and the last term, given by Eq. (2.1) approaches zero as δA. Since δV decreases to zero more rapidly then δA, in the limit it must be required that $\mathbf{S} = 0$ and the coefficient of δA in (2.1) must vanish identically, requiring components of the surface stress to satisfy

$$\Sigma_i(\mathbf{n}) = [a_j \Sigma_i(\mathbf{a}) + b_j \Sigma_i(\mathbf{b}) + c_j \Sigma_i(\mathbf{c})] n_j \tag{2.3}$$

The quantity in brackets is a tensor which can be represented as σ_{ij}, yielding

$$\Sigma_i(\mathbf{n}) = \sigma_{ij} n_j = \boldsymbol{\sigma} \cdot \mathbf{n} \tag{2.4}$$

The quantity $\boldsymbol{\sigma} = \sigma_{ij}$ is called the *stress tensor* (second order), and represents the ith-component of stress exerted on a surface element normal to the jth-direction. We can show that the stress tensor is symmetrical. For this purpose we consider the moments of forces on an arbitrary shaped fluid volume V, enclosed by a surface A shown below:

The equation for angular momentum for the fluid element can be derived from (2.2):

$$\int_V \rho \mathbf{r} \times \mathbf{a}\, dV = \int_V \rho \mathbf{r} \times \mathbf{F}\, dV + \int_A \mathbf{r} \times \boldsymbol{\Sigma}(\mathbf{n})\, dA \tag{2.5}$$

where $\mathbf{r}$ is the position vector and the last term

$$\mathbf{M} = \int_A \mathbf{r} \times \boldsymbol{\Sigma}(\mathbf{n})\, dA \tag{2.6}$$

represents the total moment created by the surface forces. By virtue of (1.9) and (2.4) the i-th component of this moment can be written as

$$M_i = \int_A \epsilon_{ijk} r_j \Sigma_k\, dA = \int_A \epsilon_{ijk}\, r_j\, \sigma_{kl}\, n_l\, dA \tag{2.7}$$

By making use of the divergence theorem (1.35) and denoting $P_{il} = \epsilon_{ijk} r_j \sigma_{kl}$, the surface integral (2.7) can be written as a volume integral

$$\begin{aligned} M_i &= \int_S P_{il} n_l \, dA = \int_S \mathbf{P} \cdot \mathbf{n} \, dA = \int_V \nabla \cdot \mathbf{P} \, dV \\ &= \int_V \frac{\partial}{\partial r_l} \epsilon_{ijk} r_j \sigma_{kl} \, dV = \int_V \epsilon_{ijk} \left(\sigma_{kj} + r_j \frac{\partial \sigma_{kl}}{\partial r_l} \right) dV \end{aligned} \tag{2.8}$$

Now, if the volume V is made arbitrarily small ($V \to 0$) it is observed that the first term of (2.8) approaches zero as $V = L^3$ (L = linear dimension), whereas the second term approaches zero as $V^{4/3} = L^4$. Likewise, the inertia and body force terms (l.h.s. and the first term on r.h.s) of (2.5) approach zero as $V^{4/3} = L^4$. Therefore the first term of (2.8) must be identically equal to zero:

$$\int_V \epsilon_{ijk} \sigma_{kj} \, dV = 0 \tag{2.9}$$

Since the volume element is arbitrary, the integrand must be zero ; $\epsilon_{ijk} \sigma_{kj} = 0$. The above result proves that the stress tensor is symmetrical, i.e.

$$\sigma_{ij} = \sigma_{ji} \tag{2.10}$$

and therefore only 6 of the 9 components of the stress tensor are independent of each other.

The three diagonal components of the stress tensor (σ_{ij}, $i = j$) are called *normal stresses*, since they represent the ith-component of surface force on the surface whose normal is in the i-direction. The off-diagonal components of the tensor (σ_{ij}, $i \neq j$) are called *tangential (or shear) stresses*, since they represent the ith-component of surface force on the elemental surface whose normal points in the j-direction.

It is always possible to choose the reference coordinates such that the off-diagonal elements of a symmetrical second order tensor are made to vanish (ref. Sect. 1.1). The stress tensor referred to these *principal coordinates* has only diagonal components $\sigma'_{11}, \sigma'_{22}, \sigma'_{33}$, called *principal stresses*.

With the coordinate transformation, the stress tensor in principal coordinates is modified such that

$$\begin{pmatrix} \sigma_{11} & \sigma_{12} & \sigma_{13} \\ \sigma_{21} & \sigma_{22} & \sigma_{23} \\ \sigma_{31} & \sigma_{32} & \sigma_{33} \end{pmatrix} \to \begin{pmatrix} \sigma'_{11} & 0 & 0 \\ 0 & \sigma'_{22} & 0 \\ 0 & 0 & \sigma'_{33} \end{pmatrix} \tag{2.11}$$

Therefore the general state of a fluid can be described by the three principal stresses σ'_{ii} on the diagonal of the transformed matrix, where σ'_{11} represents tension (compression if negative valued) in the x_1 direction; similarly for σ'_{22} and σ'_{33}.

It is a well known property of second order tensors that the sum of the diagonal components is an invariant, i.e. it remains unchanged with coordinate transformations,so that

$$\sigma_{ii} = \sigma_{11} + \sigma_{22} + \sigma_{33} = \sigma'_{11} + \sigma'_{22} + \sigma'_{33} \tag{2.12}$$

The force per unit area on any surface with normal $\mathbf{n} = (n'_1, n'_2, n'_3)$ is

$$\boldsymbol{\Sigma} = (\sigma'_{11}n'_1,\ \sigma'_{22}n'_2,\ \sigma'_{33}n'_3)$$

2.4.1 Stress in a Fluid that Is at Rest

A fluid was defined as being unable to withstand applied force tendency to deform it without any change in volume (Sect. 2.1). This definition has consequences for the stress tensor. Consider a small spherical element of fluid with the reference coordinates coinciding with the principal axes. Adding and subtracting the quantity $\sigma_{ii} = \sigma_{11} + \sigma_{22} + \sigma_{33}$ to the trace of the stress tensor in principal coordinates, we can write the stress tensor as

$$\begin{bmatrix} \sigma'_{11} & 0 & 0 \\ 0 & \sigma'_{22} & 0 \\ 0 & 0 & \sigma'_{33} \end{bmatrix} = \begin{bmatrix} \frac{1}{3}\sigma_{ii} & 0 & 0 \\ 0 & \frac{1}{3}\sigma_{ii} & 0 \\ 0 & 0 & \frac{1}{3}\sigma_{ii} \end{bmatrix} + \begin{bmatrix} (\sigma'_{11} - \frac{1}{3}\sigma_{ii}) & 0 & 0 \\ 0 & (\sigma'_{22} - \frac{1}{3}\sigma_{ii}) & 0 \\ 0 & 0 & (\sigma'_{33} - \frac{1}{3}\sigma_{ii}) \end{bmatrix} \tag{2.13}$$

i.e. a superposition of an isotropic tensor (first term) and an anisotropic tensor (second term) representing the departure from isotropy. The first tensor is isotropic since it is equal to $(\frac{1}{3}\sigma_{ii})\delta_{kl}$, and remains the same under all transformations. The corresponding contribution to the force per unit area of the sphere at a point with normal $\mathbf{n}$ is $\frac{1}{3}\sigma_{ii}\mathbf{n}$. The second term is anisotopic, and in view of (2.12), the sum of its diagonal components (its trace) is zero. Each component of this anisotropic tensor represents uniform tension or compression in one of the principal directions. Since the sum of these stresses is zero it tends to deform the fluid element without change in volume (pure tension in two of the principal directions must be balanced by compression in the third direction) and an initial spherical element would therefore deform into an ellipsoid. By definition, a fluid cannot withstand such tendencies, and the resulting deformation would contradict with the *state of rest*; therefore the second component of the tensor in (2.13) must vanish. For a fluid that is at rest, each component of the principal stresses is therefore equal to $\frac{1}{3}\sigma_{ii}$, i.e. the stress tensor is isotropic, and only normal stresses exist. Fluids at rest are normally in a state of compression so that it is convenient to write

$$\sigma'_{ij} = -p\delta_{ij} \tag{2.14}$$

where $p = -\frac{1}{3}\sigma_{ii}$ is called the *static fluid pressure.*

At any point in the fluid at rest, the surface force per unit area across a plane surface is equal to $-p\mathbf{n}$ and remains the same in any direction $\mathbf{n}$.

Finally, since the anisotropic part vanishes, the tangential (off-diagonal) stresses are zero in any arbitrary coordinates of reference.

2.4.2 Stress in a Moving Fluid

In a moving fluid, the tangential stresses are non-zero and can be made to vanish only when transformed to principal coordinates. The force per unit area acting on a surface depends on the direction of its normal vector $\mathbf{n}$. We have seem that in a fluid at rest, *static fluid pressure* p is the only force acting normal to any surface and its value remains unchanged with direction. In moving fluids it is still useful to define a similar component of stress although the *pressure* p in a moving fluid, is not the same as that in a static fluid. This can be defined as the average of the normal stresses for any orthogonal set of axes,

$$p = -\frac{1}{3}\sigma_{ii} \tag{2.15}$$

and is still invariant with respect to coordinate transformation (cf. 2.11). The pressure is a measure of the squeezing of the fluid. As a result of above discussions,it is convenient to express the stress tensor as the sum of isotropic and non-isotropic parts,

$$\sigma_{ij} = -p\delta_{ij} + d_{ij} \tag{2.16}$$

The first term represents the isotropic normal stresses, which exist in the same form as in static fluids (although its value is different). The second term represents the non-isotropic part of the stress tensor and contributes to tangential stresses as well as normal stresses (summing up to zero). The second term d_{ij} is called the *deviatoric stress tensor* and has the property of being entirely due to the motion of the fluid.

2.5 Specification of the Flow Field

Two different kinds of specification of the flow of a fluid is possible:

Eulerian Specification: Flow quantities are described as functions of $\mathbf{x}$ (position vector) and t (time). The primary flow quantity is the velocity of the fluid, $\mathbf{u}(\mathbf{x}, t)$. A spatial distribution of velocity, and other quantities such as pressure and density are provided at each instant of the motion.

Lagrangian Specification: Identifying a piece of matter, the flow quantities are defined as functions of time and of the choice of a material element of fluid. Conveniently, we identify the material element by its initial position $\mathbf{r}(t_0)$ at an initial instant t_0. For example, the velocity is specified as $\mathbf{u}(\mathbf{r}(t_0), t)$.

The Lagrangian specification often leads to complicated analyses, and often the Eulerian specification is more direct. At present, mainly the Eulerian specification will be used, although the concept of material elements will often be used to construct the Eulerian theory.

The following concepts will be useful in the description of fluids:

Streamline: A line in the fluid whose tangent is everywhere parallel to the velocity vector $\mathbf{u}(\mathbf{x}, t)$ at any instant is called a streamline. Clearly, a length increment $d\mathbf{r}$ along the streamline must satisfy

$$\mathbf{u} \times d\mathbf{r} = 0, \tag{2.17}$$

where $\mathbf{u} = (u, v, w)$, $d\mathbf{r} = (dx, dy, dz)$ referred to a rectangular coordinate system $Oxyz$. The family of streamlines satisfying (2.17) are obtained from the solutions of

$$\frac{dx}{u(\mathbf{x}, t)} = \frac{dy}{v(\mathbf{x}, t)} = \frac{dz}{w(\mathbf{x}, t)} \tag{2.18}$$

When the flow is *steady* (independent of t) the form of the streamlines is constant.

Stream-tube: A surface formed by all streamlines passing through a given closed curve in the fluid.

Path-line or trajectory: A path-line or trajectory is the line followed by a material element of fluid (fluid particle). It is defined by (ref. Sect. 1.4):

$$\frac{d\mathbf{r}}{dt} = \mathbf{u}(\mathbf{r}, t) \tag{2.19}$$

In general, the solution does not coincide with streamlines. Streamlines and path-lines coincide only in steady motion.

Streakline: A line connecting all fluid particles which at the some earlier time passed through a fixed point in space is called a streakline. For example, if a dye is introduced into a fluid at a fixed point, the line formed by the dye is a streakline. In the case of steady flow, streaklines also coincide with streamlines and pathlines.

2.6 Conservation of Mass (Continuity Equation)

2.6.1 Continuity Equation

The conservation of mass can be expressed for a material volume V of a fluid as (using single integral signs hereafter):

$$\frac{d}{dt}\int_V \rho dV = 0 \tag{2.20}$$

with ρ denoting density (mass per unit volume), through the use of Leibnitz' rule (1.43) and denoting $v_n = \mathbf{v} \cdot \mathbf{n} = \mathbf{u} \cdot \mathbf{n}$. Since the normal velocity at the material surface S enclosing volume V have to be equal to the fluid velocity (cf. definition of material surface, Sect. 1.4), Eq. (2.20) takes the form

$$\int_V \frac{\partial \rho}{\partial t} dV + \int_S \rho \mathbf{u} \cdot \mathbf{n} dS = 0 \tag{2.21}$$

On the other hand, the divergence theorem (1.30) can be used to rewrite the second term as a volume integral:

$$\int_V \left(\frac{\partial \rho}{\partial t} + \nabla \cdot \rho \mathbf{u} \right) dV = 0 \tag{2.22}$$

Since the initial choice of the material volume V is arbitrary, the statement (2.22) is true for any and all such volumes V. Therefore the integrand itself must satisfy

$$\frac{\partial \rho}{\partial t} + \nabla \cdot \rho \mathbf{u} = 0 \tag{2.23a}$$

which is the continuity equation expressing the mass conservation. Note that this is purely a kinematic statement and does not restrict the balance of forces. An alternative form can be written by making use of Eqs. (1.27a) and (1.37):

$$\frac{D\rho}{Dt} + \rho \nabla \cdot \mathbf{u} = 0 \tag{2.23b}$$

The continuity equation is more readily interpreted in this form. Consider the material volume V, whose capacity changes as a result of the movement of each surface element $\mathbf{n}\delta S$ ($\mathbf{n}$-outward normal) of the bounding material surface S:

$$\frac{dV}{dt} = \int_S \mathbf{u} \cdot \mathbf{n} dS = \int_V \nabla \cdot \mathbf{u} dV \tag{2.24}$$

where the last equality is obtained through the divergence theorem (1.30). Making the material volume V approach zero, the fractional change in volume in the limit is

$$\lim_{V \to 0} \frac{1}{V} \frac{dV}{dt} = \lim_{V \to 0} \frac{1}{V} \int_V \nabla \cdot \mathbf{u} \, dV = \nabla \cdot \mathbf{u} \tag{2.25}$$

and by virtue of (2.23b), the *rate of dilatation* (or the fractional *rate of expansion*) Δ is defined as

$$\Delta \equiv \nabla \cdot \mathbf{u} = \lim_{V \to 0} \frac{1}{V} \frac{dV}{dt} = -\frac{1}{\rho} \frac{D\rho}{Dt}. \tag{2.26}$$

Showing that the value change is balanced by the (negative) fractional rate of change of density within the material volume.

Definition: Incompressible Fluids

A fluid is said to be incompressible if the density of the fluid is not changed by pressure. For example, water is nearly incompressible; and for most practical purposes, it can be treated as incompressible. For an incompressible fluid the density of a material element should remain constant, so that we require

$$\frac{D\rho}{Dt} = \frac{\partial \rho}{\partial t} + \mathbf{u} \cdot \nabla \rho = 0 \tag{2.27a}$$

which implies using (2.23b),

$$\nabla \cdot \mathbf{u} = 0 \tag{2.27b}$$

i.e. the rate of expansion is zero everywhere in the fluid. Either of the Eqs. (2.27a and 2.27b) express the incompressibility condition. The first Eq. (2.27a) implies that the fluid need not be homogeneous (ρ =constant), and can support local or spatial variations in density as long as the material derivative is zero. The second Eq. (2.27b) implies that the velocity field **u** is *solenoidal* (zero divergence of velocity).

2.6.2 Stream Function–2-D, Incompressible, Steady Flow

Assuming an incompressible, steady flow to take place in two dimensions, $\mathbf{u} = (u, v, 0)$ and the flow field is independent of z, the continuity equation is

$$\frac{\partial u}{\partial x} + \frac{\partial v}{\partial y} = 0 \tag{2.28}$$

The from of Eq. (2.28) implies that the exact differential for stream function ψ satisfies

$$d\psi = udy - vdx = 0 \tag{2.29}$$

such that

$$u = \frac{\partial \psi}{dy}, \quad v = -\frac{\partial \psi}{\partial x} \tag{2.30}$$

Substitution of (2.30) into (2.28) shows that the *stream function* ψ satisfies the continuity equation automatically and from (2.29) its value at any point $\mathbf{x}$ =(x,y) can be obtained as

$$\psi - \psi_0 = \int_{(x_0, y_0)}^{(x, y)} (udy - vdx) \tag{2.31}$$

where ψ_0 is the value of ψ at $\mathbf{x}_0 = (x_0, y_0)$. Equation (2.30) can alternatively be written as

$$\mathbf{u} = \nabla \times \mathbf{B}, \quad \mathbf{B} = (0, 0, \psi) = \psi \mathbf{k} \tag{2.32}$$

where **B** is called a vector potential, and **k** is the unit vector in z-direction. Equivalently, using (1.27b) or (2.30) we have

$$\mathbf{u} = \nabla \psi \times \mathbf{k} = -\mathbf{k} \times \nabla \psi \tag{2.33}$$

It can be seen from (2.33) that the velocity is perpendicular to the gradient of the streamfunction, and therefore parallel to the streamlines (lines of constant ψ), as shown in Fig. 2.3.

The volume flux (discharge) per unit distance in the z-direction between any two points in the (x, y) plane is defined as follows:

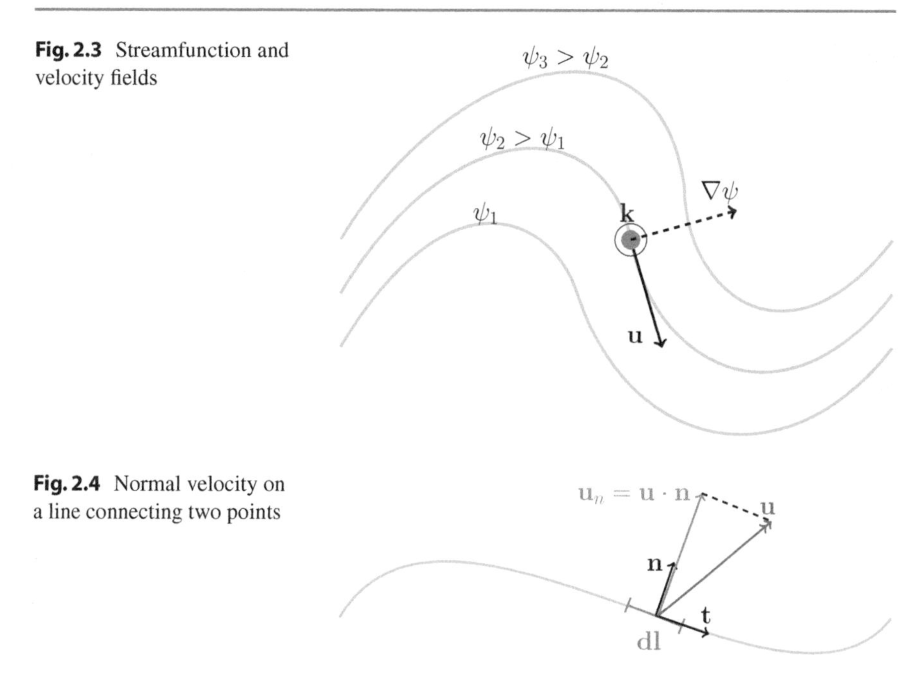

Fig. 2.3 Streamfunction and velocity fields

Fig. 2.4 Normal velocity on a line connecting two points

Let $u_n = \mathbf{u} \cdot \mathbf{n}$ be the normal velocity at any point on the line connecting the two points, $\mathbf{n}$ be the local unit normal vector to this line, $\mathbf{t}$ be the local unit tangent vector, dl be an element of distance along the same line, as in Fig. 2.4.

Then the volume flux Q between (x_0, y_0) and (x, y) can be expressed as

$$Q = \int_{(x_0,y_0)}^{(x,y)} \mathbf{u} \cdot \mathbf{n}\, dl = \int_{(x_0,y_0)}^{(x,y)} u_n\, dl \tag{2.34}$$

Substituting (2.33) and using (1.11a):

$$Q = \int_{(x_0,y_0)}^{(x,y)} (-\mathbf{k} \times \nabla\psi) \cdot \mathbf{n}\, dl = \int_{(x_0,y_0)}^{(x,y)} (\mathbf{k} \times \mathbf{n}) \cdot \nabla\psi\, dl \tag{2.35}$$

where $\mathbf{k} \times \mathbf{n} = \mathbf{t}$ gives the tangential unit vector, so that

$$\begin{aligned} Q &= \int_{(x_0,y_0)}^{(x,y)} \mathbf{t} \cdot \nabla\psi\, dl = \int_{(x_0,y_0)}^{(x,y)} \frac{\partial\psi}{\partial l}\, dl \\ &= \int_{(x_0,y_0)}^{(x,y)} d\psi = \psi - \psi_0 = \int_{(x_0,y_0)}^{(x,y)} (udy - vdx) \end{aligned} \tag{2.36}$$

by virtue of (2.31). Therefore the difference of stream function values between any two points is equal to the volume flux. It can be shown that the calculation of the volume flux is independent of the path integration.

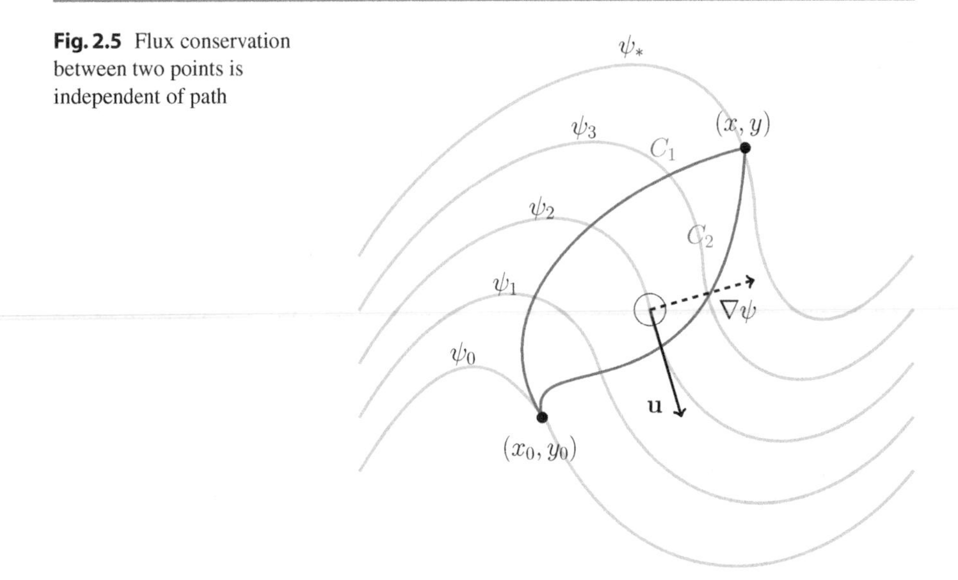

Fig. 2.5 Flux conservation between two points is independent of path

Consider two separate paths C_1 and C_2 connecting two points: We calculate the volume fluxes for each one from

$$Q_1 = \int_{C_1} \mathbf{u} \cdot \mathbf{n} \, dl, \quad Q_2 = \int_{C_2} \mathbf{u} \cdot \mathbf{n} \, dl.$$

Because the two curves make up a closed curve $C = C_1 + C_2$, the total volume flux through the closed is zero by virtue of (2.36):

$$Q_c = \oint_C \mathbf{u} \cdot \mathbf{n} \, dl = \oint_C d\psi = 0, \tag{2.37}$$

since the integration of an exact differential around a closed curve yields zero. Breaking the integral to two parts and integrating in the direction indicated by arrows in Fig. 2.5

$$Q_c = \int_{C_1} d\psi - \int_{C_2} d\psi = Q_1 - Q_2 = 0 \, ,$$

so that $Q_1 = Q_2$, i.e. the integral is independent of the path.

If we draw streamlines for a certain flow, the velocity at each point can be calculated from (2.30) or visually by dividing the volume flux (given by difference of streamfunction values) by the distance between streamlines.

2.7 Analysis of Relative Motion Near a Point

As a preliminary to dynamical considerations, it is necessary to make an analysis of the character of motion in the neighborhood of any point. This will show how fluid elements are deformed in relation to nearby fluid elements.

Let the velocity at time t and position $\mathbf{x}$ to be given by $\mathbf{u}(\mathbf{x}, t)$. Simultaneously at a neighboring point $\mathbf{x} + \mathbf{r}$, the velocity $\mathbf{u}(\mathbf{x} + \mathbf{r}, t)$ is such that

$$\mathbf{u}(\mathbf{x} + \mathbf{r}, t) = \mathbf{u}(\mathbf{x}, t) + \delta\mathbf{u}, \tag{2.38}$$

where for rectangular coordinates, $\delta\mathbf{u}$ can be approximated through Taylor series expansion. Keeping only first order terms in $\mathbf{r}$, the distance between points, we obtain

$$\delta\mathbf{u} = (\mathbf{r} \cdot \nabla)\mathbf{u}, \tag{2.39}$$

or equivalently

$$\delta u_i = r_j \frac{\partial u_i}{\partial x_j}. \tag{2.40}$$

Here, $\frac{\partial u_i}{\partial x_j}$ is a second order tensor, which can be decomposed into symmetrical and anti-symmetrical parts as follows:

$$\frac{\partial u_i}{\partial x_j} = \frac{1}{2}\left(\frac{\partial u_i}{\partial x_j} + \frac{\partial u_j}{\partial x_i}\right) + \frac{1}{2}\left(\frac{\partial u_i}{\partial x_j} - \frac{\partial u_j}{\partial x_i}\right) \tag{2.41}$$

(cf. Sect. 1.1). The symmetrical tensor is

$$e_{ij} = \frac{1}{2}\left(\frac{\partial u_i}{\partial x_j} + \frac{\partial u_j}{\partial x_i}\right) = \begin{bmatrix} \frac{\partial u_1}{\partial x_1} & \frac{1}{2}\left(\frac{\partial u_1}{\partial x_2} + \frac{\partial u_2}{\partial x_1}\right) & \frac{1}{2}\left(\frac{\partial u_1}{\partial x_3} + \frac{\partial u_3}{\partial x_1}\right) \\ \frac{1}{2}\left(\frac{\partial u_2}{\partial x_1} + \frac{\partial u_1}{\partial x_2}\right) & \frac{\partial u_2}{\partial x_2} & \frac{1}{2}\left(\frac{\partial u_2}{\partial x_3} + \frac{\partial u_3}{\partial x_2}\right) \\ \frac{1}{2}\left(\frac{\partial u_3}{\partial x_1} + \frac{\partial u_1}{\partial x_3}\right) & \frac{1}{2}\left(\frac{\partial u_3}{\partial x_2} + \frac{\partial u_2}{\partial x_3}\right) & \frac{\partial u_3}{\partial x_3} \end{bmatrix} \tag{2.42a}$$

and the anti-symmetrical tensor is given as

$$\zeta_{ij} = \frac{1}{2}\left(\frac{\partial u_i}{\partial x_j} - \frac{\partial u_j}{\partial x_i}\right) = \begin{bmatrix} 0 & \frac{1}{2}\left(\frac{\partial u_1}{\partial x_2} - \frac{\partial u_2}{\partial x_1}\right) & \frac{1}{2}\left(\frac{\partial u_1}{\partial x_3} - \frac{\partial u_3}{\partial x_1}\right) \\ -\frac{1}{2}\left(\frac{\partial u_1}{\partial x_2} - \frac{\partial u_2}{\partial x_1}\right) & 0 & \frac{1}{2}\left(\frac{\partial u_2}{\partial x_3} - \frac{\partial u_3}{\partial x_2}\right) \\ -\frac{1}{2}\left(\frac{\partial u_1}{\partial x_3} - \frac{\partial u_3}{\partial x_1}\right) & -\frac{1}{2}\left(\frac{\partial u_2}{\partial u_3} - \frac{\partial u_3}{\partial x_2}\right) & 0 \end{bmatrix} \tag{2.42b}$$

As a result, (2.40) is written as

$$\delta u_i = \delta u_i^{(s)} + \delta u_i^{(a)} \tag{2.43a}$$

$$\text{where } \delta u_i^{(s)} = r_j e_{ij} \text{ and } \delta u_i^{(a)} = r_j \zeta_{ij}\,. \tag{2.43b}$$

2.7.1 Pure-Straining Motion

The first term of (2.43a) can also be written as

$$\delta u_i^{(s)} = r_j e_{ij} = \frac{\partial \phi}{\partial r_i} = \nabla \phi \tag{2.44}$$

where

$$\phi = \frac{1}{2} r_k r_l e_{kl} \tag{2.45}$$

To verify (2.44), we substitute from (2.45), to yield

$$\begin{aligned}
\frac{\partial \phi}{\partial r_i} = \frac{\partial \frac{1}{2} r_k r_l e_{kl}}{\partial r_i} &= \frac{1}{2}\frac{\partial r_k}{\partial r_i} r_l e_{kl} + \frac{1}{2} r_k \frac{\partial r_l}{\partial r_i} e_{kl} \\
&= \frac{1}{2}\delta_{ik} r_l e_{kl} + \frac{1}{2} r_k \delta_{il} e_{kl} \\
&= \frac{1}{2} r_l e_{il} + \frac{1}{2} r_k e_{ki} \\
&= \frac{1}{2}(r_l e_{il} + r_k e_{ik}) \\
&= r_j e_{ij}
\end{aligned}$$

Note that the symetricity of e_{ij} has been used above.

The function ϕ is a function of $\mathbf{r}$ as seen from Eq. (2.45). The surfaces on which ϕ is constant form a family of quadrics:

$$\frac{1}{2} r_k r_l e_{kl} = \frac{1}{2} e_{11} r_1^2 + \frac{1}{2} e_{22} r_2^2 + \frac{1}{2} e_{33} r_3^2 + e_{12} r_1 r_2 + e_{13} r_1 r_3 + e_{23} r_2 r_3 = C \tag{2.46}$$

where C is constant.

Regarding e as locally constant, this equation defines a surface which is an ellipsoid. It is observed that the symmetric contribution to the relative velocity $\delta \mathbf{u}^{(s)}$ is thus (cf. Eq. 2.44):

$$\delta \mathbf{u}^{(s)} = \nabla \phi = |\nabla \phi|\, \mathbf{n} \tag{2.47}$$

by virtue of Eq. (1.40); i.e. this term is proportional to the gradient of the surface and *in the direction of the normal* $\mathbf{n}$ *to the surface*. The nature of $\delta \mathbf{u}^{(s)}$ becomes clear if we choose the reference coordinates to coincide with the principal axes of e. Since e is symmetric, it is always possible to make it diagonal by transforming into principal coordinates. Let $\mathbf{r}$ in these new coordinates be $r' = (r_1', r_2', r_3')$. Since off-diagonal components are zero, (2.45) takes the form

$$\phi = \frac{1}{2}\left(e_{11}' r_1'^2 + e_{22}' r_2'^2 + e_{33}' r_3'^2\right) \tag{2.48}$$

where e'_{11}, e'_{22} and e'_{33} are the diagonal components of the tensor e'_{ij} defined with respect to principal coordinates, transformed according to the formula

$$e'_{ij} = \frac{\partial r_k}{\partial r'_i}\frac{\partial r_l}{\partial r'_j} e_{kl}, \tag{2.49}$$

and $\partial r_m/\partial r'_n$ are the direction cosines of axes, which are used in the transformation.

The components of the transformed tensor e'_{ij} satisfy the invariance relation

$$\begin{aligned} e'_{11} + e'_{22} + e'_{33} &= e'_{ii} = e_{11} + e_{22} + e_{33} \\ &= e_{ii} = \frac{\partial u_i}{\partial x_i} = \nabla \cdot \mathbf{u} \equiv \Delta \end{aligned} \tag{2.50}$$

by virtue of (2.42a). The contribution $\delta \mathbf{u}^{(s)}$ then becomes (Eqs. 2.44 and 2.48):

$$\delta u_1^{(s)'} = e'_{11} r'_1, \ \delta u_2^{(s)'} = e'_{22} r'_2, \ \delta u_3^{(s)'} = e'_{33} r'_3 \tag{2.51}$$

with reference to the principal axes. Therefore any material line element near position $\mathbf{x}$ which is parallel to the r'_1 axis continues to have that direction and becomes stretched at the rate e'_{11}. Similarly material line elements in r'_2 and r'_3 directions are stretched ar rates of e'_{22} and e'_{33} respectively. The sum of these rates of stretching is equal to (Eq. 2.50) the total *fractional rate of expansion* or *dilatation* $\Delta = \nabla \cdot \mathbf{u}$ (ref. Eq. 2.26).

Due to above reasons the contribution $\delta \mathbf{u}^{(s)}$ is said to represent *pure straining motion*, and e is called the *rate of strain* tensor, completely determined by the *principal rates of strain* e'_{11}, e'_{22}, e'_{33} in the principal directions.

The relative velocity field $\delta \mathbf{u}^{(s)}$ can also be described as that part of velocity which converts a initially spherical material volume at position $\mathbf{x}$ into the shape of an ellipsoid without any rotation, as shown in Fig. 2.6.

For an incompressible fluid, the volume of the ellipsoid is constant (equal to the initial volume of the sphere), and by virtue of (2.50) and (2.27b)

$$e_{ii} = \nabla \cdot \mathbf{u} = 0 \tag{2.52}$$

For a compressible fluid $e_{ii} \neq 0$ and in general e_{ij} is non-isotropic. However, e_{ij} can be separated into isotropic and non-isotropic parts, such that

$$e_{ij} = \frac{1}{3} e_{ii} + \left(e_{ij} - \frac{1}{3} e_{ii} \right). \tag{2.53}$$

The first term $\frac{1}{3} e_{ii} = \frac{\Delta}{3}$ represents the isotropic rate of expansion leading only to a change in volume, and the second term represents the rate of strain without change of volume. Similarly the function ϕ can be separated into isotropic and non-isotropic parts as

$$\phi = \frac{1}{6} r_k r_k e_{ii} + \frac{1}{2} r_k r_l \left(e_{kl} - \frac{1}{3} e_{ii} \delta_{kl} \right). \tag{2.54}$$

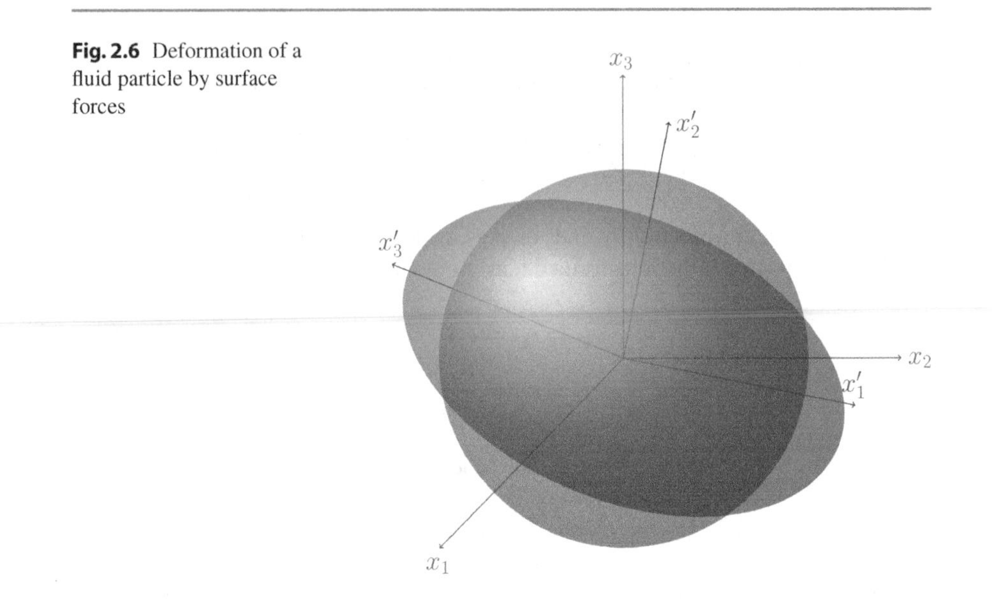

Fig. 2.6 Deformation of a fluid particle by surface forces

2.7.2 Rigid Body Rotation

Now we study the antisymmetrical contribution to relative velocity $\delta u_i{}^{(a)} = r_j \zeta_{ij}$. Defining

$$\omega_1 = \frac{\partial u_3}{\partial x_2} - \frac{\partial u_2}{\partial x_3}, \quad \omega_2 = \frac{\partial u_1}{\partial x_3} - \frac{\partial u_3}{\partial x_1}, \quad \omega_3 = \frac{\partial u_2}{\partial x_1} - \frac{\partial u_1}{\partial x_2}, \tag{2.55a}$$

as the components of the vector $\boldsymbol{\omega}$, or equivalently

$$\boldsymbol{\omega} = \nabla \times \mathbf{u}, \tag{2.55b}$$

Using Eq. (1.26) we re-write Eq. (2.42b) as

$$\zeta_{ij} = -\frac{1}{2}\epsilon_{ijk}\omega_k. \tag{2.56}$$

Then (2.43b) yields

$$\delta u_i^{(a)} = r_j \zeta_{ij} = -\frac{1}{2}\epsilon_{ijk} r_j \omega_k. \tag{2.57}$$

Equivalently (cf. Eq. 1.9), this is expected to be

$$\delta \mathbf{u}^{(a)} = -\frac{1}{2}\mathbf{r} \times \boldsymbol{\omega} = \frac{1}{2}\boldsymbol{\omega} \times \mathbf{r} \tag{2.58}$$

i.e. a velocity that would be caused by a *rigid body* rotation with an angular velocity of $\frac{1}{2}\boldsymbol{\omega}$.

The vector $\boldsymbol{\omega} = \nabla \times \mathbf{u}$ plays an important part in fluid mechanics, and is called the local *vorticity* of the fluid. Since $\frac{1}{2}\boldsymbol{\omega}$ represents rigid body rotation, fluids in which $\boldsymbol{\omega} = \nabla \times \mathbf{u} = 0$ are called *irrotational fluids.* To see why $\boldsymbol{\omega} = \nabla \times \mathbf{u}$ should represent twice the local angular velocity, consider a plane surface bounded by a circle of small radius a centered at $\mathbf{x}$ with unit normal of $\mathbf{n}$. The angular speed θ_t is calculated from the tangential velocity (averaged over the circumference) divided by the radius

$$\theta_t = \frac{\frac{1}{2\pi a}\oint \mathbf{u}\cdot d\mathbf{r}}{a} = \frac{1}{2\pi a^2}\oint \mathbf{u}\cdot d\mathbf{r}.$$

Stokes' theorem (1.37), this is equivalently written as

$$\theta_t = \frac{1}{2\pi a^2}\int_A (\nabla \times \mathbf{u})\cdot \mathbf{n}\, dA,$$

where $A = \pi a^2$ is the surface area enclosed by the circle. Since the radius of the circle is small, $\nabla \times \mathbf{u}$ may be assumed to be uniform on the surface, and it follows that

$$\theta_t \approx \frac{\pi a^2}{2\pi a^2}(\nabla \times \mathbf{u})\cdot \mathbf{n} = \frac{1}{2}(\nabla \times \mathbf{u})\cdot \mathbf{n} = \frac{1}{2}\boldsymbol{\omega}\cdot \mathbf{n} \tag{2.59}$$

i.e. the angular speed is half the normal component of vorticity.

2.7.3 Modes of Motion in a Fluid

Putting together (2.38), (2.45) and (2.57), the velocity field in the neighborhood $\mathbf{x} + \mathbf{r}$ of the point x at time t is found to consist of the following components:

$$u_i(\mathbf{x} + \mathbf{r}) = u_i(\mathbf{x}) + \frac{\partial}{\partial r_i}\left(\frac{1}{2}r_j r_k e_{jk}\right) + \frac{1}{2}\epsilon_{ijk}\omega_j r_k \tag{2.60}$$

where the time dependence has been left out, but it is be implicit in the equation. Equivalently, from (2.38), (2.47) and (2.58) we can write

$$\mathbf{u}(\mathbf{x} + \mathbf{r}) = \mathbf{u}(\mathbf{x}) + \nabla\psi + \left(\frac{1}{2}\boldsymbol{\omega}\right)\times \mathbf{r} \tag{2.61}$$

Each of the terms in Eqs. (2.60) or (2.61) represent a particular mode of motion:
(i) the first term is a *uniform translation* with velocity $\mathbf{u}(\mathbf{x})$.
(ii) the second term represents *pure straining motion* characterized by the rate of strain tensor e_{ij}. This motion itself can be decomposed into two modes (cf. Eq. 2.53): an isotropic expansion and a straining motion without change of volume, with the former of these possibilities only existing in the case of compressible fluids.
(iii) *rigid-body rotation* with angular velocity $\frac{1}{2}\boldsymbol{\omega}$.

Exercises

Exercise 1

In the last section we have seen that relative motion near a point can be a sum of three basic modes of motion, classified as *uniform translation*, *pure straining*, *rigid-body rotation*.

Based on these basic modes of fluid deformation, it is possible to construct other types of flow deformations. For instance, *simple shearing motion* refers to a particular type of velocity field which often occurs in practice, in which plane layers of fluid slide over one another. This type of motion can always be expressed as a superposition of *pure-straining* and *rigid-body rotation* modes.

To demonstrate an example for simple shearing motion in two-dimensions, consider a circle in the plane Or_1r_2. Show that the superposition of rigid body rotation with $-\frac{1}{2}\omega$ perpendicular to the $0r_1r_2$ plane, and pure straining motion (without volume change), with principal axes r_1', r_2', oriented at 45° to the r_1, r_2 axes with appropriate magnitude gives rise to a simple shearing motion in which

$$\delta\mathbf{u} = \left(r_2\frac{\partial u_1}{\partial r_2}, 0, 0\right)$$

as shown in the figure. Write analytical forms of the rigid-body rotation and pure-straining deformations that will lead to the simple shearing motion.

Exercise 2

Consider a particular type of flow in 2D,

$$\mathbf{u} = (u_\theta, u_r)$$

where the velocity components in polar coordinates (r, θ) are given as

$$u_r = -\frac{a}{r}, \quad u_\theta = -\frac{a}{r}.$$

where a is a constant. Show that the velocity field satisfies the continuity equation, which can be written as

$$\nabla\cdot\mathbf{u} = \frac{\partial}{\partial r}(ru_r) + \frac{\partial}{\partial\theta}(u_\theta) = 0.$$

A 2D stream-function ψ can be defined such that

$$\mathbf{u} = -\mathbf{k}\times\nabla\psi = -\mathbf{k}\times\left\{\mathbf{e}_r\frac{\partial\psi}{\partial r} + \mathbf{e}_\theta\frac{1}{r}\frac{\partial\psi}{\partial\theta}\right\}$$
$$= -\mathbf{e}_\theta\frac{\partial\psi}{\partial r} + \mathbf{e}_r\frac{1}{r}\frac{\partial\psi}{\partial\theta}.$$

Integrate the above to obtain the stream-function for the given velocity field. Sketch the flow by plotting the streamlines for different values of the constant a.

Then, find the flow vorticity field

$$\boldsymbol{\omega} = \nabla \times \mathbf{u} = \frac{1}{r}\left\{\frac{\partial}{\partial r}(r u_\theta) - \frac{\partial u_r}{\partial \theta}\right\},$$

to investigate its distribution, and discuss if the flow is rotational or irrotational.

3 Equations of Motion of a Fluid

3.1 Continuity Equation–Mass Conservation

We have already derived the equation expressing the conservation of mass for a fluid in Sect. 2.6. The continuity equation can be written in either of the following two forms:

$$\frac{\partial \rho}{\partial t} + \nabla \cdot \rho \mathbf{u} = 0 \tag{3.1a}$$

or

$$\frac{D\rho}{Dt} + \rho \nabla \cdot \mathbf{u} = 0 \tag{3.1b}$$

We have also noted that the continuity equation is only a kinematic statement, and does not give any information on the dynamics of the fluid, i.e. the motion of the fluid in response to applied forces.

3.2 Equation of Motion–Momentum Equation

The dynamics of the fluid is to be expressed by the statement of *Newton's second law* from classical mechanics, i.e. for a material volume of fluid V enclosed by a surface S, the applied forces should be balanced by the rate of change of momentum:

$$\frac{d\mathbf{P}}{dt} = \mathbf{F}_T \tag{3.2a}$$

where $\mathbf{P}$ is the momentum of the fluid

$$\mathbf{P} = \int_V \rho \mathbf{u} \, dV \tag{3.2b}$$

© Springer Nature Switzerland AG 2020

E. Özsoy, *Geophysical Fluid Dynamics I*, Springer Textbooks in Earth Sciences, Geography and Environment, https://doi.org/10.1007/978-3-030-16973-2_3

and the total applied force $\mathbf{F}_T$ is decomposed into body (volume) and surface forces according to Sect. 2.4:

$$\mathbf{F}_T = \int_V \rho \mathbf{F} dV + \int_S \boldsymbol{\sigma}\, dS \tag{3.2c}$$

Since V is a material volume the normal velocity on the surface is $u_n = \mathbf{u} \cdot \mathbf{n} = \mathbf{u} \cdot \mathbf{n}$, and therefore the rate of change of momentum (a left hand side of 3.2a) can be expressed by using the Leibnitz' rule (1.44c)

$$\frac{d\mathbf{P}}{dt} = \frac{d}{dt}\int_V \rho \mathbf{u}\, dV = \int_V \frac{\partial}{\partial t}\rho \mathbf{u}\, dV + \int_S \rho \mathbf{u}\mathbf{u} \cdot \mathbf{n}\, dS \tag{3.3}$$

The i-th component of the second term in Eq. (3.3) is

$$\begin{aligned}
\int_S \rho u_i \mathbf{u} \cdot \mathbf{n}\, dS &= \int_S \rho u_i u_j n_j dS = \int_S A_{ij} n_j\, dS \\
&= \int_S \mathbf{A} \cdot \mathbf{n}\, dS = \int_V \nabla \cdot \mathbf{A}\, dV \\
&= \int_V \frac{\partial A_{ij}}{\partial x_j} dV = \int_V \frac{\partial \rho u_i u_j}{\partial x_j} dV \\
&= \int_V \left(\rho u_j \frac{\partial u_i}{\partial x_j} + u_i \frac{\partial \rho u_j}{\partial x_j} \right) dV \\
&= \int_V (\rho \mathbf{u} \cdot \nabla u_i + u_i \nabla \cdot \rho \mathbf{u})\, dV
\end{aligned} \tag{3.4}$$

where use has been made of the substitution $\mathbf{A} = A_{ij} = \rho u_i u_j$ and the divergence theorem (1.34) (alternatively 1.28a). Therefore (3.3) can be expressed as

$$\begin{aligned}
\frac{d\mathbf{P}}{dt} &= \int_V \left[\frac{\partial}{\partial t}\rho \mathbf{u} + \rho(\mathbf{u} \cdot \nabla)\mathbf{u} + \mathbf{u}\nabla \cdot \rho \mathbf{u} \right] dV \\
&= \int_V \mathbf{u} \left(\frac{\partial \rho}{\partial t} + \nabla \cdot \rho \mathbf{u} \right) dV + \int_V \rho \left(\frac{\partial \mathbf{u}}{\partial t} + (\mathbf{u} \cdot \nabla)\mathbf{u} \right) dV
\end{aligned} \tag{3.5}$$

The first term of (3.5) vanishes by virtue of the continuity Eq. (3.1a). The second term which is a material derivative (cf. Eq. 1.36) survives:

$$\frac{d\mathbf{P}}{dt} = \int_V \rho \frac{D\mathbf{u}}{Dt} dV \tag{3.6}$$

The second term of (3.2c) is

$$\int_S \boldsymbol{\Sigma} dS = \int_S \boldsymbol{\sigma} \cdot \mathbf{n}\, dS = \int_V \nabla \cdot \boldsymbol{\sigma}\, dV \tag{3.7}$$

by virtue (1.34) and $\boldsymbol{\sigma} = \sigma_{ij}$ denoting the stress tensor defined in (2.4).

Substituting (3.6), (3.2c) and (3.7) into (3.2a), we obtain

$$\int_V \left(\rho \frac{D\mathbf{u}}{Dt} - \rho\mathbf{F} - \nabla\cdot\boldsymbol{\sigma}\right) dV = 0$$

Since the volume V is arbitrary, the integrand itself must vanish, i.e.:

$$\rho \frac{D\mathbf{u}}{Dt} = \rho\mathbf{F} + \nabla\cdot\boldsymbol{\sigma} \tag{3.8a}$$

which can also be written in component such that

$$\rho \frac{Du_i}{Dt} = \rho F_i + \frac{\partial \sigma_{ij}}{\partial x_j}. \tag{3.8b}$$

Equation (3.8a) or (3.8b) express the momentum balance of the fluid and therefore called *the equation of motion*. The left hand side is mass (per unit volume) multiplying the acceleration, balanced by forces (per unit volume) on the right hand side.

It is worthy of note that the surface forces contribute to the acceleration of the fluid only if the stress tensor is a function of position in the fluid, i.e. if $\nabla\cdot\boldsymbol{\sigma} = \hat{e}_i \partial\sigma_{ij}/\partial x_j \neq 0$. On the other hand, if $\nabla\cdot\boldsymbol{\sigma} = 0$, the effect of surface forces on a material element of fluid is to deform it without changing its momentum.

Equation (3.8) can not be used to determine the fluid velocity immediately, since a knowledge of the body force and stresses distribution must also be available to solve the equation.

The volume force in many instances is the gravity, $\mathbf{F} = \mathbf{g}$, and in those instances it is prescribed. However, the stress tensor is due to the internal reactions of the fluid, and therefore it is a function of the flow itself, as will be seen in the following.

3.2.1 Integral Form of the Momentum Equation

Equations (3.3) and (3.7) can be used to write Eq. (3.2) in the following form:

$$\int \frac{\partial \rho\mathbf{u}}{\partial t}\, dV + \int_S \rho\mathbf{u}\mathbf{u}\cdot\mathbf{n}\, dS = \int_V \rho\mathbf{F}\, dV + \int_S \boldsymbol{\sigma}\cdot\mathbf{n}\, dS \tag{3.9}$$

This equation will be useful in integral form if all the terms can be converted to surface integrals, since then the details of the motion in the volume enclosed in $\mathbf{S}$ would be irrelevant. For example, if the volume force $\mathbf{F}$ is a *conservative force* it can be expressed as the gradient of a scalar

$$\rho\mathbf{F} = \nabla(\rho\psi) \tag{3.10}$$

For instance, this is possible if $\mathbf{F}$ represents the gravity $\mathbf{F} = \mathbf{g}$ and ρ is uniform (the fluid is homogeneous). Therefore it follows that

$$\rho\mathbf{F} = \rho\mathbf{g} = \nabla(\rho\mathbf{g}\cdot\mathbf{x}) = \rho\nabla\mathbf{g}\cdot\mathbf{x} = \nabla\rho\psi$$

with the substitution of $\psi = \mathbf{g} \cdot \mathbf{x}$. Then making use of the volume integral in (3.9) can be expressed as

$$\int_V \rho \mathbf{F} \, dV = \int_S \rho \psi \mathbf{n} \, dS \tag{3.11}$$

In addition, if the flow is *steady*, the first term on the left hand side of (3.9) vanishes and we obtain

$$\int_S \rho \mathbf{u} \mathbf{u} \cdot \mathbf{n} \, dS = \int_S (\rho \psi \mathbf{n} + \boldsymbol{\sigma} \cdot \mathbf{n}) \, dS. \tag{3.12}$$

This equation expresses the fact that the convective flux of momentum out of the region bounded by the *control surface* S is equal to the contact force exerted at its boundary with the surrounding fluid (i.e. the momentum theorem).

3.3 Stress Tensor in a Newtonian Fluid

It now remains to show how the stress tensor σ_{ij} is related to the flow, i.e. the velocity field. In Sect. 2.4.2 we showed that the stress tensor in a moving fluid can be expressed as

$$\sigma_{ij} = -p\delta_{ij} + d_{ij} \tag{3.13}$$

where the isotopic first term represented the pressure and the anisotopic second term was called the deviatoric stress tensor, resulting solely from the motion of the fluid.

Since the deviatoric stress tensor only exists in moving fluids, it should be a function of the instantaneous fluid velocity distribution and more precisely of the departure from uniformity of the velocity distribution. The situation is analogous to heat diffusion, where the heat flux is often assumed to be proportional to temperature gradients. Here, the momentum flux due to internal friction in the sense indicated by Eq. (3.7) will be assumed to depend on the velocity gradients, from a phenomenological view point. Since internal friction can only occur in moving fluids, the deviatoric stress tensor d_{ij} will be assumed to depend on local velocity gradients, of which a typical component is $\partial u_k / \partial x_l$.

A *Newtonian fluid* is one in which a linear relation exists between the deviatoric stress and the velocity gradients, most generally stated as

$$d_{ij} = A_{ijkl} \frac{\partial u_k}{\partial x_l} \tag{3.14}$$

where A_{ijkl} is a fourth order tensor which is constant (does not depend on the velocity field). For a fluid at rest, (3.14) shows that d_{ij} is zero. In Sect. 2.7 (Eqs. 2.41, 2.42

and 2.56) we expressed the velocity gradient as

$$\frac{\partial u_k}{\partial x_l} = e_{kl} - \frac{1}{2}\epsilon_{klm}\omega_m \tag{3.15}$$

in terms of the rate of strain tensor and the angular velocity of rigid body rotation $\frac{1}{2}\omega$. Therefore (3.14) becomes

$$d_{ij} = A_{ijkl}e_{kl} - \frac{1}{2}A_{ijkl}\epsilon_{klm}\omega_m \tag{3.16}$$

The tensor coefficient A_{ijkl} takes a simpler form if the molecular structure of the fluid is isotropic, so that the stresses generated in a element of fluid are independent of the orientation. Simple fluids have isotropic structure in this sense, and would not have directional preferences for stress. If the even order tensor A_{ijkl} is isotropic, it can be expressed in terms of the basic isotropic tensor δ_{ij} (cf. Sect. 1.1):

$$A_{ijkl} = \mu\delta_{ik}\delta_{jl} + \mu'\delta_{il}\delta_{jk} + \mu''\delta_{ij}\delta_{kl} \tag{3.17}$$

where μ, μ' and μ'' are scalar coefficients. Since σ_{ij} is symmetric (cf. Sect. 2.4, Eq. 2.10), d_{ij} must also be symmetric by virtue of Eq. (2.13). On the other hand d_{ij} being symmetric in Eq. (3.14) requires that A_{ijkl} must also be symmetric with respect to the indices i and j, i.e.

$$A_{ijkl} = A_{jikl}$$

so that

$$\begin{aligned}\mu\delta_{ik}\delta_{jl} &+ \mu'\delta_{il}\delta_{ik} + \mu''\delta_{ij}\delta_{kl}\\ &= \mu\delta_{jk}\delta_{il} + \mu'\delta_{jl}\delta_{ik} + \mu''\delta_{ji}\delta_{kl}\end{aligned}$$

Since $\delta_{ij} = \delta_{ji}$ the last terms on each side is canceled leaving

$$(\mu - \mu')\delta_{ik}\delta_{jl} = (\mu - \mu')\delta_{il}\delta_{jk},$$

but this equality can only be satisfied if $\mu = \mu'$. Therefore Eq. (3.17) becomes

$$A_{ijkl} = \mu(\delta_{ik}\delta_{jl} + \delta_{il}\delta_{jk}) + \mu''\delta_{ij}\delta_{kl} \tag{3.18}$$

Now it is observed that A_{ijkl} is also symmetrical with respect to the indices k and l. Substituting (3.18) into (3.16),

$$\begin{aligned}d_{ij} &= \mu(\delta_{ik}\delta_{jl} + \delta_{il}\delta_{jk})e_{kl} + \mu''\delta_{ij}\delta_{kl}e_{kl}\\ &\quad - \frac{1}{2}\mu(\delta_{ik}\delta_{jl} + \delta_{il}\delta_{jk})\epsilon_{klm}\omega_m - \frac{1}{2}\mu''\delta_{ij}\delta_{kl}\epsilon_{klm}\omega_m\\ &= \mu(e_{ij} + e_{ji}) + \mu''\delta_{ij}e_{kk} - \frac{1}{2}\mu(\epsilon_{ijm} + \epsilon_{jim})\omega_m - \frac{1}{2}\mu''\delta_{ij}\epsilon_{kkm}\omega_m\end{aligned}$$

Noting that $e_{ij} = e_{ji}$ (e_{ij} is symmetric), $e_{kk} = \Delta = \nabla \cdot \mathbf{u}$ (cf. Eq. 2.51) and also that $\epsilon_{ijm} = -\epsilon_{jim}$, $\epsilon_{kkm} = 0$ (cf. Eq. 1.2) the last two terms are dropped to yield

$$d_{ij} = 2\mu e_{ij} + \mu'' \Delta \delta_{ij} \tag{3.19}$$

This equation can also be deduced from (3.16) where the second term vanishes because of the symmetry of A_{ijkl} with respect to indices k and l, which indicates that the deviatoric stress is only related to pure straining motion, verified in Eq. (3.19). The relation between d_{ij} and e_{ij} is a linear one and therefore *their principal axes coincide.*

Finally, since the deviatoric stress is the non-isotropic part of the stress tensor, it should make zero contribution to the mean normal stress (cf. Sect. 2.4.2, Eq. 2.15 and 2.17), i.e.

$$\begin{aligned} d_{ii} &= 2\mu e_{ii} + 3\mu'' \Delta \\ &= (2\mu + 3\mu'')\Delta \\ &= 0 \end{aligned}$$

for all values of Δ, so that $\mu'' = -\frac{2}{3}\mu$ (also known as Stokes' hypothesis) and therefore

$$d_{ij} = 2\mu \left(e_{ij} - \frac{1}{3}\Delta \delta_{ij} \right) \tag{3.20}$$

where it can be verified that $(e_{ij} - \frac{1}{3}\Delta\delta_{ij})$ is the non-isotropic part of the rate of strain tensor. Eq. (3.20) shows that the non-isotropic parts of the stress and rate of strain tensors are linearly related in *Newtonian fluids*. The constant of proportionality is referred to as *the dynamic viscosity*

The diagonal components of the stress tensor ($\sigma_{ij} with i = j$) are called *the normal stresses* and the off-diagonal components (σ_{ij}, $i \neq j$) are called *tangential (or shear) stresses* (Sect. 2.4). Sometimes it is preferred to show the shear stresses by the symbol τ_{ij}, i.e $\sigma_{ij} = \tau_{ij}$ for $i \neq j$, so that

$$\sigma_{ij} = \begin{bmatrix} \sigma_{11} & \tau_{12} & \tau_{13} \\ \tau_{21} & \sigma_{22} & \tau_{23} \\ \tau_{31} & \tau_{32} & \sigma_{33} \end{bmatrix} = -p\delta_{ij} + d_{ij} \tag{3.21}$$

by virtue of Eq. (3.13), and to show the deviatoric stress tensor as

$$d_{ij} = \begin{bmatrix} \sigma'_{11} & \tau_{12} & \tau_{13} \\ \tau_{21} & \sigma'_{22} & \tau_{23} \\ \tau_{31} & \tau_{32} & \sigma'_{33} \end{bmatrix} \tag{3.22}$$

where $\sigma'_{11} = \sigma_{11} + p$, $\sigma'_{22} = \sigma_{22} + p$, $\sigma'_{33} = \sigma_{33} + p$ are the *deviatoric normal stresses*. The shear stresses τ_{ij} only contribute to the deviatoric part of the stress tensor.

The significance of the viscosity μ is seen by considering the special case of simple shearing motion (cf. Sect. 2.7.3) which is the superposition of pure straining motion and rigid body rotation. Consider a simple shearing motion in which $\partial u_1/\partial x_2$ is the only non-zero velocity derivative, then

$$d_{12} = d_{21} = \tau_{12} = \tau_{21} = 2\mu \frac{1}{2}\left(\frac{\partial u_1}{\partial x_2} + 0\right) = \mu \frac{\partial u_1}{\partial x_2} \tag{3.23}$$

(by virtue of Eqs. 2.42 and 3.20) expresses that the shear stress is proportional to the gradient of velocity that causes fluid layers to slide over one another, and μ is the constant of proportionality.

The molecular constant μ is in general a function of temperature only. However, temperature differences in fluids are often small enough and therefore variations of μ due to temperature are often negligible.

3.4 Navier–Stokes Equation

In Sect. 3.2 the equation of motion was obtained as (Eq. 3.8b)

$$\rho \frac{Du_i}{Dt} = \rho F_i + \frac{\partial \sigma_{ij}}{\partial x_j} \tag{3.24}$$

In Sect. 3.3 we obtained an expression for the stress tensor (Eqs. 3.13 and 3.20)

$$\sigma_{ij} = -p\delta_{ij} + 2\mu\left(e_{ij} - \frac{1}{3}\Delta\delta_{ij}\right) \tag{3.25}$$

where (Eq. 2.42a and 2.50)

$$e_{ij} = \frac{1}{2}\left(\frac{\partial u_i}{\partial x_j} + \frac{\partial u_j}{\partial x_i}\right) \text{ and } \Delta = e_{kk} = i\frac{\partial u_k}{\partial x_k} \tag{3.26}$$

Substituting (3.25) into (3.24) yields *Navier–Stokes equation*

$$\rho \frac{Du_i}{Dt} = \rho F_i - \frac{\partial p}{\partial x_i} + \frac{\partial}{\partial x_j}\left\{2\mu\left(e_{ij} - \frac{1}{3}\Delta\delta_{ij}\right)\right\} \tag{3.27}$$

Taking μ = constant and substituting (3.26) into (3.27) yields

$$\rho \frac{Du_i}{Dt} = \rho F_i - \frac{\partial p}{\partial x_i} + \mu\left\{\frac{\partial^2 u_i}{\partial x_j \partial x_j} + \frac{\partial^2 u_j}{\partial x_i \partial x_j} - \frac{2}{3}\frac{\partial^2 u_k}{\partial x_i \partial x_k}\right\}$$

or

$$\rho \frac{Du_i}{Dt} = \rho F_i - \frac{\partial P}{\partial x_i} + \mu \left\{ \frac{\partial^2 u_i}{\partial x_j \partial x_j} + \frac{1}{3} \frac{\partial^2 u_k}{\partial x_i \partial x_k} \right\} \tag{3.28}$$

(since both j and k are repeated indices). In vector notation, (3.28) becomes

$$\rho \frac{D\mathbf{u}}{Dt} = \rho \mathbf{F} - \nabla p + \mu \left\{ \nabla^2 \mathbf{u} + \frac{1}{3} \nabla (\nabla \cdot \mathbf{u}) \right\} \tag{3.29}$$

In the special case of *incompressible* fluids (3.29) takes the form:

$$\rho \frac{D\mathbf{u}}{Dt} = \rho \mathbf{F} - \nabla p + \mu \nabla^2 \mathbf{u} \tag{3.30}$$

which can also be written as

$$\frac{D\mathbf{u}}{Dt} = \mathbf{F} - \frac{1}{\rho} \nabla p + \nu \nabla^2 \mathbf{u} \tag{3.31}$$

where $\nu = \mu/\rho$ is defined as the *kinematic viscosity*. Note that disregarding convective accelerations, the volume forces and the pressure gradients this equation is in the form

$$\frac{\partial \mathbf{u}}{\partial t} = \ldots + \nu \nabla^2 \mathbf{u} \tag{3.32}$$

which implies, by analogy to the heat equation, that the kinematic viscosity ν stands for *diffusivity* of the velocity $\mathbf{u}$, and has dimensions of L^2/T. Values μ and ν at 15 °C and 1 atm. pressure are listed below for air, water and mercury:

	μ (g/cm/s)	ν (cm^2/s)
air	1.8×10^{-4}	1.5×10^{-1}
water	1.1×10^{-2}	1.1×10^{-2}
mercury	1.6×10^{-2}	1.2×10^{-3}

This table shows that while the dynamic viscosity is in increasing order from air to mercury the kinematic viscosity (diffusivity) is in decreasing order. As a result shear stresses for given flow situation is larger for water and mercury as compared to air; on the other hand the diffusivity of momentum is smaller for mercury and water as compared to that for air. One must note however that this is only true for molecular transport of momentum, and the numbers for air and water are rarely of significance in the atmosphere and the ocean, where transport of momentum occurs mainly through the turbulent processes.

3.5 Pressure in a Fluid

3.5.1 Pressure in a Fluid at Rest

For a fluid at rest ($\mathbf{u} = 0$) Eq. (3.29) implies

$$\rho \mathbf{F} = \nabla p \tag{3.33}$$

i.e. the body forces are balanced by a pressure gradient in the fluid. If the body force $\mathbf{F}$ is a *conservative force*, then it can be expressed as the gradient of a scalar

$$\mathbf{F} = -\nabla \psi \tag{3.34}$$

For example, gravity is a conservative force field and can be written as

$$\mathbf{F} = \mathbf{g} = -\nabla(-\mathbf{g} \cdot \mathbf{x}) = -\nabla \psi \tag{3.35}$$

with $\psi = -\mathbf{g} \cdot \mathbf{x}$. Using (3.34), (3.33) becomes

$$-\rho \nabla \psi = \nabla p \tag{3.36}$$

Now taking the curl of both sides and using (1.27b):

$$(\nabla \rho) \times (\nabla \psi) = 0 \tag{3.37}$$

indicates that the level surfaces of ρ and ψ coincide (they are parallel everywhere). In the case of gravity, Eq. (3.33)

$$\rho \mathbf{g} = \nabla p \tag{3.38}$$

expresses *'Hydrostatic pressure'*, and surfaces of ρ, p, ψ = constant are all parallel to each other as in Fig. 3.1.

The pressure is seen to be the reaction to the body force, i.e. applied body force is internally balanced by pressure. If the fluid is homogeneous (ρ= constant), the value of the pressure at any point can be obtained by integrating (3.36)

$$p = p_0 - \rho \psi = p_0 + \rho \mathbf{g} \cdot \mathbf{x} \tag{3.39}$$

where p_0 is a constant reference value.

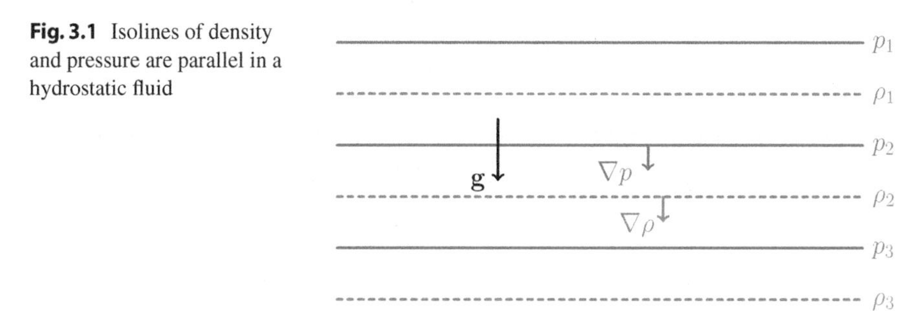

Fig. 3.1 Isolines of density and pressure are parallel in a hydrostatic fluid

3.5.2 Pressure in a Moving Homogeneous Fluid

In moving fluid pressure is generated in response to both body forces and the motion of fluid. In analogy to (3.39), the absolute pressure p can be written as

$$p = p_0 + \rho \mathbf{g} \cdot \mathbf{x} + P \tag{3.40}$$

for a homogeneous fluid. Here $p_0 + \rho \mathbf{g} \cdot \mathbf{x}$ is the pressure that would be present in a fluid at rest in response to the gravity force. The remaining part P arises due to the motion of the fluid. Substituting (3.40) into Eq. (3.29) (in which $\mathbf{F} = \mathbf{g}$ is the body force) we obtain:

$$\rho \frac{D\mathbf{u}}{Dt} = -\nabla P + \mu \left\{ \nabla^2 \mathbf{u} + \frac{1}{3} \nabla(\nabla \cdot \mathbf{u}) \right\} \tag{3.41}$$

so that the effect of gravity is completely eliminated from the equation, i.e. the flow would be unaffected by gravity (unless boundary conditions explicitly involve gravity). Since the effect of gravity does not explicitly enter the equations in most applications, we can altogether hide gravity in the equation of motion and denote P by the same symbol p in the rest of the text.

3.6 Equation of Motion Relative to Moving Frame of Reference

Consider a reference frame $O'x_1'x_2'x_3'$ which is moving relative to a Newtonian (fixed, inertial) reference frame $Ox_1x_2x_3$ as in Fig. 3.2.

Suppose that the moving coordinate system is rotating with angular velocity $\mathbf{\Omega}$ around the origin O'. Let $\mathbf{A}$ be any vector, which can be written in the moving coordinates as

$$\mathbf{A} = A_i' \hat{e}_i'$$

where $\hat{e}_i'$ are the unit vectors of the moving orthogonal coordinate system. The rate of change seen by the observer at O (fixed frame) of this vector is

$$\begin{aligned} \left(\frac{d\mathbf{A}}{dt} \right)_o &= \frac{d}{dt}(A_i' \hat{e}_i') = \frac{dA_i'}{dt} \hat{e}_i' + A_i' \frac{d\hat{e}_i'}{dt} \\ &= \frac{dA_i'}{dt} \hat{e}_i' + A_i' \mathbf{\Omega} \times \hat{e}_i' \end{aligned} \tag{3.42}$$

We can visualize the last term as follows: Seperate $\mathbf{\Omega}$ into two components $\mathbf{\Omega}_v$ and $\mathbf{\Omega}_p$ such that $\mathbf{\Omega}_v$ is perpendicular to $\hat{e}_i'$ and $\mathbf{\Omega}_p$ is parallel to $\hat{e}_i'$ as in Fig. 3.3. Then in the plane of $\hat{e}_i'$ perpendicular to $\mathbf{\Omega}_v$:

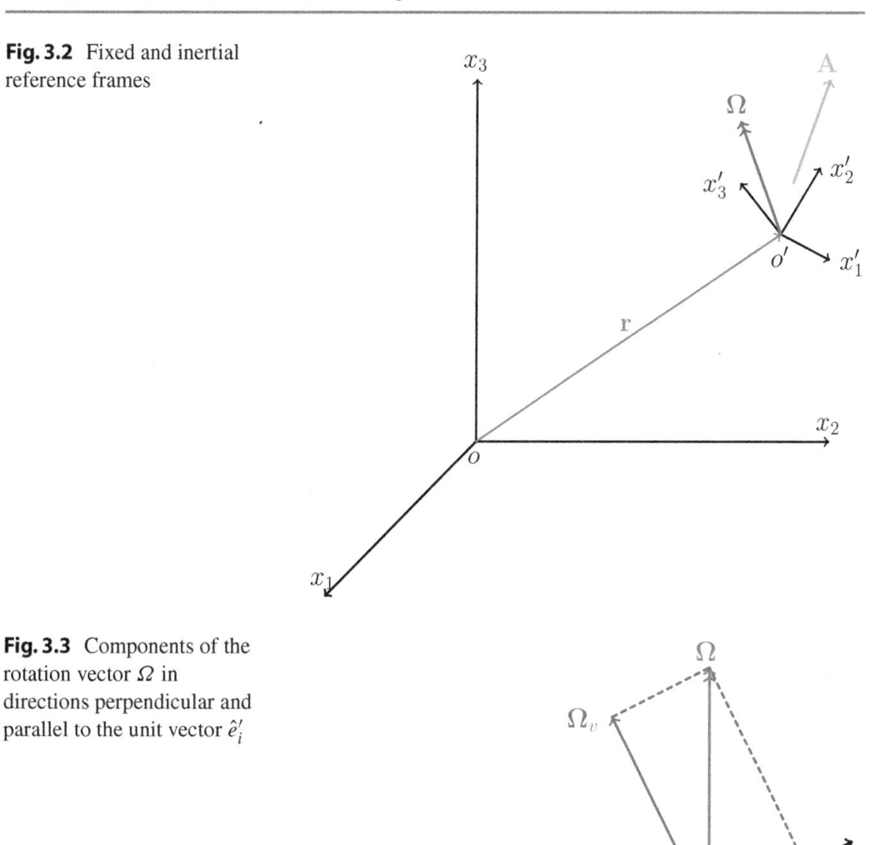

Fig. 3.2 Fixed and inertial reference frames

Fig. 3.3 Components of the rotation vector Ω in directions perpendicular and parallel to the unit vector $\hat{e}'_i$

Then,

$$\begin{aligned}\frac{d\hat{e}'_i}{dt} &= \lim_{\delta t\to 0}\frac{\delta\hat{e}'_i}{\delta t} = \lim_{\delta t\to 0}\frac{\hat{e}'_i(t+\delta t)-\hat{e}'_i(t)}{\delta t}\\ &= \mathbf{\Omega}_v\times\hat{e}'_i = (\mathbf{\Omega}-\mathbf{\Omega}_p)\times\hat{e}'_i\\ &= \mathbf{\Omega}\times\hat{e}'_i\end{aligned}$$

Since $\mathbf{\Omega}_p$ is parallel to $\hat{e}'_i$, and (3.42) follows.
Equation (3.42) is equivalent to writing

$$\left(\frac{d\mathbf{A}}{dt}\right)_o = \left(\frac{d\mathbf{A}}{dt}\right)_{o'} + \mathbf{\Omega}\times\mathbf{A} \tag{3.44}$$

where $\left(\frac{d\mathbf{A}}{dt}\right)_{o'}$ is the rate of change of appearing to an observer at o' (moving frame).

Now let $\mathbf{x}$ be the position vector of a fluid particle in the moving coordinate system, and $\mathbf{y}$ be the position vector with respect to the fixed coordinates in Fig. 3.4, so that

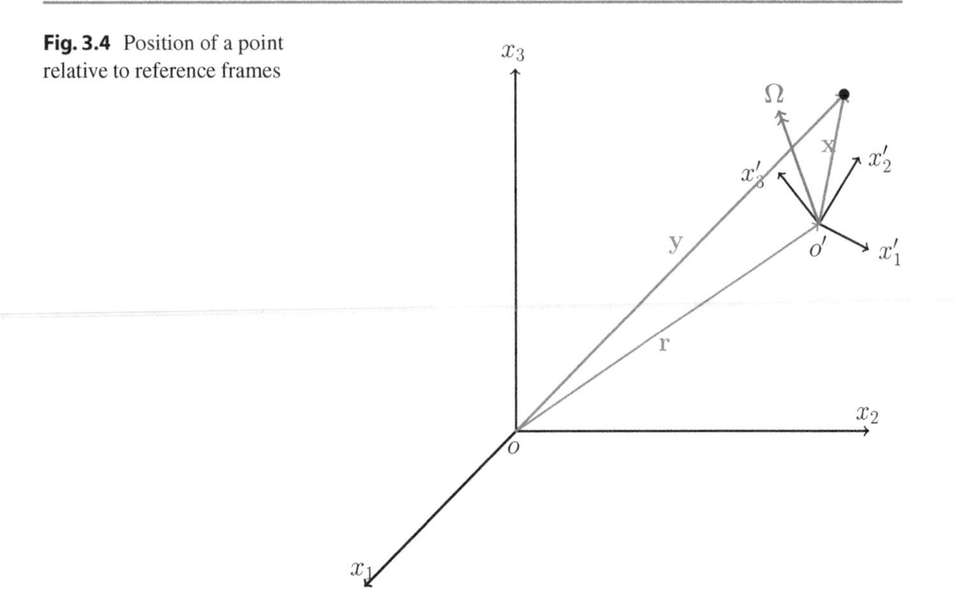

Fig. 3.4 Position of a point relative to reference frames

The fluid velocity $(\mathbf{u})_o$ with respect to the fixed system is given by virtue of (3.44) as

$$\begin{aligned}(\mathbf{u})_o &= \left(\frac{dy}{dt}\right)_o = \left(\frac{d(\mathbf{r}+\mathbf{x})}{dt}\right)_o \\ &= \left(\frac{d\mathbf{r}}{dt}\right)_o + \left(\frac{d\mathbf{x}}{dt}\right)_{o'} + \boldsymbol{\Omega}\times\mathbf{x}\end{aligned} \tag{3.45}$$

where $\left(\frac{d\mathbf{x}}{dt}\right)_{o'} = (\mathbf{u})_{o'}$ is the fluid velocity with respect to the the moving system. The fluid acceleration in the fixed system $(\mathbf{a})_o$ can be written as

$$\begin{aligned}(\mathbf{a})_o &= \left(\frac{d\mathbf{u}}{dt}\right)_o = \left(\frac{d^2\mathbf{r}}{dt^2}\right)_o + \left(\frac{d^2\mathbf{x}}{dt^2}\right)_{o'} + \\ \boldsymbol{\Omega}\times &\left(\frac{d\mathbf{x}}{dt}\right)_{o'} + \left(\frac{d(\boldsymbol{\Omega}\times\mathbf{x})}{dt}\right)_{o'} + \boldsymbol{\Omega}\times\boldsymbol{\Omega}\times\mathbf{x}\end{aligned} \tag{3.46}$$

Here $\left(\frac{d^2\mathbf{x}}{dt^2}\right)_{o'} = (\mathbf{a})_{o'}$ and $\left(\frac{d\mathbf{x}}{dt}\right)_{o'} = (\mathbf{u})_{o'}$ are respectively the fluid acceleration and velocity in the moving frame of reference. Expanding (3.46) we have

$$(\mathbf{a})_o = (\mathbf{a})_{o'} + 2\boldsymbol{\Omega}\times(\mathbf{u})_{o'} + \boldsymbol{\Omega}\times\boldsymbol{\Omega}\times\mathbf{x} + \left(\frac{d^2\mathbf{r}}{dt^2}\right)_o + \left(\frac{d\boldsymbol{\Omega}}{dt}\right)_{o'}\times\mathbf{x} \tag{3.47}$$

which shows that the accelerations in the two coordinate systems differ by the amount shown by the latter terms on the right hand side of (3.47). The second term is called the *Coriolis acceleration* and the third term is due to *Centrifugal acceleration.* The fourth term arises because of the translational motion of the moving coordinates and the fifth term represent accelerations due to rate of change of angular rotation. Now if we assume that there is no translational acceleration of the moving frame and that the angular speed of rotation is constant, the last two terms vanish, to yielding are zero.

$$(\mathbf{a})_o = (\mathbf{a})_{o'} + 2\boldsymbol{\Omega} \times (\mathbf{u})_{o'} + \boldsymbol{\Omega} \times \boldsymbol{\Omega} \times \mathbf{x} \tag{3.48}$$

The last equation may apply to a reference frame fixed on the earth, since earth's translational motion can often be neglected with respect to the reference frame fixed to the stars and since its angular speed is relatively constant.

The equation of motion (3.29) applies in an inertial (Newtonian) frame of reference with

$$(\mathbf{a})_o = \left(\frac{D\mathbf{u}}{Dt}\right)_o = \frac{D(\mathbf{u})_o}{Dt}$$

where $(\mathbf{a})_o$ is the fluid velocity in the inertial frame. In the moving frame, replacing the fluid acceleration by (3.48), the equation of motion becomes:

$$\rho\left\{\frac{D\mathbf{u}}{Dt} + 2\boldsymbol{\Omega} \times \mathbf{u} + \boldsymbol{\Omega} \times \boldsymbol{\Omega} \times \mathbf{x}\right\} = \rho\mathbf{F} - \nabla p + \mu\left\{\nabla^2\mathbf{u} + \frac{1}{3}\nabla(\nabla \cdot \mathbf{u})\right\} \tag{3.49}$$

Here the velocity $\mathbf{u} = (\mathbf{u})_{O'}$ is referenced to the moving frame. The second and the third terms on the left hand side can be passed to the right hand side with and could be interpreted as *fictitious body forces arising due to rotation* of the reference frame with respect to the Newtonian frame.

3.6.1 Modification of Gravity and Pressure Due to Centrifugal Acceleration

In the case of gravity being the only conservative body force we have shown that it can be written as

$$\mathbf{g} = -\nabla\psi = -\nabla(-\mathbf{g} \cdot \mathbf{x}) \tag{3.50}$$

Similarly, the fictitious body force due to the centrifugal acceleration can be written as

$$-\boldsymbol{\Omega} \times \boldsymbol{\Omega} \times \mathbf{x} = -\nabla\psi' = -\nabla\left(\frac{1}{2}(\boldsymbol{\Omega} \times \mathbf{x}) \cdot (\boldsymbol{\Omega} \times \mathbf{x})\right) \tag{3.51}$$

by making use of vector identities. Therefore the centrifugal force is also shown to be a conservative force field. *The modified gravity* or *gravitation* vector is defined as

$$\mathbf{g}' = \mathbf{g} - \boldsymbol{\Omega} \times \boldsymbol{\Omega} \times \mathbf{x} \tag{3.52}$$

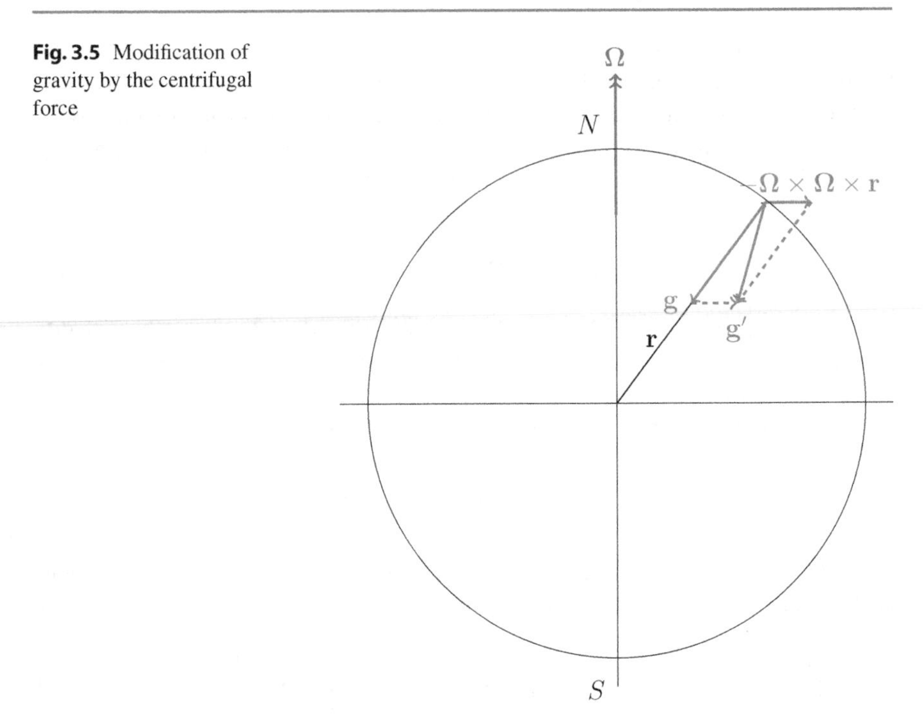

Fig. 3.5 Modification of gravity by the centrifugal force

and includes the effect of earth's centrifugal force modifying the gravity as a function of latitude: (taking O at the center of earth)

Combining Eqs. (3.50) and (3.51),

$$\mathbf{g}' = -\nabla(\psi + \psi') = -\nabla\psi'' \tag{3.53}$$

where

$$\psi'' = -\mathbf{g} \cdot \mathbf{x} - \frac{1}{2}(\boldsymbol{\Omega} \times \mathbf{x}) \cdot (\boldsymbol{\Omega} \times \mathbf{x}) \tag{3.54}$$

is used to redefine gravity, interpreted as in Fig. 3.5.

If the density ρ is constant (homogeneous fluid) then we can define a *modified pressure* P such that

$$p = p_0 + \rho\mathbf{g} \cdot \mathbf{x} + \frac{1}{2}(\boldsymbol{\Omega} \times \mathbf{x}) \cdot (\boldsymbol{\Omega} \times \mathbf{x}) + P \tag{3.55}$$

in analogy to (3.40) and the equation of motion (3.49) can thus be written as

$$\rho\left\{\frac{D\mathbf{u}}{Dt} + 2\boldsymbol{\Omega} \times \mathbf{u}\right\} = -\nabla P + \mu\left\{\nabla^2\mathbf{u} + \frac{1}{3}\nabla(\nabla \cdot \mathbf{u})\right\} \tag{3.56}$$

so that the effect of gravity and centrifugal forces (conservative body forces) are completely eliminated from the equation. The modified pressure P represents the response of the fluid to the motion (excluding gravity and centrifugal forces), and therefore can be shown by the symbol p without reference to what it represents. The only remaining fictitious force (per unit mass) is the Coriolis force $-2\boldsymbol{\Omega} \times \mathbf{u}$. We also need to consider the thermodynamics of the flow. The conservation of heat energy is expressed by the *thermodynamic equation* (the first law of thermodynamics):

$$T\frac{Ds}{Dt} = \frac{De}{Dt} - \frac{p}{\rho^2}\frac{D\rho}{Dt} = \frac{1}{\rho}\nabla \cdot K\nabla T + Q + \Phi \tag{3.57}$$

where S is the specific entropy per unit mass, e is the internal energy per unit mass K the thermal conductivity, T the temperature, Q represents the heat sources in the fluid and Φ the generation of heat by internal frictional dissipation of mechanical energy. Neglecting the last two terms (or assuming they are specified), and assuming that either entropy or internal energy can be expressed in terms of the pressure p and temperature T (for fluids of fixed composition), Eq. (3.57) provides an additional relation between the unknowns, but also increases the number of unknowns by one, by introducing T, the temperature.

We may now complement the above equation by the *equation of state*, which relates the thermodynamic quantities p, ρ, and T in a unique manner for any fluid of fixed composition and determines the thermodynamic state:

$$f(p, \rho, T) = 0 \tag{3.58}$$

where f represents a functional form that is appropriate for the fluid considered.

In principle, Eqs. (3.2b), (3.56), (3.57) and (3.58) complete the set of four equations necessary to solve for the four unknowns $\mathbf{u}$, p, ρ and T.

In oceanography and meteorology, compressibility is often negligible, and in the case of the atmosphere can often be treated by approximations that make the equations of closed form. However, geophysical fluids are often inhomogeneous and therefore stratification has to be included. We will inhomogeneous fluids later in the 'Stratified Fluids' volume, where we have to investigate thermodynamics in more detail. For the time being, we assume fluids are incompressible and homogeneous to study a simplified set of equations.

3.7 Complete Set of Governing Equations

Up to the present, we have derived the continuity and momentum equations:

$$\frac{D\rho}{Dt} + \rho\nabla \cdot \mathbf{u} = 0 \tag{3.1.b}$$

and

$$\frac{D\mathbf{u}}{Dt} + 2\boldsymbol{\Omega} \times \mathbf{u} = -\frac{1}{\rho}\nabla p + \nu\left\{\nabla^2\mathbf{u} + \frac{1}{3}\nabla(\nabla \cdot \mathbf{u})\right\}. \tag{3.59}$$

The three unknowns these equations are ρ, $\mathbf{u}$, and p. The general case of compressible flows, we see that the number of equations is insufficient.

In the case of incompressible and homogeneous fluids, the density ρ is then a specified property of the fluid, and the two equations (continuity and momentum equations) are sufficient to determine the two unknowns $\mathbf{u}$ and p.

3.8 Vorticity Dynamics

The *vorticity* defined in Eq. (2.55b)

$$\omega = \nabla \times \mathbf{u} \tag{3.60}$$

will hereafter be called *relative vorticity*, since it represents the vorticity irrespective of the reference frame, in terms of the velocity field $\mathbf{u}$ on this frame. The velocity with respect to the absolute (inertial) reference frame is (cf. Eq. 3.45)

$$\mathbf{u}_A = \mathbf{u} + \boldsymbol{\Omega} \times \mathbf{x} \tag{3.61}$$

where $\mathbf{u}_A$ is the absolute velocity (in the inertial frame) $\mathbf{u}$ the velocity in the moving frame, $\boldsymbol{\Omega}$ the angular velocity of the moving frame, and $\mathbf{x}$ the position vector with respect to the moving frame. We can define *absolute vorticity*

$$\begin{aligned}\omega_A &= \nabla \times \mathbf{u}_A = \nabla \times \mathbf{u} + \nabla \times \boldsymbol{\Omega} \times \mathbf{x}\\ &= \omega + 2\boldsymbol{\Omega}\end{aligned}$$

by virtue of Eq. (1.27d). A conservation equation for vorticity can be obtained by manipulating the momentum equation. First, using the identity

$$(\mathbf{u} \cdot \nabla)\mathbf{u} = \nabla\left(\frac{1}{2}\mathbf{u} \cdot \mathbf{u}\right) - \mathbf{u} \times \nabla \times \mathbf{u} \tag{3.62}$$

(cf. Eq. 1.27e), Eq. (3.56) is written as

$$\frac{\partial \mathbf{u}}{\partial t} + \frac{1}{2}\nabla\mathbf{u} \cdot \mathbf{u} - \mathbf{u} \times (\nabla \times \mathbf{u}) + 2\boldsymbol{\Omega} \times \mathbf{u} = -\frac{1}{\rho}\nabla p + \nu\left\{\nabla^2\mathbf{u} + \frac{1}{3}\nabla(\nabla \cdot \mathbf{u})\right\} \tag{3.63}$$

Then, taking the curl of Eq. (3.63) yields

$$\frac{\partial \nabla \times \mathbf{u}}{\partial t} + \frac{1}{2}\nabla \times \nabla \mathbf{u} \cdot \mathbf{u} + \nabla \times [(2\boldsymbol{\Omega} + \nabla \times \mathbf{u}) \times \mathbf{u}]$$
$$= -\nabla \times \frac{1}{\rho}\nabla p + \nu \left\{\nabla^2 \nabla \times \mathbf{u} + \frac{1}{3}\nabla \times [\nabla(\nabla \cdot \mathbf{u})]\right\}$$

Substituting (3.60), (3.61) and making use of the vector identity (1.27i), the above equation becomes

$$\frac{\partial \boldsymbol{\omega}}{\partial t} + \nabla \times (\boldsymbol{\omega}_A \times \mathbf{u}) = -\nabla \times \frac{1}{\rho}\nabla p + \nu \nabla^2 \boldsymbol{\omega} \tag{3.64}$$

The above equation is called the *general vorticity equation.* To put it in a more useful form, we expand

$$\nabla \times (\boldsymbol{\omega}_A \times \mathbf{u}) = (\mathbf{u} \cdot \nabla)\boldsymbol{\omega}_A - (\boldsymbol{\omega}_A \cdot \nabla)\mathbf{u} + \boldsymbol{\omega}_A(\nabla \cdot \mathbf{u}) - \mathbf{u}(\nabla \cdot \boldsymbol{\omega}_A) \tag{3.65}$$

(cf. identity 1.27d), and note that the last term

$$\nabla \cdot \boldsymbol{\omega}_A = \nabla \cdot (\boldsymbol{\omega} + 2\boldsymbol{\Omega}) = \nabla \cdot \nabla \times \mathbf{u} + \nabla \cdot 2\boldsymbol{\Omega} = 0 \tag{3.66}$$

by virtue of Eq. (1.27j) and since $\boldsymbol{\Omega}$ is constant. Substituting (3.65) and (3.66), (3.64) becomes

$$\frac{D\boldsymbol{\omega}_A}{Dt} = \frac{\partial \boldsymbol{\omega}_A}{\partial t} + \mathbf{u} \cdot \nabla \boldsymbol{\omega}_A = \boldsymbol{\omega}_A \cdot \nabla \mathbf{u} - \boldsymbol{\omega}_A \nabla \cdot \mathbf{u} - \nabla \times \frac{1}{\rho}\nabla p + \nu \nabla^2 \boldsymbol{\omega} \tag{3.67}$$

where it is also noted that $\frac{\partial \boldsymbol{\omega}}{\partial t} = \frac{\partial \boldsymbol{\omega}_A}{\partial t}$, since $\boldsymbol{\Omega}$ is a constant vector. The third term on the right hand side is (cf. 1.27b):

$$-\nabla \times \frac{1}{\rho}\nabla p = -\left(\frac{1}{\rho}\nabla \times \nabla p + \nabla\frac{1}{\rho} \times \nabla p\right) = +\frac{1}{\rho^2}\nabla \rho \times \nabla p$$

by virtue of (1.27i) and the last term on the right hand side is

$$\nu \nabla^2 \boldsymbol{\omega} = \nu \nabla^2 (\boldsymbol{\omega} + 2\boldsymbol{\Omega}) = \nu \nabla^2 \boldsymbol{\omega}_A$$

since $\boldsymbol{\Omega}$ is constant. With these results substituted, (3.67) takes the form:

$$\frac{D\boldsymbol{\omega}_A}{Dt} = \boldsymbol{\omega}_A \cdot \nabla \mathbf{u} - \boldsymbol{\omega}_A \nabla \cdot \mathbf{u} + \frac{1}{\rho^2}\nabla \rho \times \nabla p + \nu \nabla^2 \boldsymbol{\omega}_A \tag{3.68}$$

which states the conservation of absolute vorticity, and hence called the general vorticity equation. It remains to explain each term on the right hand side which contributes to the rate of change of absolute vorticity on the left hand side. First of all, if all terms on the right hand side vanish, then (3.68) states that absolute vorticity should be conserved. Next, we consider each term separately to explain their meaning.

3.8.1 The Vorticity Diffusion Term $\nu\nabla^2\omega_A$

Considering only the last term on the right hand side and neglecting convective terms on the left hand side of (3.68) the equation is similar to the heat equation

$$\frac{\partial \omega_A}{\partial t} = \cdots \nu\nabla^2\omega_A. \tag{3.69}$$

with diffusivity ν. Therefore, the last term represents vorticity diffusion due to internal (viscous) friction. This term is the most obvious contribution to the vorticity change since it represents the transport of vorticity in the fluid through viscous action, in analogy to viscous transport of momentum in the equation of motion.

3.8.2 The Tipping and Stretching Term $\omega_A \cdot \nabla\mathbf{u}$

This term can be explained better when it is written as

$$(\omega_A \cdot \nabla)\mathbf{u} = |\omega_A| \lim_{\delta s \to 0} \left(\frac{\delta \mathbf{u}}{\delta s}\right)$$

where δs is a line element along the vector ω_A, and $\delta\mathbf{u}$ is the change in velocity along the line element, as in Fig. 3.6.

Then the contribution to the fractional rate of change of vorticity on the left hand side of Eq. (3.68) becomes

$$\frac{1}{|\omega_A|}\frac{D\omega_A}{Dt} = \lim_{\delta s \to 0} \frac{\delta \mathbf{u}}{\delta s} + \cdots$$

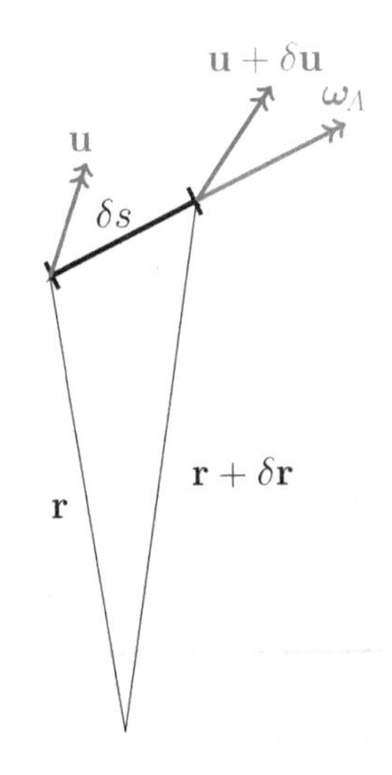

Fig. 3.6 Vorticity component along a line element oriented along velocity

A portion of this fractional rate of change is due to *tipping* i.e. rigid rotation of the vortex line element due to the normal component (δu_n) of $\delta \mathbf{u}$, and a portion of the rate of change result from *stretching*, i.e. extension or contraction of the vortex line in the direction of δu_s , the parallel component of $\delta \mathbf{u}$.

3.8.3 The Divergence Term $-\omega_A(\nabla \cdot \mathbf{u})$

It is obvious that this term contributes to the vorticity balance only if the fluid is compressible $(\nabla \cdot \mathbf{u} \neq 0)$. Replacing the divergence of velocity from the continuity Eq. (3.1b), this term reads

$$-\omega_A(\nabla \cdot \mathbf{u}) = \left(\frac{1}{\rho}\frac{D\rho}{Dt}\right)\ \omega_A$$

If the density in a material volume of fluid increases, its moment of inertia increases, so that its angular velocity (or vorticity) must also increase. This would present us an analogy to a 'compressible ballerina' rotating about herself. By increasing her body density (i.e. increasing the concentration of mass within a given volume), the ballerina would manage to rotate faster.

3.8.4 The Solenoidal Term $\frac{1}{\rho^2}\nabla\rho \times \nabla p$

This term contributes to the vorticity equation (i) only when the fluid is inhomogeneous $\left(\frac{D\rho}{Dt} \neq 0\right)$, and furthermore ($ii$) only if the level surfaces of equal density and equal pressure (i.e. isopycnals and isobars) are not parallel to each other. In fact, an inhomogeneous fluid is classified as *baroclinic*, if the isopycnals and isobars are not parallel to each other, as in Fig. 3.7.

Therefore the solenoidal term contributes to the vorticity only in *inhomogeneous, baroclinic* fluids. An inhomogeneous fluid can still be barotropic or baroclinic.

The term $-\frac{1}{\rho^2}\nabla\rho \times \nabla p$, effective for the situation in Fig. 3.7 can be interpreted as the *overturning tendency* in baroclinic fluids, in analogy with a circular disc with non-uniform mass distribution suspended at the center. In this analogous situation, gravity tends to over-turn the object to a stable position (center of gravity located below the point of suspension). If we remember that gravity is included in the pressure gradient term of the equation of motion (cf. Eq. 3.55), this tendency in a fluid arises at least partially as a result of gravity, and in general can be expressed as vorticity (rotation) attempting to orient the fluid to align its mass distribution with the conservative force field ∇p (i.e. so that $\nabla\rho \times \nabla p = 0$).

On the other hand, if pressure is a function of density in an inhomogeneous fluid, i.e. $p = f(\rho)$, so that isopycnals and isobars are parallel, as in Fig. 3.8, the fluid is termed *barotropic*.

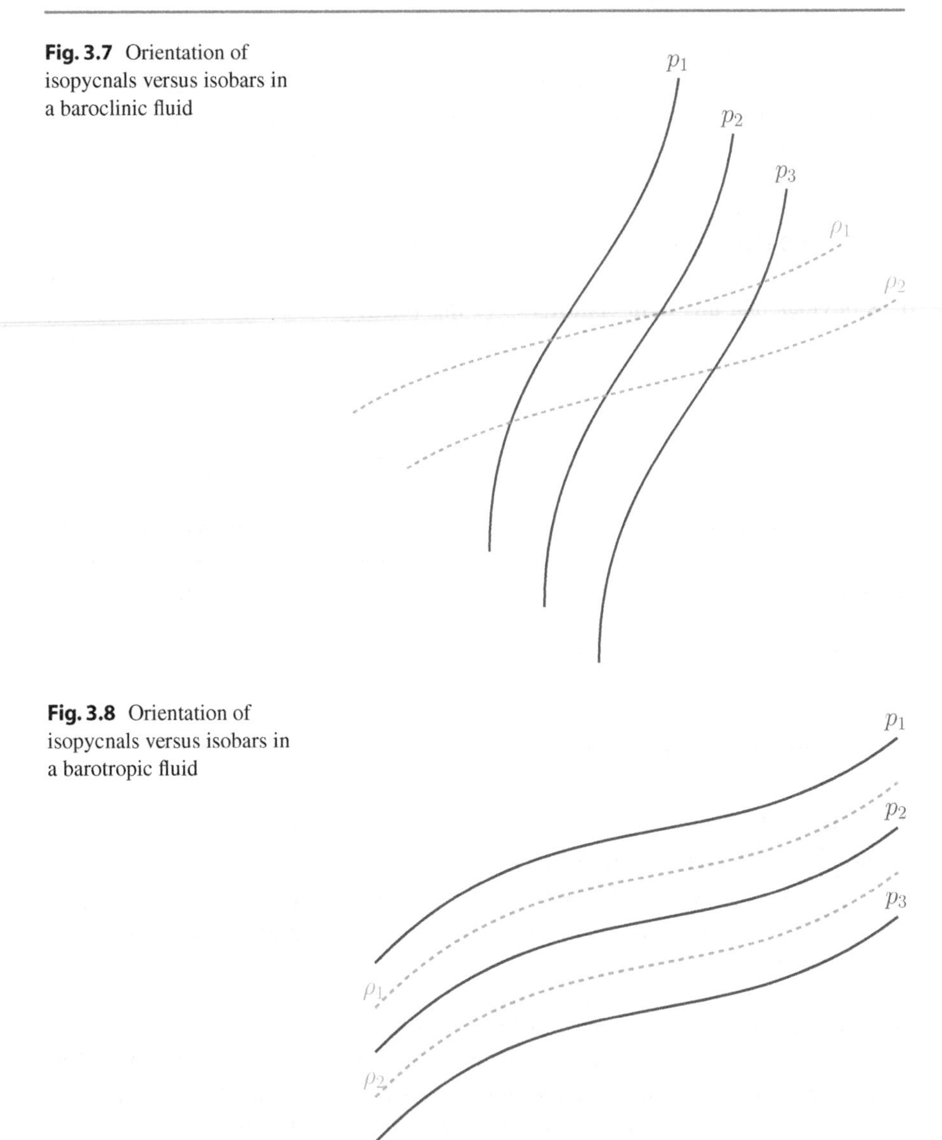

Fig. 3.7 Orientation of isopycnals versus isobars in a baroclinic fluid

Fig. 3.8 Orientation of isopycnals versus isobars in a barotropic fluid

3.9 Kelvin's Circulation Theorem

Consider an open material surface S, enclosed by a closed material curve C in a fluid. The quantity

$$\Gamma(t) = \oint_C \mathbf{u} \cdot d\mathbf{r} \tag{3.70}$$

is defined as the circulation. By virtue of the Stokes' theorem (Eq. 1.35) this is equivalent to

$$\Gamma(t) = \int_S \nabla \times \mathbf{u} \cdot \mathbf{n} dS = \int_S \omega \cdot \mathbf{n} dS \tag{3.71}$$

which expresses the vorticity flux through the surface S.

If the definition (3.70) is made for motion in a moving coordinate system, it is also possible to define the circulation with respect to the inertial (absolute) frame of reference:

$$\Gamma_A(t) = \oint_C \mathbf{u}_A \cdot d\mathbf{r} = \Gamma(t) + \oint_C \mathbf{\Omega} \times \mathbf{r} \cdot d\mathbf{r} \tag{3.72}$$

by virtue of (3.60) and (3.70). Through the of Stokes' theorem (1.35) and vector identity (1.27d) the second term can be shown to be

$$\oint_C \mathbf{\Omega} \times \mathbf{r} \cdot d\mathbf{r} = \int_S \nabla \times (\mathbf{\Omega} \times \mathbf{r}) \cdot \mathbf{n} dS = \int_S 2\mathbf{\Omega} \cdot \mathbf{n} dS \tag{3.73}$$

so that (3.72) becomes

$$\begin{aligned} \Gamma_A(t) &= \Gamma(t) + \int_S 2\mathbf{\Omega} \cdot \mathbf{n} dS \\ &= \int_S (\omega + 2\mathbf{\Omega}) \cdot \mathbf{n} dS = \int_S \omega_A \cdot \mathbf{n} dS \end{aligned} \tag{3.74}$$

where ω_A is defined as of (3.61). The integrand of the second term of (3.74) is the component of $2\mathbf{\Omega}$ normal to the surface S, or alternatively

$$\int_S 2\mathbf{\Omega} \cdot \mathbf{n} dS = 2\Omega S_p \tag{3.75}$$

since $\mathbf{\Omega}$ is a vector constant. Here $\Omega = |\mathbf{\Omega}|$ is the magnitude of the angular velocity and S_p is the projection of the surface S on the plane that is perpendicular to $\mathbf{\Omega}$. The rate of change of Γ_A is

$$\frac{d\Gamma_A}{dt} = \frac{d}{dt} \oint_C \mathbf{u}_A \cdot d\mathbf{r} = \oint_C \frac{D\mathbf{u}_A}{Dt} \cdot d\mathbf{r} + \oint_C \mathbf{u}_A \cdot \frac{D}{Dt}(d\mathbf{r}) \tag{3.76}$$

where the second term can be shown to vanish:

$$\oint_C \mathbf{u}_A \cdot d\left(\frac{D\mathbf{r}_A}{Dt}\right) = \frac{1}{2} \oint_C d(\mathbf{u}_A \cdot \mathbf{u}_A) = 0 \tag{3.77}$$

since the integral is taken for a closed curve. Utilizing (3.74), (3.75) and (3.76), it can be shown that

$$\frac{d\Gamma}{dt} = \oint_C \frac{D\mathbf{u}_A}{Dt} \cdot d\mathbf{r} - 2\Omega \frac{dS_p}{dt} \tag{3.78}$$

Now, we can substitute from the equation of motion (3.41) in the inertial frame upon which (3.78) becomes

$$\frac{d\Gamma}{dt} = -\oint_C \frac{1}{\rho}\nabla p \cdot d\mathbf{r} + \nu \oint_C \left\{\nabla^2\mathbf{u} + \frac{1}{3}\nabla(\nabla\cdot\mathbf{u})\right\} \cdot d\mathbf{r} - 2\Omega \frac{dS_p}{dt} \tag{3.79}$$

Using Stokes' theorem (1.35), and making use of (1.27b), (1.27i), this is equivalent to

$$\frac{d\Gamma}{dt} = \int_S \frac{1}{\rho^2}(\nabla\rho \times \nabla p) \cdot \mathbf{n}\, dS + \nu \int_S \nabla^2\boldsymbol{\omega} \cdot \mathbf{n}\, ds - 2\Omega \frac{dS_p}{dt} \tag{3.80}$$

which is known as *Kelvin's circulation theorem.*

Kelvin's circulation theorem (Eq. 3.80) requires that for any changes to occur in the circulation of a fluid, either the fluid must be *baroclinic* (inhomogeneous, first term on the right hand side) or there must be *diffusion of vorticity* through viscous effects (second term on the right hand side), or the *projected area must change in a rotating fluid* (third term).

If the fluid id (i) either *homogeneous* ($\rho = constant$) or *inhomogeneous barotropic* ($p = p(\rho)$) and (ii) either *inviscid* ($\nu = 0$) or *irrotational* ($\boldsymbol{\omega} = 0$), and (iii) either *non-rotating* ($\Omega = 0$) or one in which there is no change of the projected area S_p ($dS_p/dt = 0$), then we have

$$\frac{d\Gamma}{dt} = \frac{d}{dt}\int_s \boldsymbol{\omega} \cdot \mathbf{n} dS = 0 \tag{3.81}$$

which indicates that there will be no change in the circulation (*i.e.* circulation is conserved for all material surfaces S). This can also be expressed by the fact that the flux of vorticity across all arbitrarily selected open material surfaces will remain constant. If the fluid is initially irrotational ($\boldsymbol{\omega} = 0$), it will remain irrotational. If initially there is a vortex-line or a vortex-tube in the fluid, they will move with the fluid, and their strength will remain constant (since there is no diffusion). These results summarize the Helmholtz vorticity laws.

3.10 Bernoulli's Theorem

The equation of motion with the convective terms modified in Eq. (3.56) reads

$$\frac{\partial\mathbf{u}}{\partial t} - \mathbf{u} \times (\boldsymbol{\omega} + 2\boldsymbol{\Omega}) + \nabla\left(\frac{1}{2}\mathbf{u}\cdot\mathbf{u}\right) = -\frac{1}{\rho}\nabla P + \nu\left\{\nabla^2 + \frac{1}{3}\nabla(\nabla\cdot\mathbf{u})\right\} \tag{3.63}$$

where $\boldsymbol{\omega} = \nabla \times \mathbf{u}$ is the relative vorticity.

Now, if the flow is steady ($\partial \mathbf{u}/\partial t = 0$) and homogeneous ($\rho$= constant), then (3.63) becomes

$$\nabla \left(\frac{1}{2}\mathbf{u} \cdot \mathbf{u} + \frac{P}{\rho} \right) = \mathbf{u} \times \boldsymbol{\omega}_A + \nu \left\{ \nabla^2 \mathbf{u} + \frac{1}{3} \nabla (\nabla \cdot \mathbf{u}) \right\} \tag{3.82}$$

Here, $\boldsymbol{\omega}_A = \boldsymbol{\omega} + 2\boldsymbol{\Omega}$ is the absolute vorticity and the pressure p actually stands for the modified pressure P defined in Eq. (3.55). Therefore, replacing P from (3.55), (3.82) becomes

$$\nabla H = \mathbf{u} \times \boldsymbol{\omega}_A + \nu \left\{ \nabla^2 \mathbf{u} + \frac{1}{3} \nabla (\nabla \cdot \mathbf{u}) \right\} \tag{3.83a}$$

where H is defined as

$$H = \frac{1}{2}\mathbf{u} \cdot \mathbf{u} + \frac{p}{\rho} - \mathbf{g} \cdot \mathbf{x} - \frac{1}{2} (\boldsymbol{\Omega} \times \mathbf{x}) \cdot (\boldsymbol{\Omega} \times \mathbf{x}) \tag{3.83b}$$

If the fluid is also *inviscid* ($\nu = 0$) , then

$$\nabla H = \mathbf{u} \times \boldsymbol{\omega}_A \tag{3.84}$$

from which the *Bernoulli theorem* follows: the function H is constant along the *streamlines* and *vortex-lines*. This result follows since (3.84) implies that

$$\mathbf{u} \cdot \nabla H = 0 \text{ and } \boldsymbol{\omega}_A \cdot \nabla H = 0. \tag{3.85}$$

Exercises

Exercise 1

Determine the lateral deflection (distance and direction relative to the planned course) of a cannonball shot in Paris ($48°52'N$), which flies for 25 s at an average horizontal speed of 120 m/s. What would be the lateral deflection if the same shot was made in Quito ($0°13'S$) or Johannesburg ($26°12'S$)?

Exercise 2

The Great Red Spot of Jupiter centered at 22° and spanning 12° in latitude and 25° in longitude, exhibits some typical wind speeds of about 100 m/s. The planet's equatorial radius and rotation speeds respectively are $71,400$ km and 1.763×10^{-4} s^{-1}. To what degree would one estimate the Great Red Spot to be influenced by Jupiter's rotation. Would the effect be larger or smaller at a comparable latitude on earth?

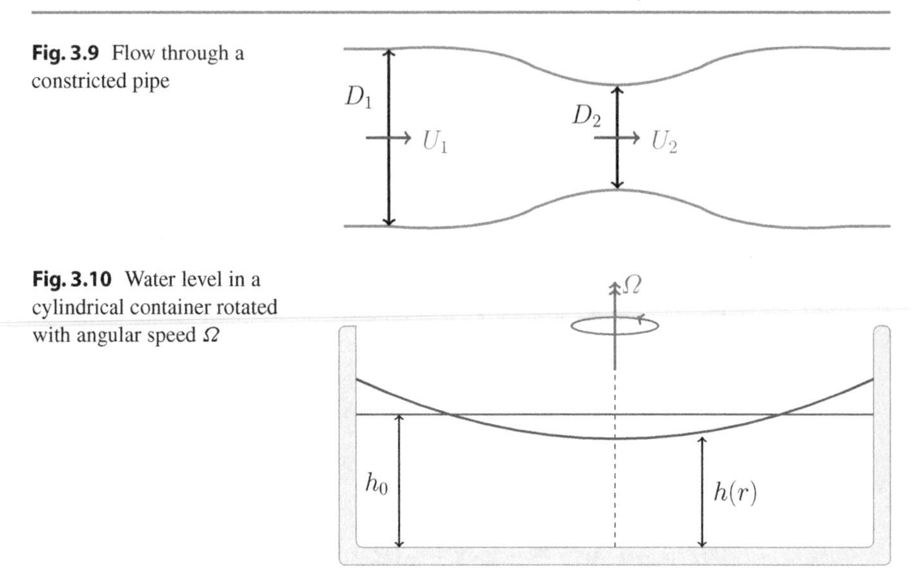

Fig. 3.9 Flow through a constricted pipe

Fig. 3.10 Water level in a cylindrical container rotated with angular speed Ω

Exercise 3

Consider the flow of an inviscid incompressible fluid through a pipe with a constriction shown in Fig. 3.9.

Assume the flow velocity is uniform across any section of the pipe. If the pressure at sections (1) and (2) could be measured, can these measurements be used to calculate the discharge Q through the pipe? Can this configuration be used as an instrument to measure the total discharge through the pipe.

Exercise 4

A cylindrical tank of radius r filled with a fluid of density ρ to a depth of h_0 is rotated with steady angular velocity Ω around its center as shown in Fig. 3.10. Assume that steady state has been reached long time after the motion has been started from rest.

By making use of the Bernoulli equation, investigate which shape would the free surface take under the effect of rotation. Calculate the difference of height at any point with respect to the depth h_0 at rest.

Flow of a Homogeneous Incompressible Fluid

4

4.1 Governing Equations

For a homogeneous and incompressible fluid, the (i) continuity (ii) momentum equations, are sufficient to determine the flow. Since the fluid is incompressible, the density of a fluid particle should remain unchanged

$$\frac{D\rho}{Dt} = 0 \tag{4.1}$$

so that the continuity Eq. (3.1b) takes the form indicates that the divergence of velocity vanishes.

$$\nabla \cdot \mathbf{u} = 0. \tag{4.2}$$

Substituting (4.2) in (3.56), the momentum equation becomes

$$\frac{D\mathbf{u}}{Dt} + 2\boldsymbol{\Omega} \times \mathbf{u} = -\frac{1}{\rho}\nabla p + \nu\nabla^2\mathbf{u} \tag{4.3}$$

Therefore, Eqs. (4.2) and (4.3) govern the flow (can be used to determine the unknowns $\mathbf{u}$ and p, providing we have appropriate boundary conditions).

If we scale the variables such that

$$|\mathbf{u}| \sim u_0, \quad |\boldsymbol{\Omega}| \sim \Omega_0, \quad |\mathbf{x}| \sim L_0, \quad p \sim \rho U_0^2, \quad t \sim L_0/U_0$$

then the dimensionless form of the momentum Eq. (4.3) is

$$\frac{D\mathbf{u}}{Dt} + \frac{1}{Ro}2\boldsymbol{\Omega} \times \mathbf{u} = -\nabla p + \frac{1}{Re}\nabla^2\mathbf{u} \tag{4.4}$$

© Springer Nature Switzerland AG 2020

E. Özsoy, *Geophysical Fluid Dynamics I*, Springer Textbooks in Earth Sciences, Geography and Environment, https://doi.org/10.1007/978-3-030-16973-2_4

where

$$Ro = \frac{U_0}{\Omega_0 L_0}, \quad \text{and} \quad Re = \frac{U_0 L_0}{\nu} \tag{4.5}$$

are dimensionless parameters, as the *Rossby number* (Ro) and the *Reynolds Number* (Re).

The quantities U_0 and L_0 are the characteristic velocity and length scales of the motion. It may be noted that the pressure is scaled as $\rho U_0{}^2$ since ∇p is a reaction to the motion note that the hydrostatic and centrifugal effects were removed from the pressure as per Sect. (3.6.1).

We noted that the inverse of the Rossby Number ($1/Ro$) measures the relative importance of Coriolis terms as compared to the inertial terms. On the other hand, the inverse of the Reynolds number ($1/Ro$) measures the relative importance of viscous terms as compared to the inertial terms.

The Coriolis terms are significant if $Ro = O(1)$, i.e. if the time scale of the motion $L_0/U_0 = O(\Omega_0{}^{-1})$. For the earth, $\Omega_0{}^{-1} = 1$ day. Therefore, for the Coriolis terms to be significant, the length scale should be larger than $L_0 \cong 100$ km for a typical velocity scale of $U_0 = 1$ m/s, and larger than $L_0 \cong 1$ km for $U_0 = 1$ cm/s. That is, Coriolis terms are significant only in large scale motions (for example: geophysical fluids). However, we also note that for smaller scale motion (say $L_0 = 1$ m), the coriolis terms can be important if the velocity scale is sufficiently small, (e.g. $U_0 = 10^{-3}$cm/s.).

In the following sections, we will neglect Coriolis terms for the sake of simplicity, promising to take up on the rotational aspects later.

The parameter controlling the characteristics of the motion in the absence of rotation is then the Reynolds' Number. We will first assume that $Re = O(1)$ and study viscous flow. The limiting cases $Re \gg O(1)$ and $Re \ll O(1)$ will be studied separately in later sections.

4.2 Steady Unidirectional Flow

The solutions of the governing Eqs. (4.2) and (4.3) constitutes a large portion of fluid dynamics. However, the main difficulty in solving these equations arises from the nonlinear (convective) terms in the material derivative of Eq. (4.3)

$$\frac{D\mathbf{u}}{Dt} = \frac{\partial \mathbf{u}}{\partial t} + \mathbf{u} \cdot \nabla \mathbf{u} \tag{4.6}$$

(the last term). These terms can be neglected for certain flow conditions, and the resulting linear equations are simpler to solve.

One such simple case arises if the velocity vector has the same direction everywhere (unidirectional flow) and is independent of distance in the flow direction. The later condition is implied by unidirectional flow. If we let u be the only nonzero component of the velocity $\mathbf{u} = (u, v, w)$, i.e. if $v = w = 0$, then the continuity Eq. (4.2) implies

$$\frac{\partial u}{\partial x} = -\left(\frac{\partial v}{\partial y} + \frac{\partial w}{\partial z}\right) = 0 \tag{4.7}$$

So that u is independent of the distance x along the flow direction, and the only component contributing to the nonlinear term $(\mathbf{u} \cdot \nabla)\mathbf{u}$ also and vanishes identically due to (4.7).

$$u\frac{\partial u}{\partial x} = 0 \tag{4.8}$$

Since the nonlinear term vanishes completely, the x component of the equation of motion (4.3) takes the form

$$\rho\frac{\partial u}{\partial t} = -\frac{\partial p}{\partial x} + \mu\left(\frac{\partial^2 u}{\partial y^2} + \frac{\partial^2 u}{\partial z^2}\right) \tag{4.9}$$

The y and z components of Eq. (4.3) reduce to

$$\frac{\partial p}{\partial y} = 0\,, \; \frac{\partial p}{\partial z} = 0 \tag{4.10}$$

The first and last terms of Eq. (4.9) do not depend on x, since u is not a function of x. Therefore the first term on the right hand side ($vis.$ pressure) is independent of x. The pressure can not be a function of y and z also by virtue of (4.10). The only functional dependence allowed for pressure is then a dependence on time, so that we can replace

$$\frac{\partial p}{\partial x} = -G(t) \tag{4.11}$$

in (4.9). When G is positive, the pressure gradient represents a uniform body force in the positive x-direction.

In the case of *steady* motion G = constant, and $\partial u/\partial t = 0$, yielding

$$\frac{\partial^2 u}{\partial y^2} + \frac{\partial^2 u}{\partial z^2} = -\frac{G}{\mu} \tag{4.12}$$

which can be solved subject to boundary conditions. The steady Eq. (4.12) expresses a balance between shear stresses and the normal stress due to pressure alone.

4.2.1 Poiseuille Flow

We can study the flow in a long pipe with length l and radius a, generated by differential pressures p_1 and p_2 applied at the two ends. The pressure gradient is

$$-G = (p_1 - p_2)/l \tag{4.13}$$

The Eq. (4.12) can be written in polar coordinates (r, θ) as

$$\frac{1}{r}\frac{\partial}{\partial r}\left(r\frac{\partial u}{\partial r}\right)+\frac{1}{r^2}\frac{\partial^2 u}{\partial \theta^2}=-\frac{G}{\mu} \tag{4.14a}$$

(refer to Appendix 2, Batchelor, 1970 for coordinate transformations). *The no-slip boundary condition*, which implies that the fluid particles on the boundary are attached to the boundary, is expressed as

$$u(r=a,\theta)=0 \tag{4.14b}$$

Since the boundary condition (4.14b) has radial symmetry, the solution to (4.14b) should also have radial symmetry. Therefore the θ-derivatives vanish, leaving

$$r^2\frac{d^2u}{dr^2}+r\frac{du}{dr}=-\frac{G}{\mu}r^2 \tag{4.15}$$

where $u=u(r)$ only. The above ordinary differential equation is the well known Euler's equation, the general solution of which is

$$u=c_1+c_2\ln|r|-\frac{G}{4\mu^2}r^2\,. \tag{4.16}$$

where the first two terms represent the homogeneous solution and the last term represents the particular solution. The logarithmic term gives a singularity at the center of the pipe ($\ln(0)\rightarrow-\infty$) requiring that we choose c_2=0. The remaining constant can be obtained by using the boundary condition (4.14b), so that the solution becomes

$$u=\frac{G}{4\mu}(a^2-r^2)\,. \tag{4.17}$$

The solution corresponds to a parabolic velocity profile cross the Section of the pipe (Fig. 4.1):

The flux of volume Q volume across any Section of the pipe is calculated from

$$Q=2\pi\int_0^a rudr=\frac{\pi Ga^4}{8\mu}=\frac{\pi a^4(p_2-p_1)}{8\mu l}\,. \tag{4.18}$$

This result shows that the volume flux is directly proportional to the pressure drop along the pipe, inversely proportional to the length and viscosity, and proportional to the fourth power of the pipe radius.

The tangential stress at any point is obtained from

$$\tau_{rx}=\tau_{xr}=\mu\frac{\partial u}{\partial r}=-\frac{1}{2}Gr \tag{4.19}$$

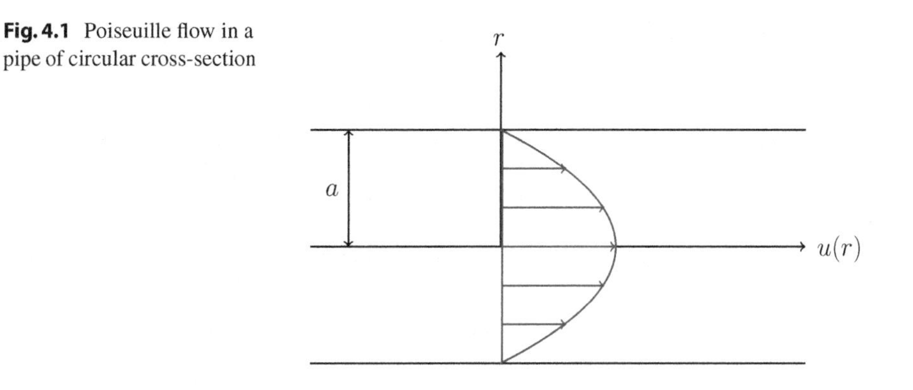

Fig. 4.1 Poiseuille flow in a pipe of circular cross-section

i.e. it increases linearly from zero at the center to a maximum of

$$\tau_{max} = \tau_{rx}(r = a) = -\frac{1}{2}Ga \tag{4.20}$$

at the pipe boundary. The total frictional force acting on the fluid at the pipe boundary is (substituting 4.20, 4.13)

$$F = 2\pi a l \tau_{max} = -\pi a^2 l G = \pi a^2 (p_1 - p_2) \tag{4.21}$$

i.e. equal and opposite to the differential pressure force applied to the pipe cross section.

4.2.2 Flow Between Plates

Consider the unidirectional flow between two infinite plates. If we assume the plates are perpendicular to the y-axis, Eq. (4.12) becomes

$$\frac{d^2u}{dy^2} = -\frac{G}{\mu} \,. \tag{4.22}$$

Let the two plates be located at $y = 0$ and $y = d$, with the plate at $y = 0$ fixed and the plate at $y = d$ moving with velocity U (Fig. 4.2).

The solution to (4.22) with no-slip boundary conditions

$$u(0) = 0 \,,\; u(d) = U \tag{4.23}$$

is

$$u = \frac{G}{2\mu} y(d - y) + \frac{Uy}{d} \tag{4.24}$$

When the upper plate is not moving ($U = 0$), the velocity profile is parabolic. When the applied pressure gradient is zero ($G = 0$), the profile is linear with maximum velocity at the top. In general, the flow is a superposition of the two cases.

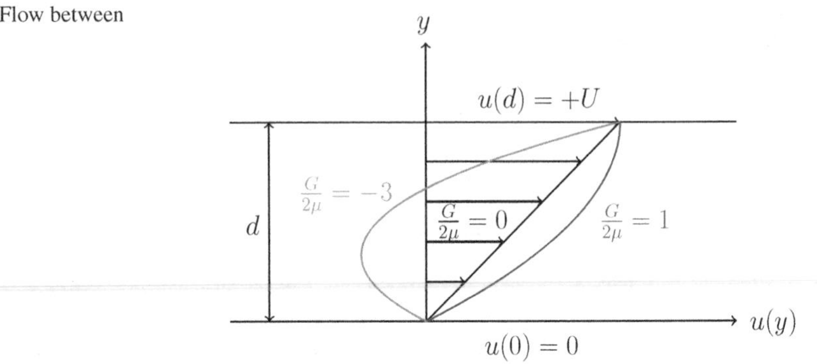

Fig. 4.2 Flow between plates

4.3 Unsteady Unidirectional Flow

It is shown in Sect. 4.2, that if a pressure gradient exists in unidirectional flow, it can only be a function of time. The flow is described by Eq. (4.9) where the functional form of the pressure gradient supplied by (4.11), is used as the forcing function.

The unsteady motion can also force a flow to be generated. Taking only this kind of forcing into account and assuming that the pressure gradient forcing does not exist ($G = 0$), Eq. (4.9) is simplified into

$$\frac{\partial u}{\partial t} = \nu \left(\frac{\partial^2 u}{\partial y^2} + \frac{\partial^2 u}{\partial z^2} \right) \tag{4.25}$$

This equation is analogous to the so-called 'heat equation' (or diffusion equations), which describes the diffusion of the flow velocity in the $y - z$ plane with diffusivity ν. Solutions to the diffusion equation under different boundary and initial conditions can be found in the literature (A good reference is the book by Carslaw and Jaeger, Oxford, 1947).

An elementary solution to (4.25) is that corresponding to the case of an infinite medium with concentrated head at a point $y = y', z = z'$ and zero elsewhere at initial time t = 0. This problem is equivalent to the solution of (4.25) with the initial condition.

$$u_0(y, z) \equiv u(y, z, 0) = \delta(y - y', z - z') \tag{4.26}$$

where the right hand side is defined as the Dirac delta 'function'. The important properties of this 'function' are that

$$\int\int_{-\infty}^{\infty} \delta(y - y', z - z')\, dydz \equiv 1 \tag{4.27a}$$

(i.e. the 'function' has a total area of unity), and that

$$\int\int_{-\infty}^{\infty} \delta(y - y', z - z') f(y, z)\, dydz \equiv f(y', z') \tag{4.27b}$$

(i.e. it takes on the value at y', z' of a function with which it is integrated).

The reason that the 'function' is written in quotes is that while it was introduced in 1926 by Dirac as a function, it was letter defined rigorously by Schwartz (1950) as a generalized function (or distribution). It can be visualized as the limiting case of a set of functions which attain large values at the origin (y', z') and decay much rapidly outside with the radial distance $r = [(y - y')^2 + (z - z')^2]^{1/2}$, so that in the limit these functions yield zero values everywhere except (y', z').

The solution to (4.25) with the initial condition (4.26) is expected to have radial symmetry with respect to the radial distance r measured from the source (y', z'). Writing (4.25) in polar coordinates ($\frac{\partial}{\partial\theta} = 0$ due to radial symmetry) we have

$$\frac{\partial u}{\partial t} = \frac{1}{r}\frac{\partial}{\partial r} r \frac{\partial u}{\partial r} \tag{4.28}$$

Equation (4.28) can be solved by a *similarity transform* of the following form

$$u(r, t) = u(y, z, t) = f(\eta)/\zeta \tag{4.29a}$$

where η and ζ are the variables transformations defended as selected to be

$$\begin{aligned} \eta &= \tfrac{r^2}{4\nu t} \\ \zeta &= 4\nu t \end{aligned} \tag{4.29b}$$

The purpose of the similarity transform is to reduce the governing partial differential equation to an ordinary differential equation. To see whether this is possible with the choice of (4.29a,b) we substitute these forms into (4.28). Noting first that

$$\begin{aligned} \frac{\partial u}{\partial t} &= \frac{\partial u}{\partial \zeta}\frac{\partial \zeta}{\partial t} + \frac{\partial u}{\partial \eta}\frac{\partial \eta}{\partial t} \\ &= -\frac{f}{\zeta^2}\frac{\partial \zeta}{\partial t} + \frac{\partial f}{\partial \eta}\frac{1}{\zeta}\frac{\partial \eta}{\partial t} \\ &= -\frac{f}{\zeta^2}4\nu - \frac{r^2}{4\nu t^2}\frac{1}{\zeta}\frac{\partial f}{\partial \eta} \\ &= -\frac{4\nu}{\zeta^2}\left[\eta f_\eta + f\right] \end{aligned}$$

and

$$\begin{aligned} \frac{1}{r}\frac{\partial}{\partial r} r \frac{\partial u}{\partial r} &= \frac{1}{r}\frac{\partial \eta}{\partial r}\frac{\partial}{\partial \eta}\left(r\frac{\partial \eta}{\partial r}\frac{\partial u}{\partial \eta}\right) \\ &= \frac{1}{r}\frac{2r}{4\nu t}\frac{\partial}{\partial \eta}\left(r\frac{2r}{4\nu t}\frac{1}{\zeta}\frac{\partial f}{\partial \eta}\right) \\ &= \frac{2}{\zeta}\frac{\partial}{\partial \eta}\left(2\eta\frac{1}{\zeta}\frac{\partial f}{\partial \eta}\right) \\ &= \frac{4}{\zeta^2}\left(\eta\frac{\partial^2 f}{\partial \eta^2} + \frac{\partial f}{\partial \eta}\right), \end{aligned}$$

Equation (4.28) takes the form of

$$\eta \frac{\partial^2 f}{\partial \eta^2} + (1+\eta)\frac{\partial f}{\partial \eta} + f = 0\,, \tag{4.30}$$

which can equivalently be written as

$$\frac{\partial}{\partial \eta}\left[\eta\left(\frac{\partial f}{\partial \eta} + f\right)\right] = 0 \tag{4.31}$$

The ordinary differential Eq. (4.31) can now be solved to determine the similarity function $f(\eta)$. Integrating once yields

$$\eta\left(\frac{\partial f}{\partial \eta} + f\right) = C = \text{Constant} \tag{4.32}$$

Since this should be valid for all η, the only possible constant on the right hand side of (4.32) should be $C = 0$. Integrating (4.32) once more yields

$$f = Ae^{-\eta} \tag{4.33}$$

as the solution. The constant A is obtained from the initial condition. Integrating (4.26) in the $y - z$ plane, yields

$$\int\int_{-\infty}^{+\infty} u(y,z,0)dydz = \int\int_{-\infty}^{+\infty} \delta(y-y', z-z')dydz \equiv 1 \tag{4.34}$$

by virtue of (4.27a). The left hand side of (4.34) can be evaluated as

$$\begin{aligned}
\int\int_{-\infty}^{+\infty} u(y,z,0)dydz &= 2\pi \int_0^{+\infty} u(r,0)rdr \\
&= 2\pi\left[\int_0^{\infty} u(r,t)rdr\right]_{t=0} \\
&= 2\pi\left[\int_0^{\infty} \frac{f(\eta)}{\zeta}\cdot\frac{\zeta}{2}d\eta\right]_{\zeta=0} \\
&= \pi\int_0^{\infty} f(\eta)d\eta \\
&= \pi A\int_0^{\infty} e^{-\eta}d\eta \\
&= \pi A
\end{aligned}$$

So that comparison with (4.34) yields $A = 1/\pi$.

The solution to the initial value problem (4.25) and (4.26) is therefore (cf. 4.29a, 4.29b and 4.33):

$$u(y,z,t) = u(r,t) = \frac{1}{4\pi\nu t}e^{-\frac{r^2}{4\nu t}} \tag{4.35}$$

where

$$r^2 = (y - y')^2 + (z - z')^2 \tag{4.36}$$

The advantage of the *principal solution* (4.35) with a delta function initial condition lies in the fact that a solution for arbitrary initial conditions can be constructed from the principal solution. Consider the problem specified as

$$L(u^\star) = 0 \tag{4.37a}$$

$$u^\star(y, z, 0) = F(y, z) \tag{4.37b}$$

where

$$L = \frac{\partial}{\partial t} - \nu \left(\frac{\partial^2}{\partial y^2} + \frac{\partial^2}{\partial z^2} \right) \tag{4.37c}$$

is the diffusion operator. Comparing this problem with (4.25 + 4.26), the solution is constructed from

$$u^\star(y, z, t) = \int \int_{-\infty}^{+\infty} F(y', z'u(y - y', z - z', t)dy'dz' \,. \tag{4.38}$$

The proof is given as follows. Operating on (4.38) yields

$$L(u^\star) = \int \int_{-\infty}^{+\infty} F(y', z')L(u)dy'dz' = 0$$

Since L operates on y, z, t only. This proves that the solution (4.38) satisfies the Eq. (4.37a). To see that the boundary conditions are also satisfied, we evaluate (4.38) at $t = 0$:

$$\begin{aligned} u^\star(y, z, 0) &= \int \int_{-\infty}^{+\infty} F(y', z')u(y - y', z - z', 0)dy'dz' \\ &= \int \int_{-\infty}^{+\infty} F(y', z')\delta(y - y', z - z')dy'dz' \\ &= F(y, z) \end{aligned}$$

by virtue of (4.26) and (4.27b).

The general solution of (4.27a,b) in an infinite plane is given by (4.38). Substituting (4.35), (4.36) and (4.37b), it takes the following form:

$$u^\star(y, z, t) = \frac{1}{4\pi\nu t} \int \int_{-\infty}^{\infty} u^\star(y', z', 0)\exp\left\{ -\frac{(y - y')^2 + (z - z')^2}{4\nu t} \right\} dy'dz' \,. \tag{4.39}$$

4.3.1 Example—Smoothing of a Velocity Discontinuity

Consider the following initial conditions for one dimensional undirectional flow of two streams in $y - z$ plane each with velocity U flowing opposite directions.

$$u(y, z, 0) = \begin{cases} +U & y>0 \\ -U & y<0. \end{cases} \tag{4.40}$$

The initial flow is sketched below (Fig. 4.3).

The flow is equivalent to a sheet vortex at $y = 0$. The velocity discontinuity will be smoothed-out with time according to the diffusion Eq. (4.25). The solution can be obtained through (4.39):

$$u(y, z, t) = \frac{1}{4\pi\nu t}\int_{-\infty}^{+\infty}\left\{\int_{-\infty}^{0}(-U)e^{-\frac{(y-y')^2+(z-z')^2}{4\nu t}}\,dy' + \int_{0}^{+\infty}(+U)e^{-\frac{(y-y')^2+(z-z')^2}{4\nu t}}\,dy'\right\}dz' . \tag{4.41}$$

Making the substitutions

$$\eta = \frac{y}{(4\nu t)^{1/2}} \ , \ \eta' = \frac{y - y'}{(4\nu t)^{1/2}} \ , \ d\eta' = -\frac{dy'}{(4\nu t)^{1/2}}$$

$$\zeta = \frac{z}{(4\nu t)^{1/2}} \ , \ \zeta' = \frac{z - z'}{(4\nu t)^{1/2}} \ , \ d\zeta' = -\frac{dz'}{(4\nu t)^{1/2}}$$

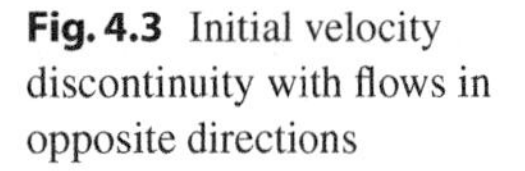

Fig. 4.3 Initial velocity discontinuity with flows in opposite directions

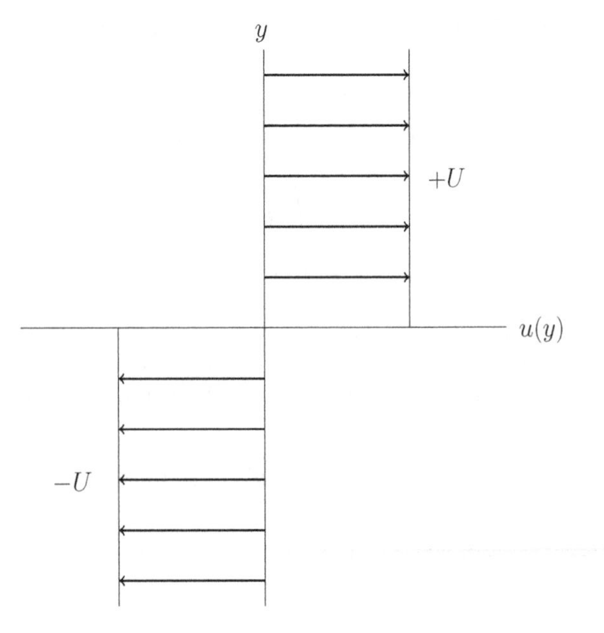

in the above integral yields

$$u(y,z,t) = u(\eta,\zeta) = \frac{1}{\pi}\left(-\int_{+\infty}^{-\infty} e^{-\zeta'^2} d\zeta'\right)\left(-\int_{+\infty}^{\eta} (-U)e^{-\eta'^2} d\eta' - \int_{\eta}^{-\infty} (+U)e^{-\eta'^2} d\eta'\right) \tag{4.42}$$

Here we define the *error function*

$$\mathrm{erf}(x) = \frac{2}{\sqrt{\pi}} \int_0^x e^{-x'^2} dx' \tag{4.43}$$

which is sketched below (Fig. 4.4).

The integral (4.43) is then written as

$$\begin{aligned} u(\eta,\zeta) &= \frac{U}{2}\left(\frac{2}{\sqrt{\pi}} 2\int_0^{\infty} e^{-\zeta'^2} d\zeta'\right)\frac{1}{\sqrt{\pi}}\left(-\int_{\eta}^{\infty} e^{-\eta'^2} d\eta' + \int_{-\infty}^{\eta} e^{-\eta'^2} d\eta'\right) \\ &= \frac{U}{2\sqrt{\pi}}\, 2\,\mathrm{erf}(\infty)\left(\left\{-\left[\int_0^{\infty} - \int_0^{\eta}\right] + \left[\int_{-\infty}^{0} + \int_0^{\eta}\right]\right\} e^{-\eta'^2} d\eta'\right) \\ &= U\frac{2}{\sqrt{\pi}} \int_0^{\eta} e^{\eta'^2} d\eta' \\ &= U\,\mathrm{erf}(\eta) \end{aligned} \tag{4.44}$$

Alternatively,

$$\begin{aligned} u(y,t) &= U\,\mathrm{erf}\left[\frac{y}{\sqrt{4\nu t}}\right] \\ &= U\frac{2}{\sqrt{\pi}} \int_0^{\frac{y}{\sqrt{4\nu t}}} e^{-\frac{y'^2}{4\nu t}} d\left(\frac{y'}{(4\nu t)^{1/2}}\right) \\ &= \frac{U}{\sqrt{\pi\nu t}} \int_0^{y} e^{-\frac{y'^2}{4\nu t}} dy' \end{aligned} \tag{4.45}$$

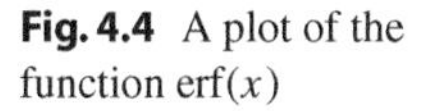
Fig. 4.4 A plot of the function erf(x)

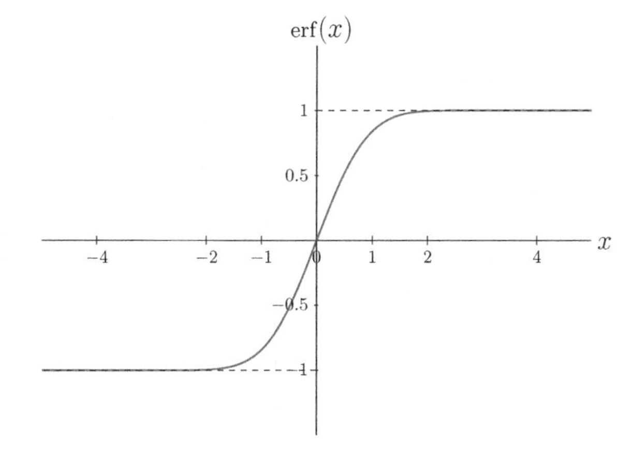

4.3.2 Flow Due to an Oscillating Plane Boundary

Consider the unsteady flow generated by an oscillating plane boundary in a semi-infinite domain. The governing equation is

$$\frac{\partial u}{\partial t} = \nu \frac{\partial^2 u}{\partial y^2} \tag{4.46}$$

The plane boundary located at $y = 0$ has a velocity predescribed as by

The no-slip boundary condition requires that the velocity at the oscillating plate is specified as

$$u(0, t) = Re\left\{U e^{i\omega t}\right\} \tag{4.47}$$

where U is an appropriate constant specifying the amplitude of the oscillating velocity at the plate boundary.

The solution in the half space $y > 0$ must have a similar sinusoidal time dependence

$$u(y, t) = Re\left\{F(y) e^{i\omega t}\right\} \tag{4.48}$$

fluctuating with frequency ω, and amplitude $F(y)$ which depends on distance y from the boundary. An additional boundary condition is obtained by requiring that the velocity remains finite at large distance away from the boundary:

$$u(y \to \infty, t) \to \infty \tag{4.49}$$

Substituting (4.48) into (4.46) we have

$$\frac{d^2 F}{dy^2} - i\frac{\omega}{\nu} F = 0 \tag{4.50}$$

whose characteristic equation is

$$m^2 = i\left(\frac{\omega}{\nu}\right) \tag{4.51}$$

The solution is

$$F = Ae^{m_1 x} + Be^{m_2 x} \tag{4.52}$$

where m_1, m_2 are obtained from (4.51):

$$\begin{aligned} m_{1,2} &= \left(i\frac{\omega}{\nu}\right)^{1/2} = \left[e^{i(\frac{\pi}{2}+2n\pi)}\frac{\omega}{\nu}\right]^{1/2} \\ &= \left(\frac{\omega}{\nu}\right)^{1/2} e^{i(\frac{\pi}{4}+n\pi)} \\ &= \begin{cases} \left(\frac{\omega}{\nu}\right)^{1/2} & \left[\frac{(1+i)}{\sqrt{2}}\right] \\ \left(\frac{\omega}{\nu}\right)^{1/2} & \left[-\frac{(1+i)}{\sqrt{2}}\right] \end{cases} \end{aligned} \tag{4.53}$$

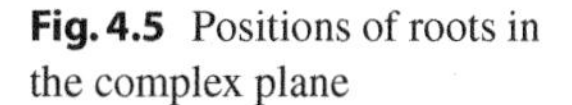
Fig. 4.5 Positions of roots in the complex plane

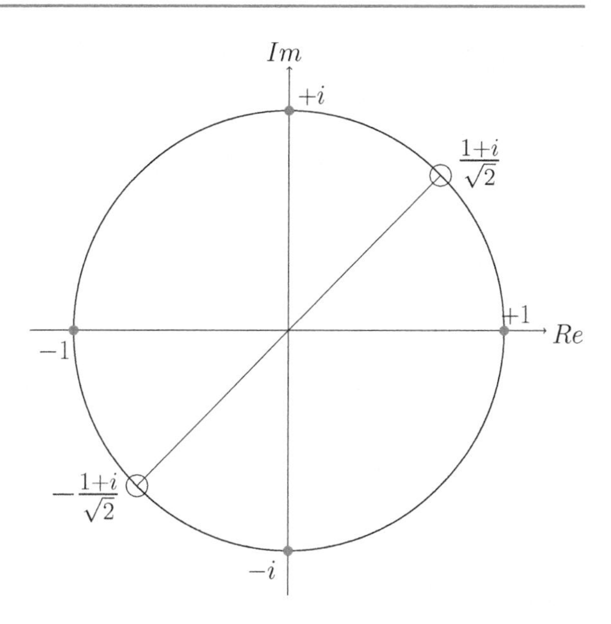

We must have ($A = 0$) in (4.52) in order to have a finite solution as $y \to \infty$ (cf. 4.49) (Fig. 4.5). Then, the solution is

$$F(y) = Be^{m_2 x} = Be^{-\sqrt{\frac{\omega}{2\nu}}(1+i)y} \tag{4.54}$$

(4.47) requires that $B = U$, so that we can write the solution (4.48) as

$$\begin{aligned} U(y,t) &= U\,Re\left\{e^{-\sqrt{\frac{\omega}{2\nu}}y} e^{i\left(\omega t - \sqrt{\frac{\omega}{2\nu}}y\right)}\right\} \\ &= Ue^{-\sqrt{\frac{\omega}{2\nu}}y} \cos\left(\omega t - \sqrt{\frac{\omega}{2\nu}}y\right) \end{aligned} \tag{4.55}$$

This solution is a decaying transverse "wave" with wavelength $2\pi(2\nu/\omega)^{1/2}$ and e-folding distance of $(2\nu/\omega)^{1/2}$) (Fig. 4.6).

Note that the solution decays very rapidly away from the boundary. For example the ratio of the velocities at two points separated by one wavelength is

$$\frac{u(y + 2\pi(2\nu/\omega)^{1/2}, t)}{u(y,t)} = \frac{e^{-[(\frac{\omega}{2\pi})y+2\pi]}}{e^{-\frac{\omega}{2\nu}y}} = e^{-2\pi} \cong 0.002$$

Near the boundary the motion is confined near the boundary to a 'penetration depth' of order $(\nu/\omega)^{1/2}$ which is smaller than the wavelength. When the frequency of oscillation is large as compared to the diffusivity of momentum penetration depth is very small. This means that if the time scale of the oscillation is smaller than the

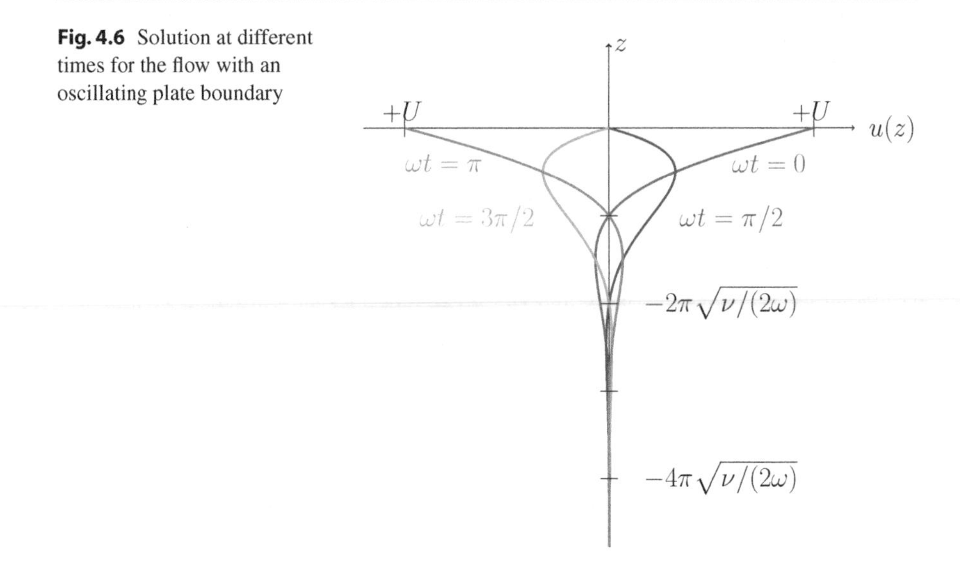

Fig. 4.6 Solution at different times for the flow with an oscillating plate boundary

time scale of diffusion, momentum can not find sufficient time to be diffused before a change occurs in the velocity of the boundary. Since the motion is confined in a thin layer near the boundary, we call this region a 'boundary layer', in this case an 'oscillatory boundary layer'.

4.4 Unidirectional Flow Including Coriolis Effects

4.4.1 The Steady Ekman Layer at the Surface

A special case of unidirectional flow occurs if we include the Coriolis terms in the equation of motion. We assume $Ro = O(1)$(cf. Sect. 4.1). We suppose the fluid is bounded by free surface on top where a constant wind stress is applied. The pressure (modified to incorporate effects of gravity and centrifugal force as in Eq. 4.4) is also assumed to be uniform in a horizontal plane (no horizontal pressure gradient). We take z-axis perpendicular to the free surface (positive upwards). We further assume that the horizontal velocity $\mathbf{u}_h = (u, v)$ is much larger than the vertical velocity w, so that essentially we take $w = 0$. It follows that the vertical (modified) pressure gradient also vanishes. The non-linear terms also vanish because the motion is driven by uniform wind stress, resulting in uniform velocity. Therefore must be uniform horizontally. Therefore the motion is described by (cf. Eq. 4.3)

$$2\boldsymbol{\Omega} \times \mathbf{u}_h = \nu \nabla^2 \mathbf{u}_h \tag{4.56}$$

where $\mathbf{u}_h$ is the horizontal velocity vector which is uniform in the horizontal, but can be a function of the vertical coordinate $\mathbf{u}_n = \mathbf{u}_n(z)$. The only component of $\boldsymbol{\Omega}$

contributing to the left hand side of (4.56) is $\boldsymbol{\Omega}_v = (\boldsymbol{\Omega} \cdot \mathbf{k})\mathbf{k}$, the vertical component of the earth's angular velocity. The only component of the right hand side is $\nu \frac{d^2\mathbf{u}_h}{dz^2}$, since $\mathbf{u}_h = \mathbf{u}_h(z)$ only, so that (4.56) becomes

$$f\mathbf{k} \times \mathbf{u}_h = \nu \frac{d^2\mathbf{u}_h}{dz^2} \tag{4.57}$$

where

$$f = 2|\boldsymbol{\Omega}_v| = 2\boldsymbol{\Omega} \cdot \mathbf{k} = 2|\boldsymbol{\Omega}| \sin \phi \tag{4.58}$$

is the *Coriolis parameter*, calculated in terms of the latitude angle ϕ of any point on the earth's surface. We further assume $f =$ constant, for a limited region on the earth's surface where $\phi \approx$ constant.

Equation (4.57) has the following components:

$$-fv = \nu \frac{d^2u}{dz^2} \tag{4.59a}$$

$$fu = \nu \frac{d^2v}{dz^2} \tag{4.59b}$$

The surface boundary conditions are supplied by the continuity of shear stress across the surface. Without loss of generality we assume that the wind stress τ_w is applied in the x-direction. Therefore the surface stresses must be

$$\mu \frac{du}{dz} = \tau_w \ , \ \mu \frac{dv}{dz} = 0 \ \text{ on } \ z = 0 \tag{4.60a,b}$$

Since the motion is generated by the surface stress it must decay at large distances away from the surface:

$$u \to 0 \ , \ \ v \to 0 \ \text{ as } \ z \to -\infty \tag{4.61a,b}$$

Equations (4.59) can now be solved under the boundary conditions (4.60a,b) and (4.61a,b). For convenience, we define a new complex variable

$$w^\star \equiv u + iv \tag{4.62}$$

so that Eqs. (4.59), (4.60a,b) and (4.61a,b) can be combined as

$$\frac{d^2w^\star}{dz^2} - i\left(\frac{f}{\nu}\right) w^\star = 0 \tag{4.63a}$$

$$\mu \frac{dw^\star}{dz} = \tau_w \ \text{ on } \ z = 0 \tag{4.63b}$$

$$w^{\star} \rightarrow 0 \text{ as } z \rightarrow -\infty \tag{4.63c}$$

Equation (4.63a) is similar to Eq. (4.50) solved in the previous section. The solution satisfying (4.63c) is given as

$$w^{\star} = Ae^{+\sqrt{\frac{f}{2\nu}}(1+i)z} \tag{4.64}$$

and the constant A is found from (4.63b) as

$$A = \frac{\tau_w}{\mu}\sqrt{\frac{\nu}{2f}}(1-i) = (1-i)\tau_w/\rho\sqrt{2f\nu}$$

so that

$$w^{\star} = \frac{(1-i)\tau_w}{\rho\sqrt{2f\nu}}e^{\sqrt{\frac{f}{2\nu}}(1+i)z} \tag{4.65}$$

The velocity components can be found as (cf. 4.62):

$$u = Re(w^{\star}) = \frac{\tau_w}{\rho\sqrt{\nu f}}e^{\sqrt{\frac{f}{2\nu}}z}\cos\left(\sqrt{\frac{f}{2\nu}}z - \frac{\pi}{4}\right) \tag{4.66a}$$

$$v = Im(w^{\star}) = \frac{\tau_w}{\rho\sqrt{\nu f}}e^{\sqrt{\frac{f}{2\nu}}z}\sin\left(\sqrt{\frac{f}{2\nu}}z - \frac{\pi}{4}\right) \tag{4.66b}$$

Note that the velocity components decay in the z-direction. Therefore we have a boundary layer of thickness $\sim \sqrt{\frac{2\nu}{f}}$ near the surface. The magnitude and direction of the velocity vector can be calculated as

$$q \equiv |\mathbf{u}_n| \equiv \sqrt{u^2 + v^2} = \frac{\tau_w}{\rho\sqrt{\nu f}}e^{\sqrt{\frac{f}{2\nu}}z} \tag{4.67a}$$

$$\alpha \equiv \tan^{-1}\left(\frac{v}{u}\right) = \sqrt{\frac{f}{2\nu}}z - \frac{\pi}{4} \tag{4.67b}$$

where α is the angle that the velocity vector makes with the x-axis. The magnitude decreases exponentially with depth. The direction angle of the current starts with $\frac{\pi}{4}$ at the surface and decreases by rotating in the clockwise direction with depth-z. The surface current is at into the wind stress direction. The flow is uni-directional everywhere in the $x - y$ plane for any given depth-z.

Ekman spiral: (projection of velocity vector on $x - y$ plane)

In three dimensions the flow can be sketched for each of the flow components (u, v) in the (x, y) directions (Fig. 4.7):

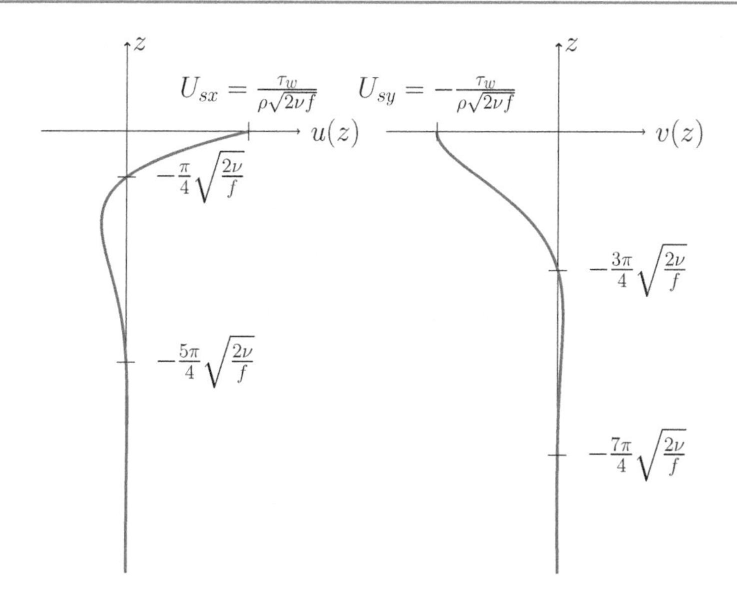

Fig. 4.7 Ekman drift current components $\mathbf{u}(z) = (u(z), v(z))$ variations with depth

The penetration depth $d \equiv \sqrt{\frac{2\nu}{f}}$ can be calculated for water with $\nu = 1 \times 10^{-2}$ cm/s (at 20 °C) and $f = 1.36 \times 10^{-5}\,\text{sec}^{-1}$ (at 36 °N latitude) as $d \equiv \sqrt{\frac{2\nu}{f}} \cong$ 40 cm. However, this penetration depth is calculated using molecular viscosity. The effective viscosity due to turbulent mixing in the sea surface is usually much larger, sometimes by as much as factor of 10^5, and the penetration depth is greater in actual conditions.

The net flux of volume can also be calculated as

$$Q \equiv \int_{-\infty}^{0} w^{\star} dz = \int_{-\infty}^{0} (u + iv) dz = -i\frac{\tau_w}{\rho f} \tag{4.68}$$

which has a magnitude of $\tau_w/\rho f$ in the $-y$ direction. Therefore the net flux of volume is at right angles to the wind stress. The coriolis force by the net flux balances the applied wind stress.

4.4.2 The Steady Ekman Layer at the Bottom

In the previous surface flow case the viscous forces arise due to applied wind stress. We now consider the flow on a rigid bottom where the viscous forces arise as a result of bottom friction.

We consider a large body of water set into steady motion by a uniform gradient of modified pressure which is balanced by the coriolis force. If the uniform pressure

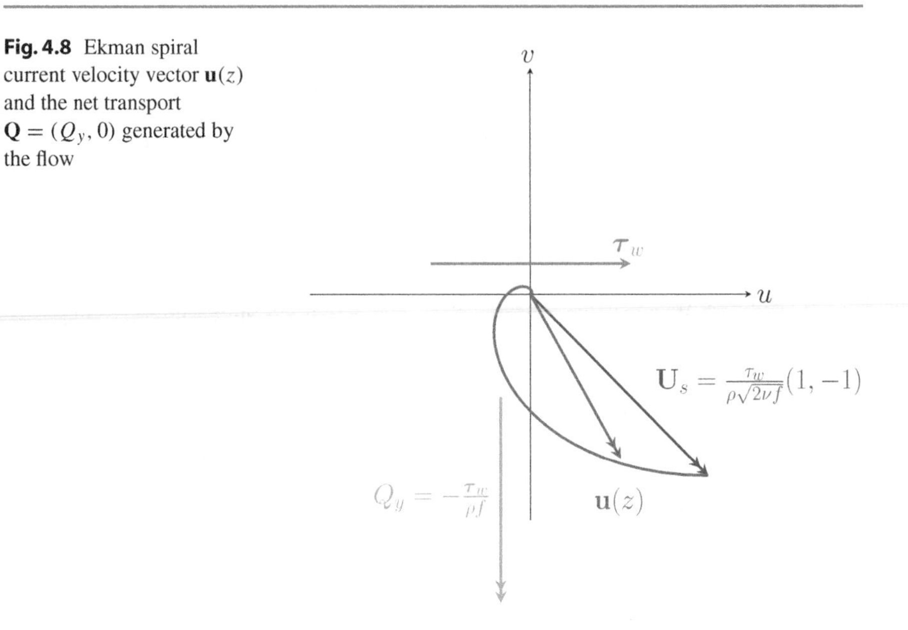

Fig. 4.8 Ekman spiral current velocity vector $\mathbf{u}(z)$ and the net transport $\mathbf{Q} = (Q_y, 0)$ generated by the flow

gradient lies in the horizontal plane (x, y) with components $(0, -G)$, the motion with velocity components (U, V) satisfies

$$-fV = 0 \tag{4.69a}$$

$$fU = \frac{1}{\rho}G \tag{4.69b}$$

So that the velocity component V in the y-direction vanishes and U is the constant velocity in the x-direction which balances the pressure gradient in the y-direction through Coriolis forces. The bulk of the motion obeys (4.69). However, the flow must be modified near the bottom through viscous stresses. Near the bottom, the velocity components (u, v) must satisfy (Fig. 4.8)

$$-fv = \nu\frac{d^2u}{dz^2} \tag{4.70a}$$

$$fu = \frac{1}{\rho}G + \nu\frac{d^2v}{dz^2} \tag{4.70b}$$

As compared to (4.69a,b) these equations are modified by the inclusion of viscous terms. By substituting from (4.69) and noting that $U = U_0 = \frac{1}{\rho}G$ =constant, Eq. (4.70) can be written as

$$-fv = \nu\frac{d^2(u - U_0)}{dz^2} \tag{4.71a}$$

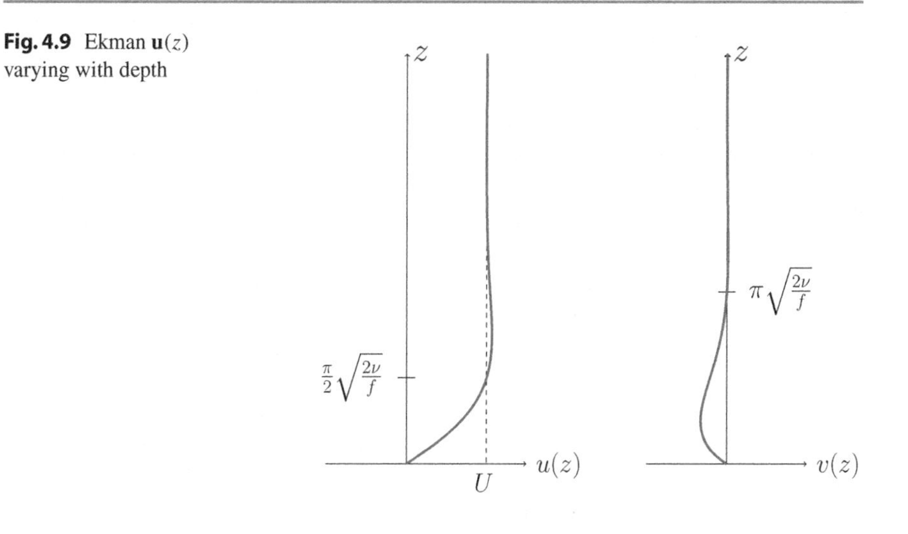

Fig. 4.9 Ekman $\mathbf{u}(z)$ varying with depth

$$f(u - U_0) = \nu \frac{d^2 v}{dz^2} \tag{4.71b}$$

The last equations are the same as Eq. (4.59) considered in the previous Section if u is replaced by $u' = u - U_0$. Thus the solutions obtained in Sect. 4.4.1 can be used. The boundary conditions are different, however. At the bottom ($z = 0$) the no-slip boundary condition must be satisfied (Fig. 4.9).

$$u = 0 \; (u' = -u) \;\; , \;\; v = 0 \;\; \text{on} \;\; z = 0 \tag{4.72a}$$

At large distance from the bottom boundary, the flow must approach the bulk motion

$$u \to U_0 \; (u' \to 0) \;\; , \;\; v \to 0 \;\; \text{as} \;\; z \to \infty \tag{4.72b}$$

The solution of (4.59) or (4.71) satisfying (4.72b) is

$$w^\star \equiv u' + iv = u - U_0 + iv = Ae^{-\sqrt{\frac{f}{2\nu}}(1+i)z} \tag{4.73}$$

and the constant A can be obtained from (4.72a) as A$= -$U. The velocity components are obtained as

$$u = U_0 + u' = U_0 + Re(w^\star) = U_0\left(1 - e^{-\sqrt{\frac{f}{2\nu}}z}\cos\sqrt{\frac{f}{2\nu}}z\right) \tag{4.74a}$$

$$v = Im(w^\star) = U_0 e^{-\sqrt{\frac{f}{2\nu}}z}\sin\sqrt{\frac{f}{2\nu}}z \tag{4.74b}$$

The Ekman spiral has characteristics similar to the surface flow:
In three dimensions the flow may be sketched as follows:

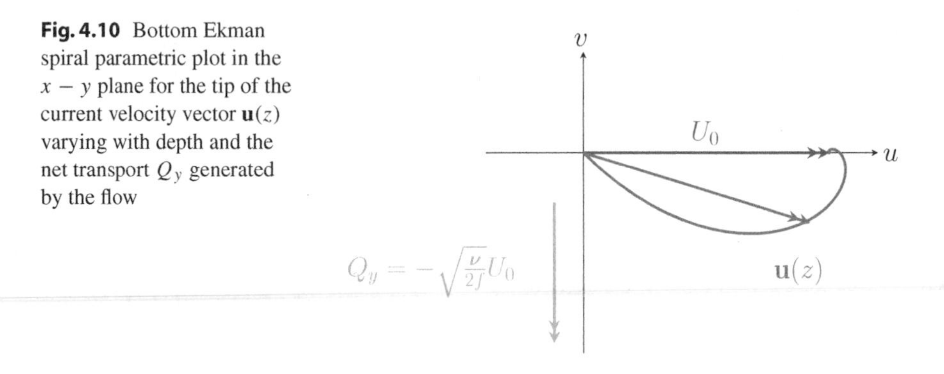

Fig. 4.10 Bottom Ekman spiral parametric plot in the $x - y$ plane for the tip of the current velocity vector $\mathbf{u}(z)$ varying with depth and the net transport Q_y generated by the flow

Near the bottom, the current vectors sum up to a net flow in the y-direction (Fig. 4.10).

The volume flux in the y-direction (normal to the uniform U velocity) can be calculated to be

$$Q_y = \int_0^\infty v dz = \sqrt{\frac{\nu}{2f}}U_0 \tag{4.75}$$

4.4.3 The Oscillatory Ekman Boundary Layer at the Surface

In previous sections (4.4.1 and 4.4.2) we considered steady Ekman layers. We now considered the unsteady case, with the rest of the previous assumptions (undirectional flow). The governing equations for the horizontal flow is

$$\frac{\partial \mathbf{u}}{\partial t} + f\mathbf{k} \times \mathbf{u} = \nu \frac{\partial^2 \mathbf{u}}{\partial z^2} \tag{4.76}$$

where $\mathbf{u}$ is the horizontal velocity with components (u, v) in the (x, y) coordinates. Let the motion be forced by wind stress $\boldsymbol{\tau}_w = ({\tau_x}^w, {\tau_y}^w)$ at the surface. The surface boundary condition is then

$$\mu \frac{\partial \mathbf{u}}{\partial z} = \boldsymbol{\tau}_w \text{ at } z = 0 \tag{4.77a}$$

At large distances away from the surface the fluid velocity must vanish

$$\mathbf{u} \to 0 \text{ as } z \to -\infty \tag{4.77b}$$

The components of Eqs. (4.76), (4.77a,b) can be written as

$$\frac{\partial u}{\partial t} - fv = \nu \frac{\partial^2 u}{\partial z^2} \tag{4.78a}$$

$$\frac{\partial v}{\partial t} + fu = \nu \frac{\partial^2 v}{\partial z^2} \tag{4.78b}$$

$$\mu \frac{\partial u}{\partial z} = \tau_{wx}, \quad \mu \frac{\partial v}{\partial z} = \tau_{wy} \text{ on } z = 0 \tag{4.79a,b}$$

$$u \to 0 \ , \ v \to 0 \text{ as } z \to -\infty \tag{4.80a,b}$$

We can define the following complex variables as we have done in the previous sections

$$w^\star = u + iv \tag{4.81a}$$

$$\tau^\star = \tau_{wx} + i\tau_{wy} \tag{4.81b}$$

Combining (4.78a,b), (4.79a,b) and (4.80a,b) yields

$$\frac{\partial w^\star}{\partial t} + ifw^\star = \nu \frac{\partial^2 w^\star}{\partial z^2} \tag{4.82}$$

$$\mu \frac{\partial w^\star}{\partial z} = \tau^\star \text{ on } z = 0 \tag{4.83}$$

$$w^\star \to 0 \text{ as } z \to -\infty \tag{4.84}$$

We will consider the flow to be driven by an oscillatory wind stress. Most generally an oscillatory wind stress can be expressed as

$$\tau^\star = \Pi_+ e^{i\omega t} + \Pi_- e^{-i\omega t} \tag{4.85}$$

where the complex numbers Π_+ and Π_- represent the anti-clockwise and clockwise rotating components of wind-stress respectively. By choosing appropriate values for these constants, the wind stress vector oscillation can be described according to its sense of rotation. In general, Eq. (4.85) describes an ellipse followed by the tip of the wind stress vector with parametric oscillatory dependence on the time t.

Special cases with suitable special values are listed below as examples

(i) $\Pi_+ = \Pi_- = \frac{1}{2}\Pi_x$ (real)
$\tau^\star = \Pi_x \cos wt$ (real)

(ii) $\Pi_+ = -\Pi_- = \frac{1}{2}\Pi_y$ (real)
$\tau^\star = i\Pi_y \sin wt$ (imaginary)

(iii) $\Pi_+ = \Pi_p \ , \ \Pi_- = 0$ (complex)
$\tau^\star = \Pi_p(\cos \omega t + i \sin \omega t)$ (complex)

(iv) $\Pi_+ = 0 \ , \ \Pi_- = \Pi_n$ (complex)
$\tau^\star = \Pi_n(\cos \omega t - i \sin \omega t)$ (complex)

The solution to (4.82) must also be oscillatory with the same general form. We therefore express

$$w^{\star} = U_{+}(z)e^{i\omega t} + U_{-}(z)e^{-i\omega t} \tag{4.86}$$

where U_{+} and U_{-} are allowed to be functions of depth. Since the problem is linear, solutions in response to anti-clockwise and clockwise rotating components of wind stress can be obtained separately and then combined. Substituting (4.85) and (4.86) upon the corresponding sets of equations become

$$i(f+\omega)U_{+} = \nu\frac{d^2U_{+}}{dz^2} \tag{4.87a}$$

$$\mu\frac{dU_{+}}{dz} = \Pi_{+} \text{ on } z=0 \tag{4.87b}$$

$$U_{+} \to 0 \text{ as } z \to -\infty \tag{4.87c}$$

$$i(f-\omega)U_{-} = \nu\frac{d^2U_{-}}{dz^2} \tag{4.88a}$$

$$\mu\frac{dU_{-}}{dz} = \Pi_{-} \text{ on } z=0 \tag{4.88b}$$

$$U_{-} \to 0 \text{ as } z \to -\infty \tag{4.88c}$$

Let m_1, m_2, be the roots of the characteristic equation

$$m^2 - i\left(\frac{f+\omega}{\nu}\right) = 0 \tag{4.89a}$$

for the first Eq. (4.87a) and n_1, n_2, be the roots of the characteristic equation

$$n^2 - i\left(\frac{f-\omega}{\nu}\right) = 0 \tag{4.89b}$$

The corresponding roots of (4.89a) are

$$m_{1,2} = \mp\left(\frac{f+\omega}{2\nu}\right)^{1/2}(1+i) \tag{4.90a}$$

whereas the roots of (4.89b) are

$$m_{1,2} = \begin{cases} \mp\left(\frac{f-\omega}{2\nu}\right)^{1/2}(1+i) \quad if \ \omega < f \\ \mp\left(\frac{\omega-f}{2\nu}\right)^{1/2}(1-i) \quad if \ \omega > f \end{cases} \tag{4.90b}$$

Therefore the solutions to be accepted with regard to the boundary conditions (4.87c) and (4.88c) differ according to the value of the frequency? We accept the solutions

$$w^{\star} = A_{+}e^{\sqrt{\frac{f+\omega}{2\nu}}z}e^{i\left(\sqrt{\frac{f+\omega}{2\nu}}z+\omega t\right)} + A_{-}e^{\sqrt{\frac{f-\omega}{2\nu}}z}e^{i\left(\sqrt{\frac{f-\omega}{2\nu}}z-wt\right)} \quad \text{if } \omega < f, \tag{4.91a}$$

and

$$w^{\star} = A_{+}e^{\sqrt{\frac{f+\omega}{2\nu}}z}e^{i\left(\sqrt{\frac{f+\omega}{2\nu}}z+\omega t\right)} + A_{-}e^{\sqrt{\frac{\omega-f}{2\nu}}z}e^{i\left(\sqrt{\frac{\omega-f}{2\nu}}z+\omega t\right)} \quad \text{if } \omega > f, \tag{4.91b}$$

The constant can be evaluated through boundary conditions (4.87b) and (4.88b):

$$A_{+} = \frac{(1-i)\Pi_{+}}{\rho\sqrt{2(f+\omega)\nu}} \tag{4.92a}$$

$$A_{-} = \begin{cases} \frac{(1-i)\Pi_{-}}{\rho\sqrt{2(f-\omega)\nu}}, & \text{if } \omega < f \\ \frac{(1+i)\Pi_{-}}{\rho\sqrt{2(\omega-f)\nu}}, & \text{if } \omega > f \end{cases} \tag{4.92b}$$

The components of velocity can then be obtained Eq. (4.81a).

Exercises

Exercise 1

An L-shaped tube is initially filled with a fluid of density ρ. The lower end (A) is closed and the upper end (C) is left open to atmospheric pressure p_a.

Describe the pressure distribution along the tube (Fig. 4.11).

At time $t = t_0$, the lower end at point A is opened. Immediately after opening the lower end, the fluid is accelerated in the tube. Assuming that the fluid velocity is uniform along the tube, calculate the fluid acceleration at time $t = t_{0+} > t_0$ immediately after opening of the lower end.

Sketch the pressure distribution before ($t = t_0$) and immediately after ($t = t_{0+}$) opening the lower end of the tube.

Exercise 2

It is observed that fragments of tea leaves at the bottom of a stirred tea cup conglomerate toward the center. Can this phenomenon be exlained with boundary layer dynamics? Why do the tea leaves accumulate at the center irrespective of the direction of stirring (clockwise or counterclockwise)?

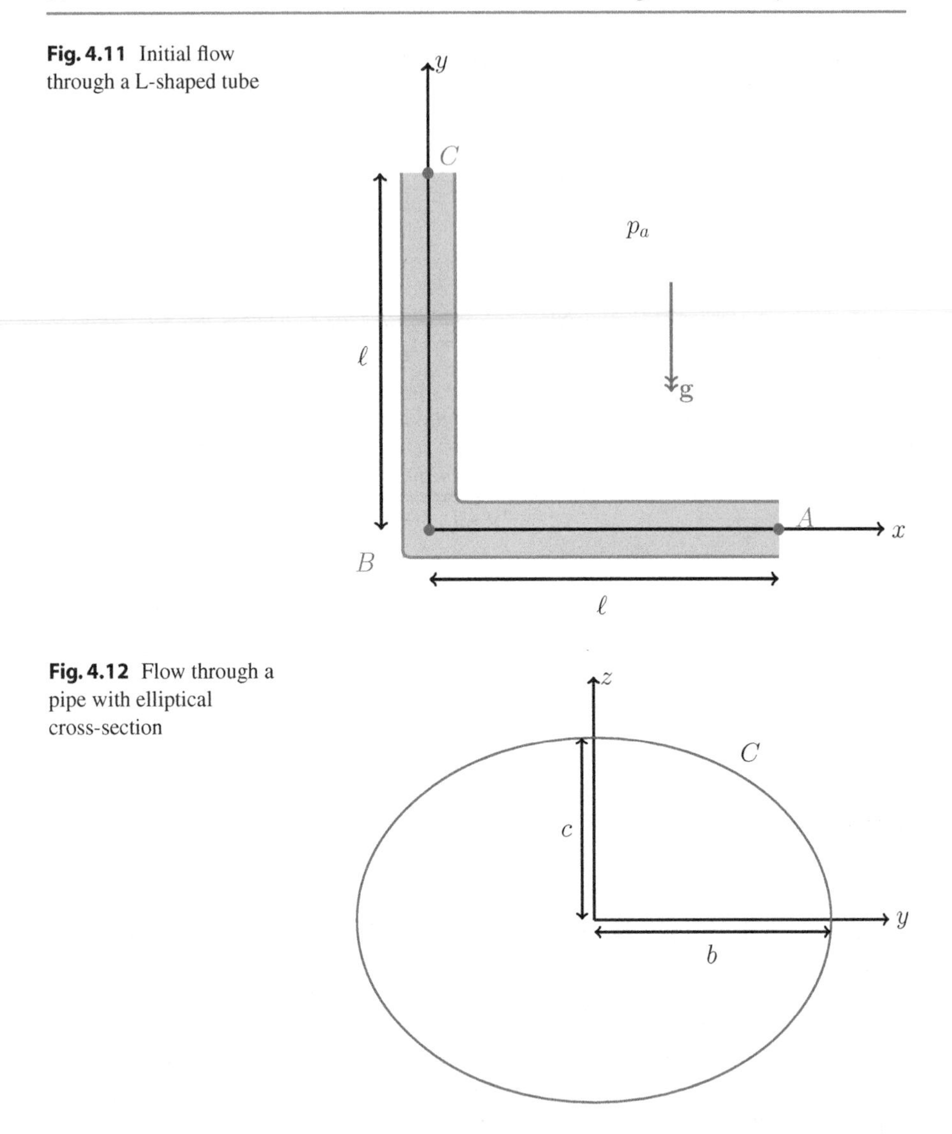

Fig. 4.11 Initial flow through a L-shaped tube

Fig. 4.12 Flow through a pipe with elliptical cross-section

Exercise 3

Consider a pipe with elliptical cross-section (Fig. 4.12), whose boundary C in the (y, z)-plane is given by

$$\frac{y_0^2}{b^2} + \frac{z_0^2}{c^2} = 1.$$

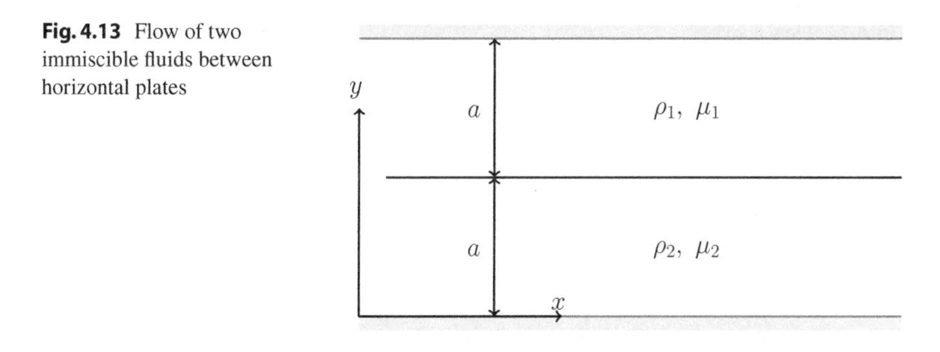

Fig. 4.13 Flow of two immiscible fluids between horizontal plates

The unidirectional viscous flow of a fluid with dynamic viscosity μ through the pipe, subject to a pressure gradient $G = -\frac{dp}{dx}$ is described by the following equation

$$\frac{\partial^2 u}{\partial y^2} + \frac{\partial^2 u}{\partial z^2} = -\frac{G}{\mu},$$

with the boundary condition

$$u(\text{on boundary C} : (\text{y}_0, \text{z}_0)) = 0.$$

Test if the trial solution

$$u = A + By + Cz + Dy^2 + Ez^2$$

would satisfy the above viscous flow equation and boundary conditions, with appropriate coefficients A, B, C, D, E.

`Exercise 4`

Consider the steady flow of two immiscible and incompressible fluids of equal depth a, between horizontal plates separated by a distance of $2a$. The lighter density fluid of density ρ_1 and dynamic viscosity μ_1 lies on top of the other with respective properties ρ_2 and μ_2 (Fig. 4.13).

The flow is driven by a constant pressure gradient $\frac{\partial p}{\partial x}$. The conditions required at the interface is that the velocity and shear stress be continuous. Determine the velocity distribution.

`Exercise 5`

Consider the uniform, steady flow of a viscous fluid above a horizontal plate. The plate has small perforations through which fluid is sucked out with uniform velocity V_0. The flow is independent of x, with a free stream velocity of U_0 (as $z- > \infty$), and without any pressure gradients ($\frac{\partial p}{\partial x} = 0$).

Determine the velocity components for this flow. Also determine the momentum flux across the plate and the force applied per unit area of the plate.

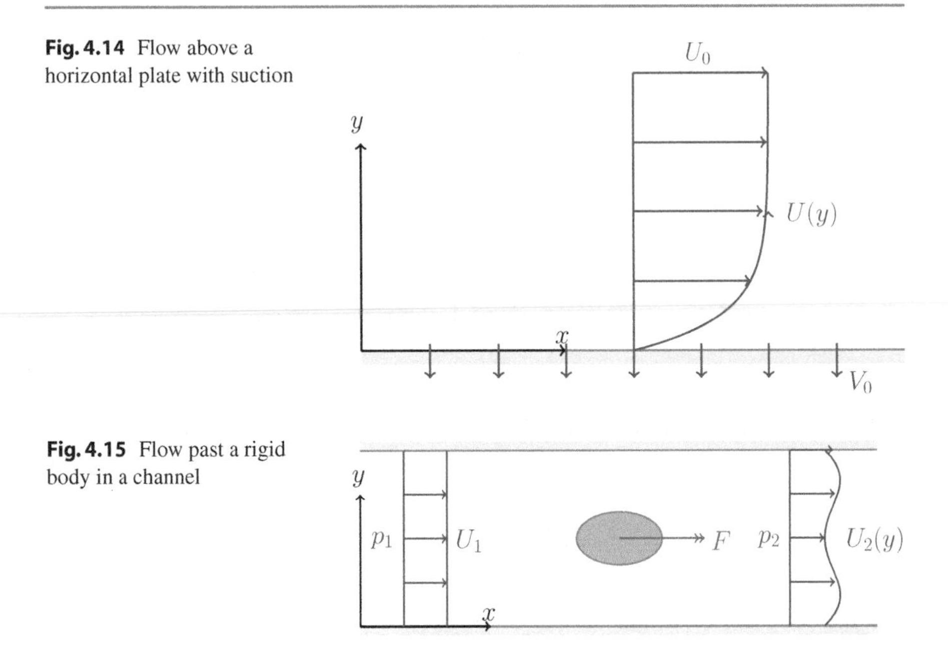

Fig. 4.14 Flow above a horizontal plate with suction

Fig. 4.15 Flow past a rigid body in a channel

Exercise 6

A rigid body is placed in a horizontal channel with cross sectional area A, where a uniform flow with velocity U_1 is approaching at it in the x-direction (Fig. 4.14).

The flow is disturbed past the object, to produce a variable velocity profile $U_2(y, z)$ in the $y - z$-plane. Assume that friction is negligible at the channel boundaries and that the pressure field is uniform in cross section, only having a gradient in the x-direction, defined by the pressure variation $p(x)$ (Fig. 4.15).

How can one determine the force applied on the rigid body if pressure $p(x)$ and velocity profiles $U(y, z)$ were to be measured at sections before and after the rigid body?

Exercise 7

Discuss coastal upwelling in the case of a semi-infinite ocean adjacent to a coast. Try to describe the phenomenon based on Ekman layer dynamics in a homogeneous ocean. Under which conditions do you expect (a) upwelling or (b) downwelling. How do you expect the surface properties to change in a stratified ocean with water properties changing with depth. What is the reason for the property changes at the surface? Do you expect changes in air-sea fluxes of heat and water mass in an upwelling region? Do you expect changes in the productivity of the ocean?

Rotating, Homogeneous, Incompressible Fluids

5

5.1 Equations of Motion for an Incompressible Homogeneous Fluid

In Chap. 3, the governing equations applied to a homogeneous ($\rho = constant$), incompressible ($D\rho/Dt = 0$) fluid in a rotating, non-inertial frame of reference were derived. For an incompressible, homogeneous fluid these Eqs. (3.1b) and (3.56) are simplified to become

$$\nabla \cdot \mathbf{u} = 0 \tag{5.1)[3.1b]}$$

and

$$\frac{D\mathbf{u}}{Dt} + 2\boldsymbol{\Omega} \times \mathbf{u} = -\frac{1}{\rho}\nabla p + \nu\nabla^2\mathbf{u}. \tag{5.2)[3.56]}$$

As a side product of these governing equations, an equation for the vorticity balance was also derived. Simplifying again for homogeneous, incompressible fluids, the vorticity Eq. (3.68) reads:

$$\frac{D\boldsymbol{\omega}_A}{Dt} = \boldsymbol{\omega}_A \cdot \nabla\mathbf{u} + \nu\nabla^2\boldsymbol{\omega}_A \tag{5.3)[3.68]}$$

where $\boldsymbol{\omega}_A = \boldsymbol{\omega} + 2\boldsymbol{\Omega}$ has been referred to as absolute vorticity. In addition, an equation governing the circulation around a closed curve was also derived, expressing the balance of vorticity. For a homogeneous fluid, we have

$$\frac{d\Gamma}{dt} = \nu \int_S \nabla^2\boldsymbol{\omega} \cdot \mathbf{n}\, dS - 2\Omega\frac{dS_p}{dt} \tag{5.4)[3.80]}$$

where Γ is the circulation around the closed curve C enclosing the surface S, and S_p is the projection of the surface S on the plane perpendicular to the angular velocity vector $\boldsymbol{\Omega}$. We recall the above result as *Kelvin's circulation theorem.*

© Springer Nature Switzerland AG 2020

E. Özsoy, *Geophysical Fluid Dynamics I*, Springer Textbooks in Earth Sciences, Geography and Environment, https://doi.org/10.1007/978-3-030-16973-2_5

Finally, *Bernoulli's theorem* was stated for steady flow of a homogeneous incompressible fluid, based on

$$\nabla H = \mathbf{u} \times \omega_A + \nu \nabla^2 \mathbf{u} \tag{5.5a)[3.83a]}$$

where

$$H = \frac{1}{2}\mathbf{u} \cdot \mathbf{u} + \frac{p'}{\rho} - \mathbf{g} \cdot \mathbf{x} - \frac{1}{2}(\boldsymbol{\Omega} \times \mathbf{x}) \cdot (\boldsymbol{\Omega} \times \mathbf{x}). \tag{5.5b)[3.83b]}$$

Note that p' represents the *fluid pressure* in (5.5b), while the notation p in Eq. (5.2) represents the *modified pressure*, replacing the last three terms of (5.5b).

These basic equations will be considered in this volume, since our subject will only cover the motion of homogeneous, incompressible fluids with respect to a rotating frame of reference. We will elaborate the effects of rotation, leading to important new types of behaviour of *geophysical fluids* (i.e. in fluid systems on a rotating earth). The effects due to density inhomogeneity of the fluid (stratification) are deliberately omitted in this volume, and will be studied under the title of Stratified Fluids, later in this series.

5.2 Inviscid Rotating Flows

In this section, we first study the relatively simpler class of motions where the viscous effects are to be ignored, i.e. by setting viscosity $\nu = 0$ in Eqs. (5.1) and (5.2). Without loss of generality, we assume the angular velocity vector is aligned with the z-axis in a Cartesian coordinate system (x, y, z) with unit vectors $(\imath, \jmath, \mathbf{k})$, i.e., $\boldsymbol{\Omega} = \Omega \mathbf{k}$.

$$\nabla \cdot \mathbf{u} = 0 \tag{5.6a}$$

$$\frac{D\mathbf{u}}{Dt} + 2\Omega \mathbf{k} \times \mathbf{u} = -\frac{1}{\rho} \nabla p \tag{5.6b}$$

We consider flows bounded by a container with solid boundary B with normal vector $\mathbf{n}$, where the normal velocity component should vanish:

$$\mathbf{u} \cdot \mathbf{n} = 0 \text{ on } B \tag{5.6c}$$

The effects of rotation are expressed by the Coriolis term (second term in Eq. 1.2), since the centrifugal force has already been included in the modified pressure p of Eq. (5.2). We will now demonstrate novel effects in rotating fluid motion arising due to this apparently minor modification of the governing equations. One of the most important of these effects is the elasticity created in rotating fluid motion. This effect is important, because the presence of a restoring mechanism allows particular types of wave motions to be supported.

In particular, if we seek oscillatory solutions in time, where we assume

$$\mathbf{u} = \mathbf{U}e^{i\omega t}$$
$$p = \phi e^{i\omega t},$$

we obtain

$$\nabla \cdot \mathbf{U} = 0 \tag{5.7a}$$

$$i\omega\mathbf{U} + 2\Omega\mathbf{k} \times \mathbf{U} = -\frac{1}{\rho}\nabla\phi. \tag{5.7b}$$

$$\mathbf{U} \cdot \mathbf{n} = 0 \text{ on } B \tag{5.7c}$$

We can then transpose these equations in better manageable form by first taking the divergence and later the curl of (5.7b). Firstly, the divergence results in

$$i\omega\nabla \cdot \mathbf{U} + 2\Omega\nabla \cdot (\mathbf{k} \times \mathbf{U}) = -\frac{1}{\rho}\nabla^2\phi$$

and making use of (5.7a) and vector identity (1.27c) yields

$$\nabla^2\phi - 2\Omega\rho\mathbf{k} \cdot \nabla \times \mathbf{U} = 0 \tag{5.8a}$$

Secondly, the curl gives

$$i\omega\nabla \times \mathbf{U} + 2\Omega\nabla \times (\mathbf{k} \times \mathbf{U}) = -\frac{1}{\rho}\nabla \times \nabla\phi \equiv 0$$

by vector identity (1.27i). Then, by using vector identity (1.27d), one obtains

$$i\omega\nabla \times \mathbf{U} - 2\Omega(\mathbf{k} \cdot \nabla)\mathbf{U} = 0. \tag{5.8b}$$

Multiplying (5.8a) by $i\omega$ and (5.8b) by $2\Omega\rho\mathbf{k}$ and adding together, with some cancellations, gives

$$i\omega\nabla^2\phi - (2\Omega)^2\rho\mathbf{k} \cdot (\mathbf{k} \cdot \nabla)\mathbf{U} = i\omega\nabla^2\phi - (2\Omega)^2\rho\mathbf{k} \cdot (\mathbf{k} \cdot \nabla)\mathbf{k} \cdot \mathbf{U} = 0. \tag{5.9}$$

Then dot product of $\mathbf{k}$ with (5.7b) yields

$$i\omega\mathbf{k} \cdot \mathbf{U} = \frac{1}{\rho}(\mathbf{k} \cdot \nabla)\phi$$

which can be substituted in (5.9) to obtain

$$\nabla^2\phi - \left(\frac{2\Omega}{\omega}\right)^2 (\mathbf{k} \cdot \nabla)^2\phi = 0. \tag{5.10}$$

Noting that $\mathbf{k} \cdot \nabla = \frac{\partial}{\partial z}$, the open form of (5.10) can now be written as

$$\frac{\partial^2 \phi}{\partial x^2} + \frac{\partial^2 \phi}{\partial y^2} + (1 - \lambda^2)\frac{\partial^2 \phi}{\partial z^2} = 0, \tag{5.11a}$$

where

$$\lambda = \frac{2\Omega}{\omega}.$$

The boundary condition (5.7c) can be transposed in terms of ϕ by first dot multiplying (5.7b) by $\mathbf{k} \times \mathbf{n}$ and making use of vector identities (1.11a) and (1.11c), then combining with the boundary condition (5.7c), to yield

$$-\lambda^2 \mathbf{n} \cdot \nabla\phi + 4(\mathbf{n} \cdot \mathbf{k})(\mathbf{k} \cdot \nabla\phi) + 2i\lambda(\mathbf{k} \times \mathbf{n}) \cdot \nabla\phi = 0 \text{ on } B. \tag{5.11b}$$

(5.11a) is the governing equation with prescribed boundary condition (5.11b) applied to the present case of inviscid, homogeneous, unsteady motions.

The striking feature that emerges from Eqs. (5.11a,b) is the possible change of regime that we may expect from the studied flows in respect to the parameter λ. In particular, the nature of the solution will depend on the value of $\lambda = \frac{2\Omega}{\omega}$, which is in the form of an inverse ratio of the frequency to planetary (inertial) frequency. It is immediately observed that the equations are of *parabolic* (Laplacian) form if $\lambda < 1$ i.e. for the *super-inertial* frequencies $\omega > 2\Omega$. On the other hand, the equations are of *hyperbolic* form if $\lambda > 1$ i.e. for the *sub-inertial* frequencies $\omega < 2\Omega$.

Based on these results, it is expected that the solutions would be smooth in parabolic regime, $\omega > 2\Omega$. On the other hand, for values of $\lambda > 1$, the flow will be in hyperbolic regime which allows wave-like solutions. For instance, an oscillatory source placed at (0, 0, 0) with small displacements aligned with the z-coordinate in the hyperbolic regime would create wave-like solutions spreading along characteristics in the form of cones

$$z = \pm\lambda(4 - \lambda)^{1/2}(x^2 + y^2)^{1/2}.$$

5.2.1 Inertial Motion—Unsteady (Periodic) Uniform Flow

To see the restoring mechanism of the Coriolis term in more detail, consider an inviscid fluid ($\nu = 0$) with vanishing pressure gradients ($\nabla p = 0$). In the absence of a pressure gradient, the flow is uniform in space, subject to the fictitious Coriolis force. Equation (5.2) becomes

$$\frac{D\mathbf{u}}{Dt} = -2\boldsymbol{\Omega} \times \mathbf{u}, \tag{5.12}$$

i.e., the fluid acceleration is balanced only by the restoring Coriolis force (per unit mass) $-2\boldsymbol{\Omega} \times \mathbf{u}$. Without loss of generality, we assume that the angular velocity

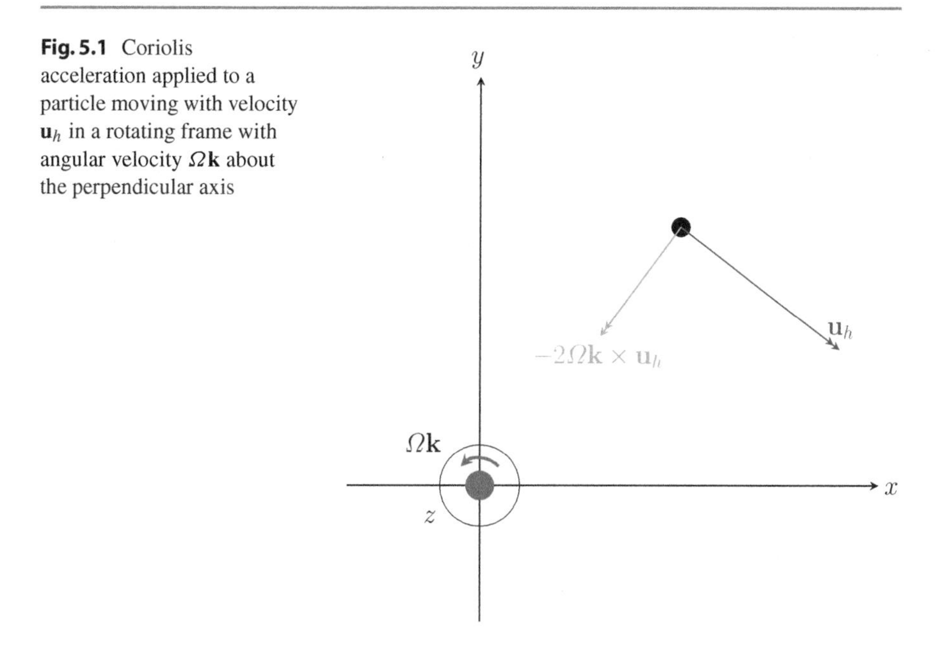

Fig. 5.1 Coriolis acceleration applied to a particle moving with velocity $\mathbf{u}_h$ in a rotating frame with angular velocity $\Omega\mathbf{k}$ about the perpendicular axis

vector is aligned with the z-axis in a Cartesian coordinate system (x, y, z) with unit vectors $(\imath, \jmath, \mathbf{k})$, i.e., $\boldsymbol{\Omega} = \Omega\mathbf{k}$. Then, the only component of velocity $\mathbf{u} = (u, v, w)$ contributing to the right hand side of (5.6) is $\mathbf{u}_h = (u, v, 0)$ in the plane perpendicular to $\boldsymbol{\Omega} = \Omega\mathbf{k}$), so that $\mathbf{u}$ can be replaced by $\mathbf{u}_h$. The direction of the restoring force is at right angles to the lateral component of fluid velocity $\mathbf{u}_h$, and its sense is to the right of this vector:

For a uniform flow in infinite domain, the nonlinear advection terms can be neglected, so that (5.12) becomes

$$\frac{\partial \mathbf{u}_h}{\partial t} + 2\Omega\mathbf{k} \times \mathbf{u}_h = 0 \tag{5.13}$$

Cross multiplying with $-2\Omega\mathbf{k}$ and adding with the time derivative of the above equation yields

$$\frac{\partial^2 \mathbf{u}_h}{\partial t^2} + (2\Omega)^2 \mathbf{u}_h = 0 \tag{5.14}$$

This equation has sinusoidal solutions (harmonic motion), analogous to a spring-mass system. For example, we can use the initial conditions (Fig. 5.1):

$$\mathbf{u}_h(0) = \mathbf{U}_0 \tag{5.15a}$$

$$\frac{\partial \boldsymbol{u}_h}{\partial t}(0) = -2\Omega\mathbf{k} \times \mathbf{U}_0 \tag{5.15b}$$

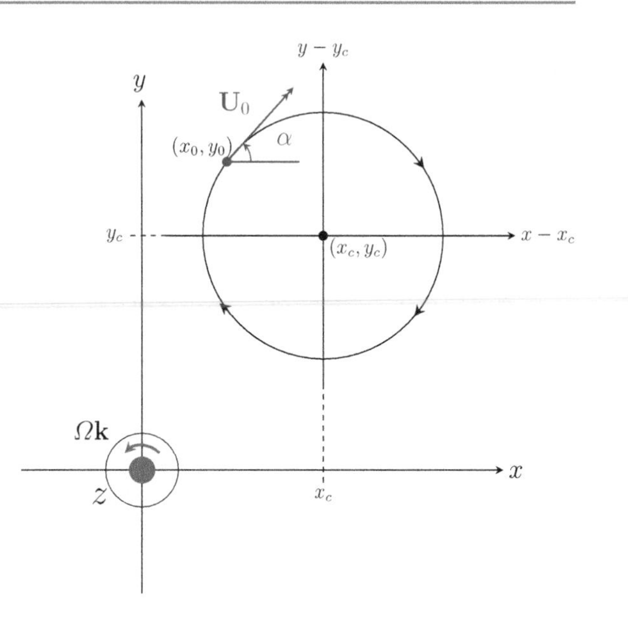

Fig. 5.2 Inertial motion

The solution follows as

$$\mathbf{u}_h = \mathbf{U}_0 \cos 2\Omega t - \mathbf{k} \times \mathbf{U}_0 \sin 2\Omega t. \tag{5.16}$$

The components (u, v), of the velocity $\mathbf{u}_h = (u, v, 0)$ are

$$u = U_0 \cos(2\Omega t - \alpha) \tag{5.17a}$$

$$v = -U_0 \sin(2\Omega t - \alpha) \tag{5.17b}$$

where α is the angle that the initial velocity vector $\mathbf{U}_0$ makes with the x-axis, and $U_0 = \|\mathbf{U}_0\|$.

For small amplitude motions, the displacements (x, y) of a material point (fluid particle) with respect to its initial position (x_0, y_0) can be obtained by integrating (5.17a,b) with respect to time (Fig. 5.2):

$$(x - x_0) = \frac{U_0}{2\Omega}[\sin(2\Omega t - \alpha) + \sin\alpha] \tag{5.18a}$$

$$(y - y_0) = \frac{U_0}{2\Omega}[\cos(2\Omega t - \alpha) - \cos\alpha] \tag{5.18b}$$

These can be combined by eliminating t between (5.18a,b), to yield

$$(x - x_c)^2 + (y - y_c)^2 = \left(\frac{U_0}{2\Omega}\right)^2 \tag{5.19}$$

where x_c, y_c are appropriate constants determined from (5.18).

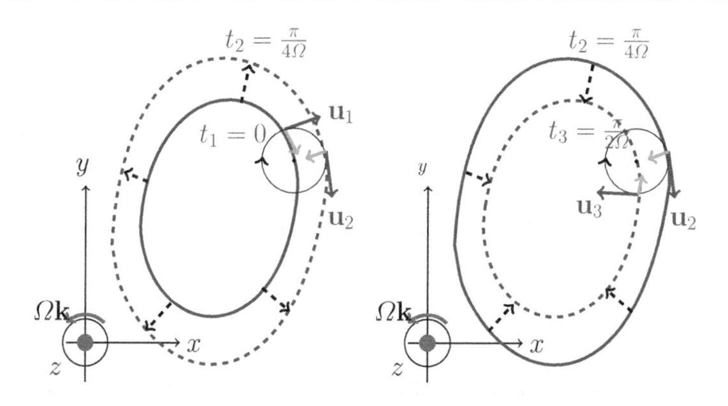

Fig. 5.3 Demonstration of elasticity in a rotating fluid by the action of Coriolis force as restoring agent

In the above solutions, both the sense of rotation of the velocity vector $\mathbf{u}_h$ and the trajectory $(x(t), (y(t))$ in (5.18) are in the *clockwise* direction. Each particle rotates clockwise, and comes to its initial position after one *inertial period* $T_I = \frac{2\pi}{2\Omega} = \frac{\pi}{\Omega}$. This *inertial motion* demonstrates the restoring effects of the Coriolis force. Because it arises due to the inertia, without any external (surface or body) forces, this motion is considered to be a free oscillation in a rotating fluid, corresponding to a natural frequency of 2Ω.

5.2.2 "Elasticity" in a Rotating Fluid—Restoring Effects of Coriolis

The previous example on inertial motion illustrates the restoring mechanism rotating fluids. Since particles displayed return to their initial positions after one characteristic (inertial) period, the fluid acts as if it has some special form of elasticity, whereby particles are forced into closed circular trajectories.

To further demonstrate the elastic behavior, consider a closed material curve C whose projection in the lateral plane ($\perp$ *to* $\boldsymbol{\Omega}$) is C_p (Fig. 5.3).

Suppose that a motion is generated in the fluid such that it will cause a positive rate of expansion in the lateral plane, i.e. with

$$\nabla_h \cdot \mathbf{u}_h = \frac{\partial u}{\partial x} + \frac{\partial v}{\partial y} > 0.$$

This outward motion along the material curve C is going to give rise to Coriolis forces in the clockwise direction along the curve since the induced force is to the right hand side of the motion (in the northern hemisphere). This is also seen exactly by Eq. (5.4) (Kelvin's theorem) since an increase in the projected area S_p enclosed by curve C leads to a negative contribution to the circulation. On the other hand, clockwise motion along the material curve will give rise to Coriolis forces in the inward direction (i.e. with $\nabla_h \cdot \mathbf{u}_h < 0$) then the material line C will then tend to contract.

Thus the fluid is seen to resist elastically to any motion that would cause displacement of fluid elements leading to a change in the projection of an are enclosed by a curve of such elements.

The relative importance of Coriolis effects is determined by the inverse of the Rossby number $R_0 = U_0/L_0\Omega_0$ measuring the ratio of Coriolis terms to other inertial terms (cf. Eq. 4.4). When $Ro \ll 1$, the elasticity effect of rotation is expected to be dominant.

5.2.3 Geostrophic Motion: Steady Flow with a Pressure Gradient

When the flow is *steady* $\left(\frac{\partial \mathbf{u}}{\partial t} = 0\right)$, *inviscid* ($\nu = 0$), and if the Rossby number $Ro \ll 1$, then the nonlinear term $\mathbf{u} \cdot \nabla \mathbf{u}$ is negligibly small compared to the Coriolis term $2\boldsymbol{\Omega} \times \mathbf{u}$). In this limit the momentum Eq. (5.2) becomes

$$2\boldsymbol{\Omega} \times \mathbf{u} = -\frac{1}{\rho}\nabla p \tag{5.20}$$

Without loss of generality we can let $\boldsymbol{\Omega} = \Omega\mathbf{k}$ be aligned with the z-axis of a Cartesian coordinate system (x, y, z) with unit vectors $(\imath, \jmath, \mathbf{k})$, such that

$$2\Omega\mathbf{k} \times \mathbf{u} = -\frac{1}{\rho}\nabla p. \tag{5.21}$$

The continuity equation

$$\nabla \cdot \mathbf{u} = 0 \tag{5.22}$$

complements the momentum equation. In principle (5.21) and (5.22) should be sufficient to solve for the unknowns $\mathbf{u}$ and p. However, it turns out that the so-called *geostrophic motion* by these equations have some very special characteristics.

First, by taking the curl of (5.21) we can show that

$$\nabla \times \mathbf{k} \times \mathbf{u} = -\frac{1}{2\Omega\rho}\nabla \times \nabla p \equiv 0 \tag{5.23}$$

by virtue of (1.27i). Then, by making use of (1.27d) the *l.h.s.* is

$$\nabla \times \mathbf{k} \times \mathbf{u} \equiv \mathbf{k}\nabla \cdot \mathbf{u} - \mathbf{k} \cdot \nabla \mathbf{u} = 0, \tag{5.24}$$

of which, the first term on the *r.h.s.* vanishes by (5.22). Then (5.24) states that

$$\mathbf{k} \cdot \nabla \mathbf{u} = \frac{\partial \mathbf{u}}{\partial z} = 0 \tag{5.25}$$

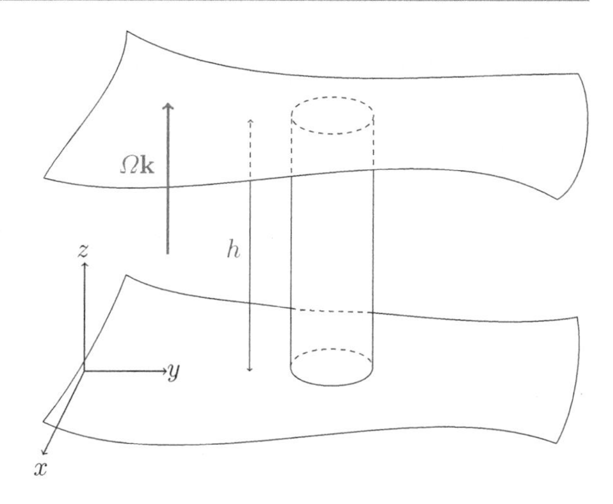

Fig. 5.4 Flow bounded by two rigid surfaces

expressing the fact that the velocity field has to be two-dimensional; $\mathbf{u} = \mathbf{u}(x, y)$ only. On the other hand, (5.21) dictates that

$$\mathbf{k} \cdot \nabla p = \frac{\partial p}{\partial z} = 2\Omega \rho \mathbf{k} \cdot (\mathbf{k} \times \mathbf{u}) = 0, \tag{5.26}$$

so that we find the pressure to be also two-dimensional, $p = p(x, y)$ only.

The above results, namely that none of the flow variables depend on the vertical coordinate z, indicates that the flow is essentially two-dimensional and occurs in the (x, y) plane (Fig. 5.4).

By virtue of the above results, (5.21) and (5.22) can equivalently be written as

$$2\Omega \mathbf{k} \times \mathbf{u}_h = -\frac{1}{\rho} \nabla_h p \tag{5.27}$$

$$\nabla_h \cdot \mathbf{u}_h = 0, \tag{5.28}$$

where $\mathbf{u}_h(x, y) = (u, v, 0)$ is the horizontal velocity vector, $p = p(x, y)$ is the pressure, such that both variables are independent of z, and $\nabla_h = \left(\frac{\partial}{\partial x}, \frac{\partial}{\partial y}, 0\right)$ is the horizontal gradient operator in the (x, y) plane.

The result, that the rotating flow at the limit $Ro \to 1$ must be two dimensional, is known as the *Taylor–Proudman theorem.* Consider flow bounded by two rigid surfaces

$$\Phi_1 = z - f_1(x, y) = 0, \Phi_2 = z - f_2(x, y) = 0. \tag{5.29a,b}$$

If a mound was placed in an otherwise constant depth motion, would bypass the mound (Fig. 5.5).

Since the flow is two-dimensional, any fluid column that is initially vertical will remain vertical. However, while moving, the net height of the column h would have

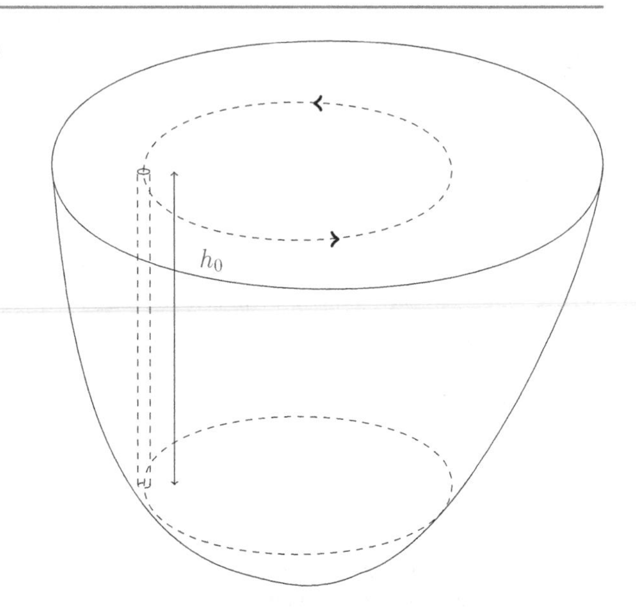

Fig. 5.5 Geostrophic motion in a basin with variable bottom topography has to follow closed contours of constant depth

to adjust itself to the distance of separation between the two surfaces, requiring that (Fig. 5.6)

$$\frac{D\Phi_1}{Dt} = 0, \frac{D\Phi_2}{Dt} = 0 \tag{5.30}$$

Since Φ_1, Φ_2 are material surfaces according the (1.41). Substituting (5.29a, b) (Fig. 5.7):

$$\frac{D\Phi_1}{Dt} = \mathbf{u} \cdot \nabla \Phi_1 = \mathbf{u}_h \cdot \nabla_h f_1 - w = 0, \text{ on } z = f_1 \tag{5.31a}$$

$$\frac{D\Phi_2}{Dt} = \mathbf{u} \cdot \nabla \Phi_2 = \mathbf{u}_h \cdot \nabla_h f_2 - w = 0, \text{ on } z = f_2 \tag{5.31b}$$

Subtracting (5.31b) from (5.31a) and since $\mathbf{u}_h = \mathbf{u}_h(x, y)$ only, we have

$$\begin{aligned} w \mid_{z=f_1} - w \mid_{z=f_2} &= \mathbf{u}_h \cdot \nabla (f_1 - f_2) \\ &= \mathbf{u}_h \cdot \nabla h \\ &= \frac{Dh}{Dt} \end{aligned} \tag{5.32}$$

On the other hand, since $w = w(x, y)$ only ($\frac{\partial w}{\partial z} = 0$), the vertical velocity w at the upper surface can not be different from that at the lower surface, i.e. the *l.h.s.* of (5.32) must vanish, so that

$$\frac{Dh}{Dt} = 0 \tag{5.33}$$

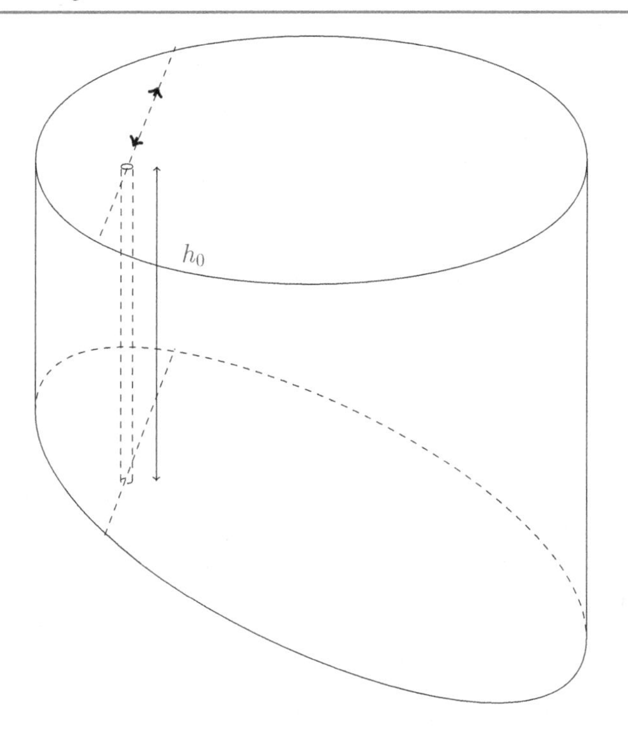

Fig. 5.6 Geostrophic motion in a container with sliced cylinder bottom topography is not possible, since the flow has to follow constant depth contours

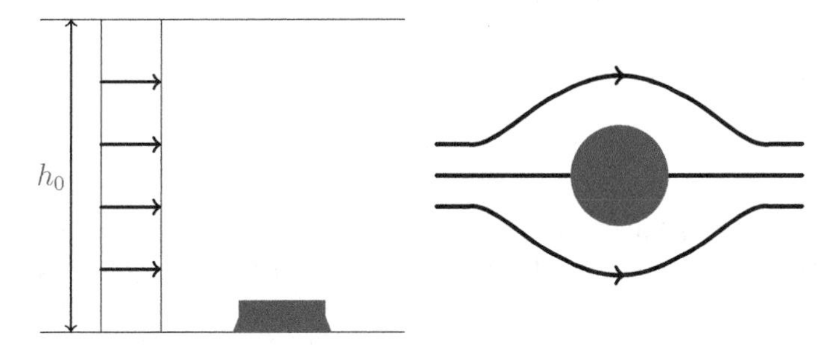

Fig. 5.7 Geostrophic flow passing a mound has to follow constant depth

This result indicates that any moving fluid column must preserve its height in geostrophic motion, i.e. the fluid column moves along a very special trajectory that would make $h =$ constant. In a closed container, this would mean that fluid columns could only move along closed contours having h = const, if such closed contours exist. If there are no such closed contours, geostrophic motion would not be possible.

Such columns which are identified with their constant thicknesses in geostrophic motion are called Taylor columns, since *G. I. Taylor* was the first to discover them. The flow modelled by Eqs. (5.21) and (5.22) [equivalently (5.27) and (5.28)] is called

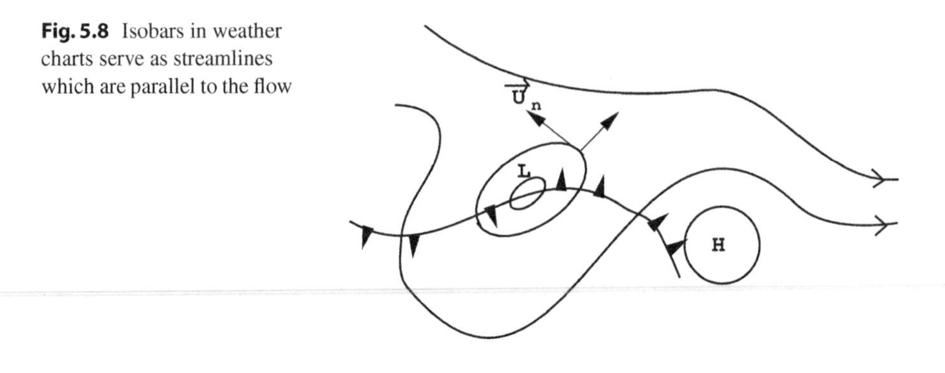

Fig. 5.8 Isobars in weather charts serve as streamlines which are parallel to the flow

geostrophic flow. It is a steady approximation to the governing equations for inviscid, homogeneous, incompressible rotating fluids in the limit $Ro \to 0$.

On the other hand, it can immediately be seen that *geostrophic flow* is in fact degenerate or indeterminate, i.e. while such a flow would exist, it is impossible to obtain a "solution" to equations. This is seen if pressure is eliminated from (5.27), by first rearranging such that

$$2\Omega \mathbf{u}_h = \frac{1}{\rho}\mathbf{k} \times \nabla_h p$$

then taking divergence of both sides

$$\nabla \cdot \mathbf{u}_h = \frac{1}{2\Omega\rho}\nabla \cdot (\mathbf{k} \times \nabla_h p).$$

Utilizing (1.27c) and (1.27i), the above equation is equivalent to

$$\nabla \cdot \mathbf{u}_h = -\frac{1}{2\Omega\rho}\mathbf{k} \cdot (\nabla_h \times \nabla_h p) \equiv 0,$$

i.e. the same thing as Eq. (5.28). This result shows that both of the statements (5.27) and (5.28) are equivalent to each other, i.e. one of the two equations is *redundant*. Because there is only one independent equation with insufficient information to solve for the two variables $\mathbf{u}$ and p, a simultaneous solution can not be obtained; which shows that *geostrophic flow is indeterminate* or *degenerate*.

Since there are two unknowns $\mathbf{u}_h$ and p in (5.27) and (5.28) it is only possible to infer one of these fields from given values of the other field. For example if pressure is given we can infer the velocity distribution, or vice versa. This *diagnostic method* is often used for interpreting weather charts or interpreting hydrographic fields in oceanography (Fig. 5.8).

Equation (5.27) can be put into the form

$$\begin{aligned} \mathbf{u}_h &= \frac{1}{2\Omega\rho}\mathbf{k} \times \nabla_h p \\ &= \mathbf{k} \times \nabla_h \left(\frac{p}{2\Omega\rho}\right), \end{aligned} \tag{5.34}$$

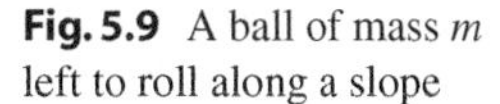
Fig. 5.9 A ball of mass m left to roll along a slope

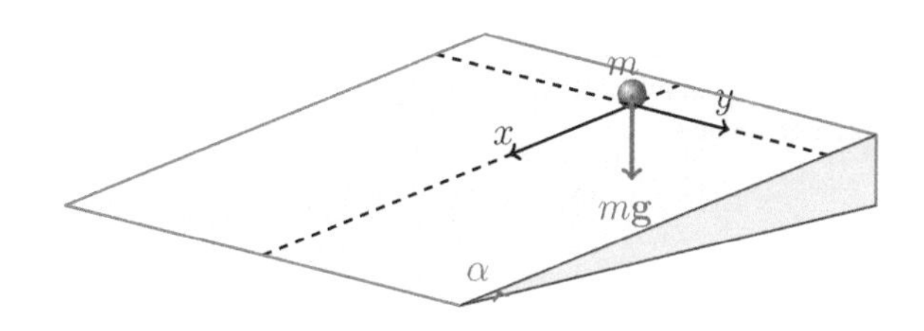

so that the velocity vector is perpendicular to the pressure gradient, and its sense is such that it takes high pressure to its right hand side. In fact, $\Psi = (\frac{p}{2\Omega\rho})$ acts as the stream function for the two dimensional flow; comparing (5.34) with (2.34). Around low-pressure centers L, the flow is *cyclonic*, i.e. it rotates in the *anti-clockwise* sense; and around high pressure centers H it is *anti-cyclonic* (i.e., rotation in clockwise sense). We must finally note that, to remove the geostrophic indeterminacy, we must include other effects in the dynamics, such as friction, unsteady variations, etc. The inclusion of these effects can be in the form of small corrections if $Ro \ll 1$, but nevertheless they would render the equations determinate.

Exercises

Exercise 1

Consider a ball of mass m released from rest at the origin ($x = 0$, $y = 0$) on an inclined plane as shown. The plane makes a small inclination angle α with the horizontal. The coordinates (x, y) are aligned with the inclined plane, where x is downward of the slope.

At the latitude ϕ where the experiment is performed, the Coriolis parameter is calculated from $f = 2\Omega \sin\phi$, reflecting the earth's rotational effects. Linear friction opposite to the direction of motion and proportional to the velocity of motion $\mathbf{u} = (u, v)$ with respect to the coordinates (x, y) is represented by the friction factor k (Fig. 5.9).

By simple mechanical arguments, we can show that the motion is governed by

$$\frac{du}{dt} - fv = g \sin\alpha - ku$$

$$\frac{dv}{dt} + fu = -kv$$

Find the position of the ball as a function of time by solving the above equations with initial condition $\mathbf{u}(t = 0) = 0$.

For realistic values of parameters, assume the experiment is done at latitude $\phi = 30\,°$N, on a slope of $\alpha = 10°$ and $k = 0.01$. What would be the space and time scales of the motion?

Sketch the motion for different values of f and k and discuss the motion for the following cases:

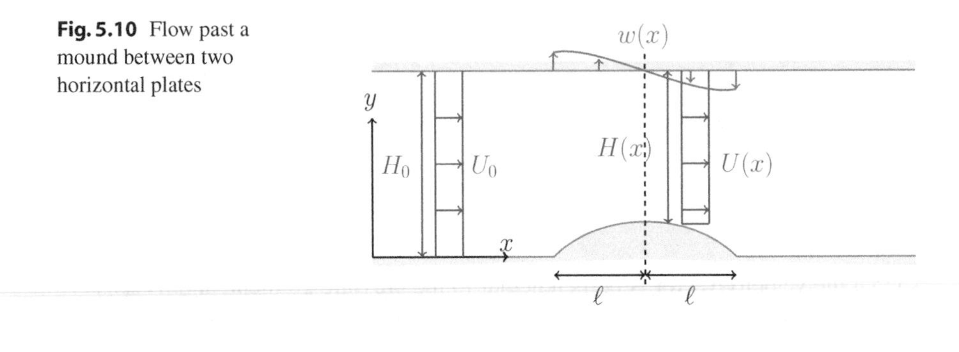

Fig. 5.10 Flow past a mound between two horizontal plates

(i) $f \neq 0,\ k \neq 0$;
(ii) $f \neq 0,\ k = 0$;
(iii) $f = 0,\ k \neq 0$;
(iv) $f = 0,\ k = 0$.

(v) What would change if the initial conditions would be changed? For instance, what would be the trajectory if an initial force was applied to give initial acceleration a_0 to the small ball?

`Exercise 2`

Consider an incompressible fluid of constant density confined between two plates of infinite horizontal extent. A mound of length 2ℓ in the x-direction and uniform in the y-direction is placed on the bottom, perpendicular to a flow with uniform speed U_0 in the x-direction. The total thickness of fluid between the solid boundaries is H_0, constant outside the mound region, and $H(x)$ at the mound (Fig. 5.10).

By making use of the governing equations with inviscid solid boundary conditions and assuming small Rossby number and negligible friction, investigate if it would be possible for the flow to pass over the mound.

Show that a flow with uniform profile $U(x)$ adjusted to the mound would only be able to pass over the mound if fluid was either sucked out or injected in at the upper boundary, with a vertical velocity

$$w(x) = -U_0 H_0 \frac{d}{dx}(ln[H(x)]) \text{ in the region } -\ell < x < \ell, \text{ at } y = H_0.$$

Where is it, along x, that the fluid is sucked out and where is it injected in?

`Exercise 3`

The linearized momentum equation for an incompressible fluid rotating with angular velocity $\boldsymbol{\Omega} = \Omega\mathbf{k}$ about the z-axis, subject to linearized viscous friction can be written as

$$\frac{\partial \mathbf{u}_h}{\partial t} + 2\Omega\mathbf{k} \times \mathbf{u}_h + \mu\mathbf{u}_h = 0.$$

This equation would represent inertial motions modified by frictional effects, due to the additional term.

(i) Solve the above equation with the initial condition for horizontal velocity, $\mathbf{u}_h(0) = \mathbf{U}_0$.

(ii) Based on an oscillatory solution, what kind of motion is expected? Which processes do terms in the governing equation represent, and how do they affect the motion?

(ii) Determine the components u, v of the complex vector variable $\mathbf{u} = (u, v) = u + iv$ as a function of time, and obtain linear displacements of a particle initially placed at (x_0, y_0) where the initial velocity is given as $\mathbf{u}_0$.

Plot the velocity components and particle trajectory as a function of time. Compare the cases $\mu \neq 0$ and $\mu = 0$.

6 Shallow Water Theory

6.1 Tangent Plane Approximation

Since we are dealing with motions on a spherical earth we need to use Eqs. (5.1) and (5.2) written in spherical coordinates. However, these are often complicated, and therefore their solutions would be difficult. In order to overcome this difficulty we often employ the "tangent plane" approximation. We envision the motions to take place on a plane that is tangent to the earth near the region of interest. This is a feasible approximation if the horizontal domain of interest is only a small portion of the earth's surface area as shown in Fig. 6.1.

The selected plane is tangent to the earth at points O, which has a latitude angle of ϕ.

The Cartesian coordinates on this tangent plane are chosen such that x, y axes point in the east and north directions, and the z-axis points in the vertical direction (perpendicular to the tangent plane). We can conveniently decompose the velocity vector $\mathbf{u} = (u, v, w)$ and the angular velocity vector $\boldsymbol{\Omega} = (\Omega_x, \Omega_y, \Omega_z)$ as follows:

$$\mathbf{u} = \mathbf{u}_h + w\mathbf{k} \tag{6.1a}$$

$$\boldsymbol{\Omega} = \boldsymbol{\Omega}_h + \Omega_z\mathbf{k} \tag{6.1b}$$

where

$$\mathbf{u}_h = (u, v) \tag{6.1c}$$

$$\boldsymbol{\Omega}_h = (\Omega_x, \Omega_y) \tag{6.1d}$$

Now the Coriolis terms in Eq. (5.2) can be written as

$$\begin{aligned} 2\boldsymbol{\Omega} \times \mathbf{u} &= 2(\boldsymbol{\Omega}_h + \Omega_z\mathbf{k}) \times (\mathbf{u}_h + w\mathbf{k}) \\ &= 2\Omega_z\mathbf{k} \times \mathbf{u}_h + 2\boldsymbol{\Omega}_h \times \mathbf{u}_h + 2\boldsymbol{\Omega}_h \times \mathbf{k}w \end{aligned} \tag{6.2}$$

© Springer Nature Switzerland AG 2020

E. Özsoy, *Geophysical Fluid Dynamics I*, Springer Textbooks in Earth Sciences, Geography and Environment, https://doi.org/10.1007/978-3-030-16973-2_6

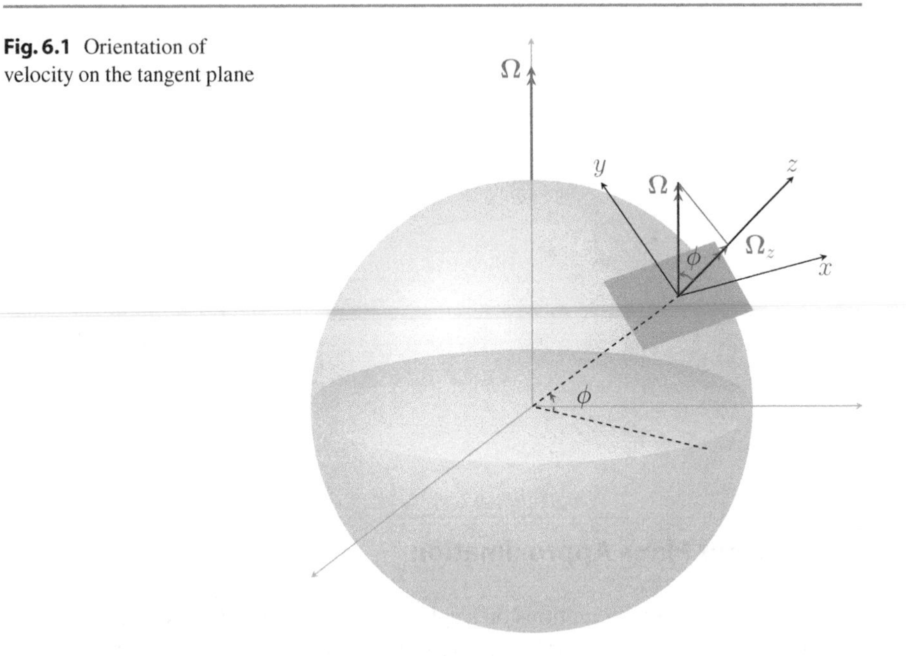

Fig. 6.1 Orientation of velocity on the tangent plane

Next, we note that because of the selected orientation of axes,

$$\mathbf{\Omega}_h = \Omega_y \mathbf{i} = \Omega(\cos\phi)\mathbf{i}, \quad \Omega_z = \Omega \sin\phi$$

where $\Omega =\mid \mathbf{\Omega} \mid$, so that (6.2) becomes

$$\begin{aligned} 2\mathbf{\Omega} \times \mathbf{u} &= 2\Omega_z \mathbf{k} \times \mathbf{u}_h + 2\Omega_y \mathbf{i} \times \mathbf{u}_h + 2\Omega_y \mathbf{i} \times \mathbf{k} w \\ &= (2\Omega \sin\phi)\mathbf{k} \times \mathbf{u}_h + (2\Omega \cos\phi)(w\mathbf{i} - u\mathbf{k}) \end{aligned} \tag{6.4}$$

We can now substitute this form into Eq. (5.2). First we utilize the definition of the "del" operator decomposed into horizontal and vertical components:

$$\nabla = \nabla_h + \mathbf{k}\frac{\partial}{\partial z} \tag{6.5}$$

where

$$\nabla_h = \left(\frac{\partial}{\partial x}, \frac{\partial}{\partial y}\right) = \mathbf{i}\frac{\partial}{\partial x} + \mathbf{j}\frac{\partial}{\partial y},$$

to write the horizontal and vertical components of (5.2):

$$\frac{\partial \mathbf{u}_h}{\partial t} + \mathbf{u}_h \cdot \nabla_h \mathbf{u}_h + w\frac{\partial \mathbf{u}_h}{\partial z} + (2\Omega \sin\phi)\mathbf{k} \times \mathbf{u}_h + (2\Omega \cos\phi) w\mathbf{i} = -\frac{1}{\rho}\nabla_h p + \nu\left(\nabla_h^2 \mathbf{u}_h + \frac{\partial^2 \mathbf{u}_h}{\partial z^2}\right) \tag{6.6a}$$

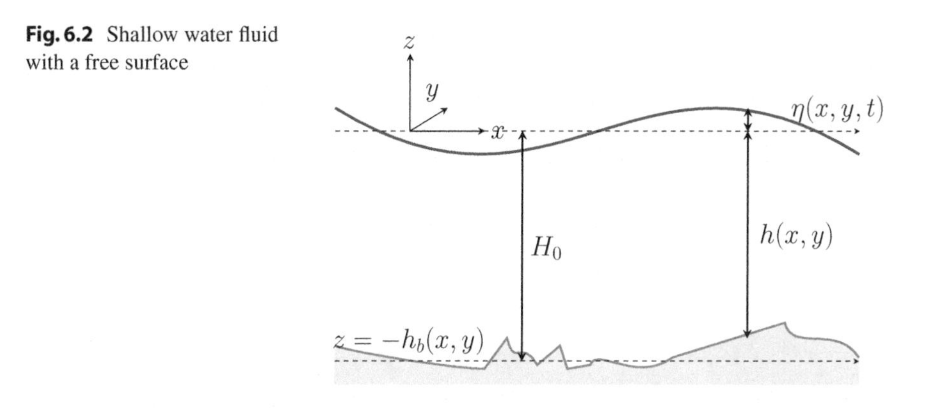

Fig. 6.2 Shallow water fluid with a free surface

and

$$\frac{\partial w}{\partial t} + \mathbf{u}_h \cdot \nabla_h w + w\frac{\partial w}{\partial z} - (2\Omega \cos\phi)u = -\frac{1}{\rho}\frac{\partial p}{\partial z} + \nu\left(\nabla_h^2 w + \frac{\partial^2 w}{\partial z^2}\right) \tag{6.6b}$$

The continuity equation (5.1) can also be written as

$$\nabla_h \cdot \mathbf{u}_h + \frac{\partial w}{\partial z} = 0 \tag{6.7}$$

The Eqs. (7.6) and (7.7) constitute the governing equations written in tangent plane coordinates.z

We can therefore make use of this fact in simplifying our equations. The so-called shallow-water approximation $(\lambda \ll 1)$ arises from the fact that the oceans and the atmosphere have essentially a small thickness as compared to the earth's radius.

We therefore consider a typical motion sketched below:

The lower surface $z = -h(x, y)$ describes the bottom topography (i.e., the ocean-bottom or the earth's surface in the case of the atmosphere) as shown in Fig. 6.2. The upper surface $z = \eta(x, y, t)$ describes the displacement of the sea-surface or an equivalent, imaginary "tropospheric upper surface" in the atmosphere. The magnitude of this displacement is characterized by the scale a_0, which is typically small, compared to the total depth H_0:

6.2 Shallow Water Approximations

6.2.1 Scaling of the Equations

A common feature of geophysical motions (i.e., the oceanic and atmospheric flows) is that they are too often characterized by horizontal length scales that are much larger than vertical scales. For example, the largest depth to be found in the world's oceans is about 10 km. Similarly the thickness of the troposphere (the lowest atmospheric layer in which most of the weather processes take place) is also of the order of 10 km. The horizontal scale of a typical domain of study extends from 100 km (meso-scale) to 1000 km (synoptic scale) or more. Therefore if L_0 represents the horizontal length scale and H_0 represents the vertical length scale, the dimensionless ratio λ (aspect ratio) is typically:

$$\lambda = \frac{H_0}{L_0} = O(10^{-1}) - O(10^{-2}) \ll 1, \tag{6.8}$$

$$\mu = \frac{a_0}{H_0} \ll 1. \tag{6.9}$$

In order to non-dimensionalize Eqs. (6.6a), (6.6b) and (6.7) for a cursory examination, we choose the following scales:

$$\begin{aligned} (x, y) &\sim L_0 \\ z &\sim H_0 \\ t &\sim T_0 \\ \mathbf{u}_h = (u, v) &\sim U_0 \\ w &\sim \left(\frac{a_0}{L_0}\right) U_0 \\ p &\sim \rho g H_0 \end{aligned} \tag{6.10a-f}$$

The vertical velocity scale is selected as $(a_0/L_0)\,U_0$ since the vertical motion should be proportional to the displacement of the upper surface $z = \eta(x, y, t)$, whose scale is a_0 To give it correct dimensions we divide by the time scale L_0/U_0. The pressure scale is selected as $\rho g H_0$, since the symbol p represents the modified pressure.

$$p = p' - \rho \mathbf{g} \cdot \mathbf{x} - \frac{1}{2}\rho \mid \mathbf{\Omega} \times \mathbf{x} \mid^2 \tag{6.11}$$

(ref. Eq. (3.55), where p' stands for fluid pressure. Since $\mathbf{g} = -g\mathbf{k}$, $p = O(\rho g z)$.

If we use the scales (6.10a-f) in Eqs. (6.6a), (6.6b) and (6.7), the non-dimensional equations can be written as

$$\begin{aligned} &Ro\left(\epsilon_T \frac{\partial \mathbf{u}_h}{\partial t} + \mathbf{u}_h \cdot \nabla \mathbf{u}_h + \mu w \frac{\partial \mathbf{u}_h}{\partial z}\right) + (2\sin\phi)\mathbf{k} \times \mathbf{u}_h + \mu\lambda(2\cos\phi)w \\ &\quad = -S\lambda \nabla_h p + E^2\left(\lambda^2 \nabla_h^2 \mathbf{u}_h + \frac{\partial^2 \mathbf{u}_h}{\partial z^2}\right) \end{aligned} \tag{6.12a}$$

$$\mu\lambda Ro\left(\epsilon_T \frac{\partial w}{\partial t} + \mathbf{u}_h \cdot \nabla_h w + \mu w \frac{\partial w}{\partial z}\right) - (2\cos\phi)u = -S\frac{\partial P}{\partial z} + \mu\lambda E^2\left(\lambda^2 \nabla_h^2 w + \frac{\partial^2 w}{\partial z^2}\right) \tag{6.12b}$$

$$\nabla_h \cdot \mathbf{u}_h + \mu \frac{\partial w}{\partial z} = 0 \tag{6.13}$$

where the variables are all non-dimensional and the following non-dimensional parameters are defined:

$$\begin{gathered} \mu = \frac{a_0}{H_0}, \quad \lambda = \frac{H_0}{L_0}, \quad \epsilon_T = \frac{L_0}{U_0 T_0}, \\ Ro = \frac{U_0}{\Omega L_0}, \quad E^2 = \frac{\nu}{\Omega H_0^2}, \quad S = \frac{g}{U_0 \Omega}. \end{gathered} \tag{6.14a-f}$$

Now we can use these non-dimensional parameters to estimate the relative orders of magnitude of the various terms. We know that $\mu \ll 1$ and $\lambda \ll 1$. The parameter ϵ_T is the ratio of the length scale L_0 to particle excursion length $U_0 T_0$ and is usually $O(1)$. We can also assume that the Rossby number $Ro = O(1)$ or smaller. The Ekman number E can also be estimated as $E = O(1)$ or smaller, by using typical values. On the other hand, the parameters can be estimated as follows:

$$\begin{aligned} S = \frac{g}{U_0 \Omega} &= \frac{g H_0}{U_0^2} \cdot \frac{U_0}{\Omega L_0} \cdot \frac{L_0}{H_0} \\ &= \frac{1}{F^2} \cdot Ro \cdot \frac{1}{\lambda} \end{aligned} \tag{6.15}$$

where $F = U_0/\sqrt{g H_0}$ is the Froude number. The Froude number is typically $O(1)$ or smaller. We therefore find that since $\lambda \ll 1$,

$$S = \frac{Ro}{F^2} \frac{1}{\lambda} \gg 1. \tag{6.16}$$

With these typical estimates of the parameters, it can be seen that in Eq. (6.12b) the dominating term is the vertical gradient of pressure. Neglecting all other terms, we therefore have

$$\frac{\partial p}{\partial z} = 0, \tag{6.17}$$

i.e. the modified pressure p is independent of the vertical coordinate, $p = p(x, y, t)$ only.

If the same scales are used for a cursory examination of Eq. (6.12a), it is first seen that the coefficient of the pressure gradient term

$$S\lambda = O(1) \tag{6.18}$$

by virtue of (6.16). All the other terms are also of $O(1)$, except the term arising due to the horizontal component of earth's angular speed which is multiplied by $\mu\lambda \ll 1$.

Therefore this term is neglected. If we also neglect the $w\frac{\partial \mathbf{u}_h}{\partial z}$ term ($\mu \ll 1$) and the $\nabla_h^2 \mathbf{u}_h$ term ($\lambda^2 \ll 1$), the equation becomes:

$$Ro\left(\epsilon_T \frac{\partial \mathbf{u}_h}{\partial t} + \mathbf{u}_h \cdot \nabla_h \mathbf{u}_h\right) + (2\sin\phi)\mathbf{k} \times \mathbf{u}_h = -S\lambda\nabla_h p + E^2 \frac{\partial^2 \mathbf{u}_h}{\partial z^2} \tag{6.19}$$

the dimensional equivalent of which is

$$\frac{\partial \mathbf{u}_h}{\partial t} + \mathbf{u}_h \cdot \nabla_h \mathbf{u}_h + (2\Omega \sin\phi)\mathbf{k} \times \mathbf{u}_h = -\frac{1}{\rho}\nabla_h p + \nu \frac{\partial^2 \mathbf{u}_h}{\partial z^2}. \tag{6.20}$$

The obvious result of Eq. (6.19) is that since $p = p(x, y, t)$ only (cf. Eq. 2.17) then $\mathbf{u}_h = \mathbf{u}_h(x, y, t)$ only (if frictional terms are neglected). The situation similar to that found in geostrophic flow, and the *flow is essentially two-dimensional.*

Finally we can observe that, to the same approximation (6.13) becomes

$$\nabla_h \cdot \mathbf{u}_h = 0 \tag{6.21}$$

which implies that

$$\frac{\partial w}{\partial z} = 0 \tag{6.22}$$

and therefore w is also independent of depth $w = w(x, y, t)$.

We have shown that the *modified* pressure is independent of z (cf. Eq. 2.17). The actual fluid pressure can be found from (6.11). First noting that we can write *modified gravity* or *gravitation* (cf. (3.52)) as

$$\mathbf{g}' = \mathbf{g} - \mathbf{\Omega} \times \Omega \times \mathbf{x} \tag{6.23}$$

we can write (6.11) as

$$\begin{aligned} p &= p' + \rho\left(\mathbf{g}\cdot\mathbf{x} + \frac{1}{2}(\mathbf{\Omega}\times\mathbf{x})\cdot(\mathbf{\Omega}\times\mathbf{x})\right) \\ &= p' + \rho\left(\mathbf{g}\cdot\mathbf{x} - \frac{1}{2}\mathbf{\Omega}\times(\mathbf{\Omega}\times\mathbf{x})\cdot\mathbf{x}\right) \\ &= p' + \rho\mathbf{g}'\cdot\mathbf{x} \end{aligned} \tag{6.24}$$

Now the tangent plane is actually perpendicular to the gravitation vector $\mathbf{g}'$. In fact the difference between the $\mathbf{g}$ and $\mathbf{g}'$ vectors is only minor, arising due to centrifugal forces. If the earth had uniform density, it would take the form of an ellipsoid where the tangent plane would be exactly perpendicular to the $\mathbf{g}'$ vector. In this case, substituting

$$\begin{aligned} \mathbf{g}' &= -g'\mathbf{k} \\ \mathbf{x} &= x\mathbf{i} + y\mathbf{j} + z\mathbf{k} \end{aligned} \tag{6.25}$$

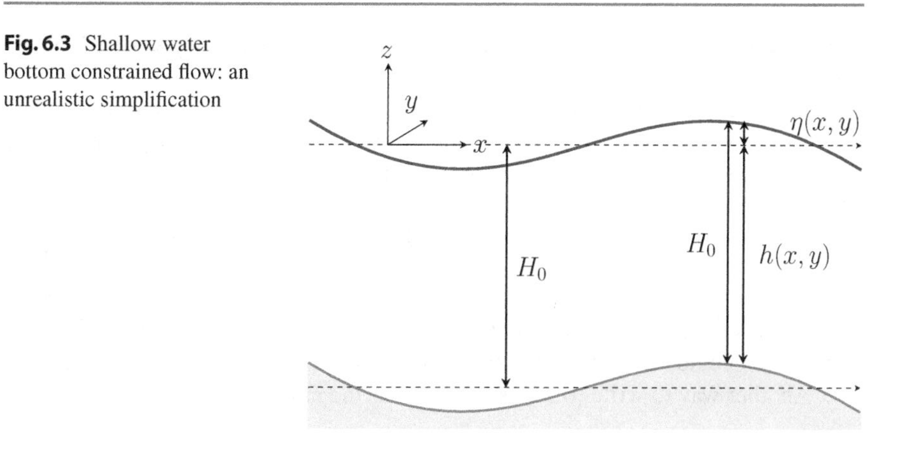

Fig. 6.3 Shallow water bottom constrained flow: an unrealistic simplification

in (6.24) gives

$$p = p' - \rho g' z. \quad (6.26)$$

By virtue of (6.17) the modified pressure $p = p(x, y, t)$, and therefore p is constant with respect to z. Equation (6.26) is said to express *hydrostatic pressure*, since the pressure *at any horizontal position* (x, y) and *at any instant t* depends on z as if it was for a static fluid. Note that in this approximation, $\mathbf{u}_h = \mathbf{u}_h(x, y, t)$, $p = p(x, y, t)$, $w = w(x, y, t)$, the fluid motion is essentially two-dimensional and bears much similarity to the characteristics of geostrophic motion. Fluid columns which are initially vertical remain vertical more like Taylor columns. Since $\frac{\partial w}{\partial z} = 0$ the total height of these columns would not change. A fluid column moving over topography (if it does) would therefore adjust its surface elevation such that

$$H \equiv h + \eta = \text{constant}$$

Since the upper surface is adjusted to depth as shown in Fig. 6.3, it seems no longer required for the fluid column to follow constant bottom depth ($h(x, y)$) contours, as it would in the case of a geostrophic flow with a rigid upper surface. However, this behavior would be too demanding to be observed in nature.

It seems that the above approximations are in fact too rigid and will be somewhat relaxed in later sections. However, in spite of the excessive rigidity of the present approximation, some of its features have been observed in the ocean. In recent years satellite altimetry methods have allowed the measurement of the ocean surface elevation from space. It has been often found that the surface of the ocean takes almost the same shape as the underlying topography especially in the mid-ocean regions. It is not necessarily true that $H = h + \eta = constant$, but nevertheless the topography is often "impressed" on the sea surface.

6.2.2 Continuity of Surface Forces (Dynamic Boundary Conditions)

We will consider an element of the upper surface at $z = \eta(x, y, t)$ and investigate the continuity of surface forces across this surface. In short, we should insist that the surface stress $\boldsymbol{\Sigma}$ (force per unit area of the upper surface) be continuous across the surface

$$\boldsymbol{\Sigma}\ |_{z=\eta^-} = \boldsymbol{\Sigma}\ |_{z=\eta^+} \tag{6.27}$$

where $\eta^- = \eta - \delta$, $\eta^+ = \eta + \delta$, such that $\delta \to 0$; i.e. the surface stress just above the surface should be balanced by that just below. We had seen in Sect.2.4 that the body forces could be neglected if the fluid volume considered as infinite small in size. Another way to write (6.27) is to state it as the jump condition

$$[\boldsymbol{\Sigma}]_{z=\eta^-}^{z=\eta^+} = 0 \tag{6.28}$$

i.e., there will be no jump in the value of $\boldsymbol{\Sigma}$ across the surface.

In Sect.2.4 the surface stress vector $\boldsymbol{\Sigma}$ was expressed as the dot product of the stress tensor with the normal vector

$$\Sigma_i(\mathbf{n}) = \sigma_{ij} n_j = \boldsymbol{\sigma} \cdot \mathbf{n} \tag{6.29}$$

(cf. Eq. 1.15). In Fig. 6.4 we take an arbitrary element of the surface oriented perpendicular to the normal vector $\mathbf{n}$, where $\Sigma_i(\mathbf{n})$ represents the ith component of the stress on this surface element.

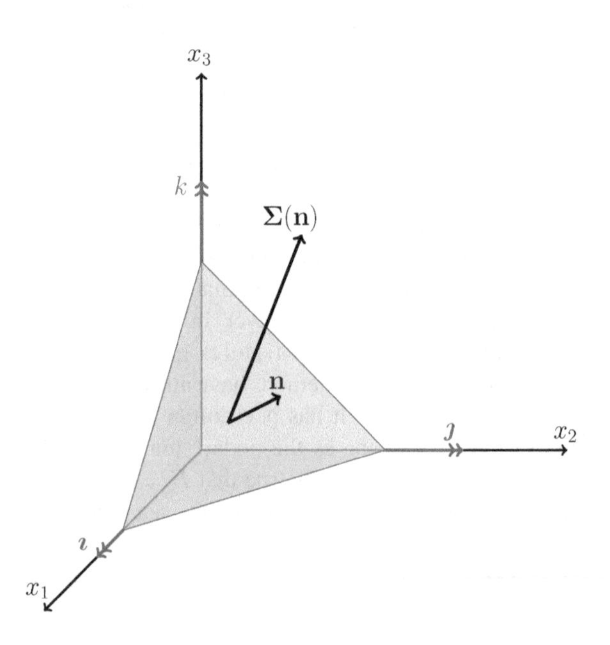

Fig. 6.4 A surface fluid element

In a moving fluid the stress tensor $\boldsymbol{\sigma}$ is expressed as (Sect. 2.4.2, Eq. 2.17)

$$\sigma_{ij} = -p\delta_{ij} + d_{ij} \tag{6.30}$$

where $\mathbf{d} = d_{ij}$ was the *deviatoric stress tensor* expressed later in Sect. 3.3, Eq. (3.22) as

$$\mathbf{d} = d_{ij} = \begin{bmatrix} \sigma'_{11} & \tau_{12} & \tau_{13} \\ \tau_{21} & \sigma'_{22} & \tau_{23} \\ \tau_{31} & \tau_{32} & \sigma'_{33} \end{bmatrix} \tag{6.31}$$

whose off-diagonal elements are shear stresses arising only due to the motion of the fluid. Now using (6.29) and (6.30) the components of the stress on the upper surface are then

$$\begin{aligned} \Sigma_i &= (-p\delta_{ij} + d_{ij})n_j \\ &= -pn_i + d_{ij}n_j \end{aligned} \tag{6.32}$$

or in vector form

$$\boldsymbol{\Sigma} = -p\mathbf{n} + \mathbf{d} \cdot \mathbf{n} \tag{6.33}$$

This vector has components in the x, y, z directions referred to 1, 2, 3 in index notation. The x and y-components of (6.33) are

$$\Sigma_x = -p\mathbf{n} \cdot \mathbf{i} + (\mathbf{d} \cdot \mathbf{n}) \cdot \mathbf{j} = -pn_1 + (d_{1j}n_j)e_1 \tag{6.34a}$$

$$\Sigma_y = -p\mathbf{n} \cdot \mathbf{j} + (\mathbf{d} \cdot \mathbf{n}) \cdot \mathbf{j} = -pn_2 + (d_{2j}n_j)e_2 \tag{6.34b}$$

Instead of writing the z-component of the vector $\boldsymbol{\Sigma}$, we choose to write its component perpendicular to the surface in direction n, since the vector $\mathbf{n}$ is a linear combination of the $(\mathbf{i}, \mathbf{j}, \mathbf{k})$ vectors. The n-component is

$$\Sigma_n = -p\mathbf{n} \cdot \mathbf{n} + (\mathbf{d} \cdot \mathbf{n}) \cdot \mathbf{n} = -p + d_{ij}n_jn_i \tag{6.34c}$$

In these equations the components of the normal vector is

$$\begin{aligned} n_1 &= (\mathbf{n} \cdot \mathbf{i}) \\ n_2 &= (\mathbf{n} \cdot \mathbf{j}) \\ n_3 &= (\mathbf{n} \cdot \mathbf{k}) \end{aligned} \tag{6.35a-c}$$

Using (6.35a-c) and (6.31), (6.34a)–(6.34c) can be written as

$$\Sigma_x = -p(\mathbf{n} \cdot \mathbf{i}) + \sigma'_{xx}(\mathbf{n} \cdot \mathbf{i}) + \tau_{xy}(\mathbf{n} \cdot \mathbf{j}) + \tau_{xz}(\mathbf{n} \cdot \mathbf{k})$$

$$\Sigma_y = -p(\mathbf{n} \cdot \mathbf{j}) + \tau_y x(\mathbf{n} \cdot \mathbf{i}) + \sigma_{yy}(\mathbf{n} \cdot \mathbf{j}) + \tau_{yz}(\mathbf{n} \cdot \mathbf{k})$$

$$\Sigma_z = -p + (\mathbf{n}\cdot\mathbf{i})^2\sigma'_{xx} + (\mathbf{n}\cdot\mathbf{j})^2\sigma'_{yy} + (\mathbf{n}\cdot\mathbf{k})^2\sigma'_{zz} + 2(\mathbf{n}\cdot\mathbf{i})(\mathbf{n}\cdot\mathbf{j})\tau_{xy} + 2(\mathbf{n}\cdot\mathbf{i})(\mathbf{n}\cdot\mathbf{k})\tau_{xz} + 2(\mathbf{n}\cdot\mathbf{j})(\mathbf{n}\cdot\mathbf{k})\tau_{yz} \quad (6.36\text{a-c})$$

where use has been made of the symmetry property of the deviatoric stress tensor, $d_{ij} = d_{ji}$.

Defining the surface $z = \eta(x, y, t)$ by the equation

$$\phi = z - \eta(x, y, t) = 0 \quad (6.37)$$

and its gradient by

$$\nabla\phi = -\frac{\partial\eta}{\partial x}\mathbf{i} - \frac{\partial\eta}{\partial y}\mathbf{j} + \mathbf{k}, \quad (6.38\text{a})$$

the normal vector $\mathbf{n}$ is found to be (cf. (1.39))

$$\mathbf{n} \equiv \frac{\nabla\phi}{|\nabla\phi|} = \frac{-\frac{\partial\eta}{\partial x}\mathbf{i} - \frac{\partial\eta}{\partial y}\mathbf{j} + \mathbf{k}}{\sqrt{\left(\frac{\partial\eta}{\partial x}\right)^2 + \left(\frac{\partial\eta}{\partial y}\right)^2 + 1}}. \quad (6.39)$$

In order to shorten the expressions, let

$$S = \sqrt{1 + \left(\frac{\partial\eta}{\partial x}\right)^2 + \left(\frac{\partial\eta}{\partial y}\right)^2}. \quad (6.40)$$

Now, the components of the unit normal vector $\mathbf{n}$ are

$$(\mathbf{n}\cdot\mathbf{i}) = -\frac{1}{S}\frac{\partial\eta}{\partial x}, \quad (\mathbf{n}\cdot\mathbf{j}) = -\frac{1}{S}\frac{\partial\eta}{\partial y}, \quad (\mathbf{n}\cdot\mathbf{k}) = \frac{1}{S} \quad (6.41\text{a-c})$$

Which, upon substituting into (6.36a-c) yield

$$\Sigma_x = \left(\frac{1}{S}\frac{\partial\eta}{\partial x}\right)p - \left(\frac{1}{S}\frac{\partial\eta}{\partial x}\right)\sigma'_{xx} - \left(\frac{1}{S}\frac{\partial\eta}{\partial y}\right)\tau_{xy} + \left(\frac{1}{S}\right)\tau_{xz} \quad (6.41\text{a})$$

$$\Sigma_y = \left(\frac{1}{S}\frac{\partial\eta}{\partial y}\right)p - \left(\frac{1}{S}\frac{\partial\eta}{\partial x}\right)\tau_{xy} - \left(\frac{1}{S}\frac{\partial\eta}{\partial y}\right)\sigma'_{yy} + \left(\frac{1}{S}\right)\tau_{yz} \quad (6.41\text{b})$$

$$\Sigma_n = -p + \left(\frac{1}{S}\frac{\partial\eta}{\partial x}\right)^2\sigma'_{xx} + \left(\frac{1}{S}\frac{\partial\eta}{\partial y}\right)^2\sigma'_{yy} + \left(\frac{1}{S}\right)^2\sigma'_{zz} + \frac{2}{S^2}\left(\frac{\partial\eta}{\partial x}\right)\left(\frac{\partial\eta}{\partial y}\right)\tau_{xy} - \frac{2}{S^2}\left(\frac{\partial\eta}{\partial x}\right)\tau_{xz} - \frac{2}{S^2}\left(\frac{\partial\eta}{\partial y}\right)\tau_{yz} \quad (6.41\text{c})$$

We can now use the scales introduced in Sect. 6.2.1, namely that $(x, y) \sim L_0$ and $\eta \sim a_0$. Then, we find that

$$\begin{aligned}
\frac{\partial \eta}{\partial x} &= O\left(\frac{a_0}{L_0}\right) = O\left(\frac{a_0}{H_0} \cdot \frac{H_0}{L_0}\right) = O(\mu\lambda) \ll 1 \\
\frac{\partial \eta}{\partial y} &= O(\mu\lambda) \ll 1 \\
S &= \sqrt{1 + |\nabla_h \eta|^2} = \sqrt{1 + 2O(\mu^2\lambda^2)} \simeq 1 = O(1)
\end{aligned} \tag{6.42a-c}$$

Since the $O(\mu\lambda)$ terms are very small, neglecting these and setting S = 1 in Eq. (6.41a-c) to the same order used in shallow water approximations yields:

$$\begin{aligned}
\Sigma_x &\simeq \tau_{xz} \\
\Sigma_y &\simeq \tau_{yz} \\
\Sigma_n &\simeq -p + \sigma'_{zz}.
\end{aligned} \tag{6.43a-c}$$

In the last Eq. (6.43c) σ'_{zz} stands for the vertical normal stress component (i.e., stress pointing in direction z on the (x, y) plane) arising only due to the motion of the fluid. In fact

$$-p + \sigma'_{zz} = \sigma_{zz} \tag{6.44}$$

is the total normal stress. This term is usually small as compared to the fluid pressure ($\sigma'_{zz} \ll p$) at the surface and can be neglected altogether:

$$\Sigma_n \approx -p \tag{6.45}$$

After these simplifications, the continuity of surface forces across the surface in (6.28) requires

$$\begin{aligned}
\left[\tau_{xz}\right]_{z=\eta^-}^{z=\eta^+} &= 0 \\
\left[\tau_{yz}\right]_{z=\eta^-}^{z=\eta^+} &= 0 \\
\left[p\right]_{z=\eta^-}^{z=\eta^+} &= 0
\end{aligned} \tag{6.46a-c}$$

i.e. the horizontal components of the vertical shear and the fluid pressure must be continuous across the surface. At the surface of the ocean the shear stress occur due to the stresses applied by wind. Assuming that the wind only applies horizontal forces (a horizontal vector $\boldsymbol{\tau}^s$) defined as

$$\boldsymbol{\tau}^s = \left(\tau_{xz}\,|_{z=\eta^+}\right)\mathbf{i} + \left(\tau_{yz}\,|_{z=\eta^+}\right)\mathbf{j} \tag{6.47}$$

and writing the fluid shear in terms of the velocity gradient a boundary condition is found:

$$\mu \frac{\partial \mathbf{u}_h}{\partial z}\,|_{z=\eta^-} = \boldsymbol{\tau}^s \tag{6.48}$$

The pressure at the sea surface is the atmospheric pressure $p^s = p\mid_{z=\eta^+}$ so that (6.46c) implies another surface boundary condition:

$$p\mid_{z=\eta^-} = p^s \tag{6.49}$$

At the sea bottom similar conditions may be applied, the mean bottom surface is assumed to be horizontal (constant depth) with slowly varying small undulations superimposed: i.e., again assuming $a_0/L_0 = \mu\lambda \ll 1$, so that the shear stresses must be continuous. Defining similarly a bottom stress vector:

$$\boldsymbol{\tau}^b = \left(\tau_{xz}\mid_{z=-h^-}\right)\mathbf{i} + \left(\tau_{yz}\mid_{z=-h^-}\right)\mathbf{j} \tag{6.50}$$

we can write

$$\mu\frac{\partial \mathbf{u}_h}{\partial z}\mid_{z=-h^+} = \boldsymbol{\tau}^b \tag{6.51}$$

It is shown in the above derivations that horizontal components of vertical shear and the pressure is transmitted into the fluid purely when the shallow water approximation $\mu\lambda \ll 1$ holds, i.e. when the surface across which these forces are transmitted is almost flat.

6.2.3 Hydrostatic Pressure

We can now combine (6.26) and (6.49) since p is constant by virtue of (6.17). At the surface

$$p\mid_{z=\eta^-} = p' - \rho g'\eta^- = p^s \tag{6.52}$$

so that

$$p'(x, y, t) = p^s_{(x,y,t)} + \rho g'\eta^-(x, y, t) \tag{6.53}$$

Substituting into (6.26) yields

$$p = p^s(x, y, t) + \rho g[\eta(x, y, t) - z] \tag{6.54}$$

where prime and minus signs have been dropped and will not be used hereafter.

The above expression (6.54) simply states *hydrostatic pressure*, i.e. that the pressure at any depth $-z$ is the weight per unit area of the overlying fluid plus the atmospheric pressure. This is true for any horizontal position (x, y) and instant t.

Furthermore, by using (6.53), the horizontal pressure gradients appearing in Eq. (6.6a) appear as

$$-\frac{1}{\rho}\nabla_h p = -\frac{1}{\rho}\nabla_h p^s + g\nabla_h \eta \tag{6.55}$$

i.e. horizontal pressure gradients are partly caused by gradients in the atmospheric pressure or may be manifested as the gradients of the undulations in the upper surface. The latter of these, i.e. the surface elevation gradients impose a horizontal pressure gradient in the fluid through the action of *gravity*.

6.2.4 Kinematic Boundary Conditions

In addition to the dynamic boundary conditions reviewed in Sect. 6.2.2, we can derive kinematic boundary conditions utilizing the fact that the upper and lower surfaces of the fluid are material surfaces, i.e.

$$\frac{D\phi_1}{Dt} = 0, \quad \frac{D\phi_2}{Dt} = 0 \tag{6.56a, b}$$

where

$$\begin{aligned} \phi_1 &= z - \eta(x, y, t) = 0 \\ \phi_2 &= z + h(x, y) = 0 \end{aligned} \tag{6.57a, b}$$

describe these surfaces. Substituting (6.57) into (6.26) gives

$$-\frac{\partial \eta}{\partial t} - \mathbf{u}_h \cdot \nabla_h \eta + w = 0 \;\; on \;\; z = \eta(x, y, t) \tag{6.58a}$$

$$\mathbf{u}_h \cdot \nabla_h h + w = 0 \;\; on \;\; z = -h(x, y) \tag{6.58b}$$

or equivalently

$$w \mid_{z=\eta} = \frac{\partial \eta}{\partial t} + \mathbf{u}_h \mid_{z=\eta} \cdot \nabla_h \eta \tag{6.59a}$$

$$w \mid_{z=-h} = -\mathbf{u}_h \mid_{z=-h} \cdot \nabla_h h \tag{6.59b}$$

6.2.5 Shallow Water Equations

In Sect. 6.2.1 we used estimated scales for a cursory examination of the governing equations. Then, making the approximations $\mu \to 0, \ \lambda \to 0$ and $S = Ro/(F^2\lambda) \to \infty, \ S\lambda \to O(1)$ resulted in purely two-dimensional equations, with all flow variables becoming independent of the vertical coordinate. However we saw that this was not very realistic, and it was concluded that the assumptions used were rather restrictive. We now relax these assumptions a little, especially with regard to vertical velocity. The vertical velocity scale chosen was

$$w \sim \frac{a_0}{L_0} U_0 = \frac{a_0}{H_0} \frac{H_0}{L_0} U_0 = \mu \lambda U_0$$

as compared to the horizontal velocity scale of

$$\mathbf{u}_h \sim U_0$$

so that

$$\frac{w}{\mid \mathbf{u}_h \mid} = O(\mu\lambda) \tag{6.60}$$

where it was assumed that $\mu \ll 1,\ \lambda \ll 1$.

Since we did not have any prior knowledge of the vertical velocity scale we choose the above scales arbitrarily. However if we re-consider Eq. (6.13)

$$\nabla_h \cdot \mathbf{u}_h + \mu \frac{\partial w}{\partial z} = 0$$

we see that if the horizontal divergence is to be balanced by the vertical gradient term, we should have $\mu = O(1)$ or that $w \sim (\lambda U_0) = O(\frac{H_0}{L_0} U_0)$, i.e. vertical velocity should be smaller than the horizontal velocity by only the ratio H_0/L_0 of depth to horizontal scale of motion. This essentially means that perhaps we should have scaled vertical velocity as $w \sim (H_0/L_0)U_0$ at the beginning. In actuality, vertical velocity is at most balanced by the horizontal divergence is to be balanced by the horizontal divergence as indicated above, or somewhat smaller, so that

$$\mu \leq O(1). \tag{6.61}$$

in other words, in the previous scaling we assumed both of the two small parameters $\mu,\ \lambda \to 0$ without stating which one of these two parameters is actually smaller. Here we assume $\lambda \ll \mu$ by virtue of (6.61). With approximation (6.61) and the other previous approximations in Sect. 2.2.1, the non dimensional Eqs. (6.12a), (6.12b) and (6.13) can be simplified accurate to $O(\lambda)$ as follows:

$$Ro\left(\epsilon_T \frac{\partial \mathbf{u}_h}{\partial t} + \mathbf{u}_h \cdot \nabla_h \mathbf{u}_h + \mu w \frac{\partial \mathbf{u}_h}{\partial z}\right) + (2\sin\phi)\mathbf{k} \times \mathbf{u}_h = -S\lambda \nabla_h p + E^2 \frac{\partial^2 \mathbf{u}_h}{\partial z^2} \tag{6.62a}$$

$$0 = -S\frac{\partial p}{\partial z} \tag{6.62b}$$

$$\nabla_h \cdot \mathbf{u}_h + \mu \frac{\partial w}{\partial z} = 0 \tag{6.63}$$

The dimensionless equivalents of the above are:

$$\frac{\partial \mathbf{u}_h}{\partial t} + \mathbf{u}_h \cdot \nabla_h \mathbf{u}_h + w \frac{\partial \mathbf{u}_h}{\partial z} + f\mathbf{k} \times \mathbf{u}_h = -\frac{1}{\rho}\nabla_h p + \nu \frac{\partial^2 \mathbf{u}_h}{\partial z^2} \tag{6.63a}$$

$$\frac{\partial p}{\partial z} = 0 \tag{6.63b}$$

$$\nabla_h \cdot \mathbf{u}_h + \frac{\partial w}{\partial z} = 0 \tag{6.64}$$

where the *Coriolis parameter* f has been defined as

$$f = 2\Omega \sin\phi. \tag{6.65}$$

Note that Eqs. (6.63a), (6.63b)and (6.64) can alternatively be written as

$$\frac{D\mathbf{u}_h}{Dt} + f\mathbf{k} \times \mathbf{u}_h \equiv \frac{\partial \mathbf{u}_h}{\partial t} + \mathbf{u} \cdot \nabla \mathbf{u}_h + f\mathbf{k} \times \mathbf{u}_h = -\frac{1}{\rho}\nabla_h p + \nu \frac{\partial^2 \mathbf{u}_h}{\partial z^2} \tag{6.66a}$$

$$\frac{\partial p}{\partial z} = 0 \tag{6.66b}$$

$$\nabla \cdot \mathbf{u} = 0 \tag{6.67}$$

Equation (6.66b) indicates that the former approximation (that modified pressure is independent of z) in Sects. 6.2.1 and 6.2.3 are still valid. That is the fluid pressure is *hydrostatic*, even with the new form of approximations. Earlier arguments have shown that the horizontal velocity $\mathbf{u}_h$ is also expected to be *approximately uniform* in the vertical:

$$\mathbf{u}_h \simeq \mathbf{u}_h(x, y, t) \tag{6.68}$$

More exactly, we can decompose horizontal velocity into two components: one having no dependence on z, and the other with a dependence on z:

$$\mathbf{u}_h(x, y, z, t) = \bar{\mathbf{u}}_h(x, y, t) + \mathbf{u}'_h(x, y, z, t) \tag{6.69}$$

where $\bar{\mathbf{u}}_h$ is the vertically averaged horizontal velocity

$$\bar{\mathbf{u}}_h \equiv \frac{1}{h+\eta} \int_{z=-h}^{\eta} \mathbf{u}_h(x, y, z, t)dz, \tag{6.70}$$

and $\mathbf{u}'_h$ is the deviation from this average velocity.

Using these approximations we can now integrate the governing Eqs. (6.66) and (6.67) vertically to derive equations for the vertically averaged component $\bar{\mathbf{u}}_n$. First, we integrate the continuity equation (6.67) or (6.64), to yield

$$\int_{-h(x,y)}^{\eta(x,y,t)} \nabla_h \cdot \mathbf{u}_h dz + w \mid_{z=\eta} - w \mid_{z=-h} = 0. \tag{6.71}$$

Since the limits of integration are functions of (x, y), the Leibnitz' rule (1.44a) is used to write

$$\nabla_h \cdot \int_{-h}^{\eta} \mathbf{u}_h dz - \mathbf{u}_h \mid_{z=\eta} \cdot \nabla_h \eta - \mathbf{u}_h \mid_{z=-h} \cdot \nabla_h(-h) + w \mid_{z=\eta} - w \mid_{z=-h} = 0.$$

We can now insert the kinematic boundary conditions (6.59a,b) and the definition (6.70) in the above, to yield

$$\nabla_h \cdot (h+\eta)\bar{\mathbf{u}}_h + \frac{\partial \eta}{\partial t} = 0.$$

Defining the total depth

$$H = h + \eta \tag{6.72}$$

this result becomes

$$\frac{\partial \eta}{\partial t} + \nabla_h \cdot (H\bar{\mathbf{u}}_h) = 0. \tag{6.73}$$

Similarly, the momentum equation is integrated. However, first we make some manipulations in Eq. (6.66a).

The identity (1.28) will be of some use:

$$\begin{aligned} \nabla \cdot (\mathbf{a} \circ \mathbf{b}) &= \mathbf{a}(\nabla \cdot \mathbf{b}) + \mathbf{b} \cdot (\nabla \circ \mathbf{a}) \\ &= \mathbf{a}(\nabla \cdot \mathbf{b}) + (\mathbf{b} \cdot \nabla)\mathbf{a} \end{aligned} \tag*{[1.28.a]}$$

Proof

$$\begin{aligned} \nabla \cdot (\mathbf{a} \circ \mathbf{b}) &= \hat{e}_i \frac{\partial}{\partial x_j} a_i b_j \\ &= \hat{e}_i a_i \frac{\partial b_j}{\partial x_j} + \hat{e}_i b_j \frac{\partial a_i}{\partial x_j} \\ &= \mathbf{a}(\nabla \cdot \mathbf{b}) + \mathbf{b} \cdot (\nabla \circ \mathbf{a}) \\ &= \hat{e}_i a_i \frac{\partial b_j}{\partial x_j} + b_j \frac{\partial a_i \hat{e}_i}{\partial x_j} \\ &= \mathbf{a}(\nabla \cdot \mathbf{b}) + (\mathbf{b} \cdot \nabla)\mathbf{a}. \end{aligned} \tag{6.73}$$

Applying this identity to the second term on the left hand side of (6.66a) gives

$$\begin{aligned} (\mathbf{u} \cdot \nabla)\mathbf{u}_h &= \nabla \cdot (\mathbf{u}_h \circ \mathbf{u}) - \mathbf{u}_h(\nabla \cdot \mathbf{u}) \\ &= \nabla \cdot (\mathbf{u}_h \circ \mathbf{u}) \end{aligned} \tag{6.74}$$

where the second term on the *r.h.s.* vanishes by virtue of (6.67). The divergence of the dyadic product $\mathbf{u}_h \circ \mathbf{u}$ in Eq. (6.74) is simply

$$\begin{aligned} \nabla \cdot (\mathbf{u}_h \circ \mathbf{u}) &= \hat{e}_i \frac{\partial}{\partial x_j} u_{hi} u_j \\ &= \mathbf{i}\left(\frac{\partial}{\partial x} u^2 + \frac{\partial}{\partial y} uv + \frac{\partial}{\partial z} uw\right) \\ &\quad + \mathbf{j}\left(\frac{\partial}{\partial x} vu + \frac{\partial}{\partial y} v^2 + \frac{\partial}{\partial z} vw\right) + \mathbf{k}\,(0) \\ &= \nabla_h \cdot (\mathbf{u}_h \circ \mathbf{u}_h) + \frac{\partial}{\partial z}(\mathbf{u}_h w). \end{aligned} \tag{6.75}$$

Substituting (6.74) and (6.75) into Eq. (6.66a) the momentum equation becomes

$$\frac{\partial \mathbf{u}_h}{\partial t} + \nabla \cdot (\mathbf{u}_h \circ \mathbf{u}) + f\mathbf{k} \times \mathbf{u}_h = -\frac{1}{\rho}\nabla_h p^s - g\nabla_h \eta + \nu \frac{\partial^2 \mathbf{u}_h}{\partial z^2}, \tag{6.76}$$

which will next be integrated vertically. Note that the variables $p^s = p^s(x, y, t)$, $\eta = \eta(x, y, t)$ are constants with respect to vertical integration. Integrating from $z = -h$ to $z = \eta$, we have

$$\int_{-h}^{\eta} \frac{\partial \mathbf{u}_h}{\partial t} dz + \int_{-h}^{\eta} \nabla \cdot (\mathbf{u}_h \circ \mathbf{u}) dz + f\mathbf{k} \times \int_{-h}^{\eta} \mathbf{u}_h dz = -\frac{H}{\rho} \nabla_h p^s - gH\nabla\eta + \nu \left[\frac{\partial \mathbf{u}_h}{\partial z} \right]_{z=-h}^{z=\eta}$$

Using the definition (6.70) and substituting dynamic boundary conditions (6.48) and (6.51) derived in Sect. 6.2.2 we have

$$\int_{-h}^{\eta} \frac{\partial \mathbf{u}_h}{\partial t} dz + \int_{-h}^{\eta} \nabla .(\mathbf{u}_h \circ \mathbf{u}) dz + fH\mathbf{k} \times \bar{\mathbf{u}}_h = -\frac{H}{\rho} \nabla_h p^s - gH\nabla\eta + \frac{1}{\rho} \left(\tau^s - \tau^b \right), \quad (6.77)$$

where, the definition $\nu = \mu/\rho$ has also been used. The integration of the first term on the left hand side is carried out, using the Leibnitz' rule (1.44a):

$$\begin{aligned} \int_{-h(x,y)}^{\eta(x,y,t)} \frac{\partial \mathbf{u}_h}{\partial t} &= \frac{\partial}{\partial t} \int_{-h}^{\eta} \mathbf{u}_h dz - u_h \mid_{z=\eta} \frac{\partial \eta}{\partial t} \\ &= \frac{\partial}{\partial t} H\bar{\mathbf{u}}_h - u_h \mid_{z=\eta} \frac{\partial \eta}{\partial t}. \end{aligned} \quad (6.78)$$

The second term on the left hand side of (6.77) can similarly be integrated, this time making use of (6.75):

$$\begin{aligned} &\int_{-h}^{\eta} \nabla \cdot (\mathbf{u}_h \circ \mathbf{u}_h) dz = \int_{-h}^{\eta} \nabla_h \cdot (\mathbf{u}_h \circ \mathbf{u}_h) dz + \int_{-h}^{\eta} \frac{\partial}{\partial z} (\mathbf{u}_h w) dz \\ &= \nabla_h \cdot \int_{-h}^{\eta} (\mathbf{u}_h \circ \mathbf{u}_h) dz - (\mathbf{u}_h \circ \mathbf{u}_h) \mid_{z=\eta} \cdot \nabla_h \eta \\ &- (\mathbf{u}_h \circ \mathbf{u}_h) \mid_{z=-h} \cdot \nabla_h h + \mathbf{u}_h w \mid_{z=\eta} - \mathbf{u}_h w \mid_{z=-h} . \end{aligned} \quad (6.79)$$

Here, we can make use of the identity

$$(\mathbf{a} \circ \mathbf{a}) \cdot \mathbf{b} = (a_i a_j) b_j = a_i (a_j b_j) = \mathbf{a}(\mathbf{a} \cdot \mathbf{b}) \quad (6.80)$$

to write (6.79) as

$$\begin{aligned} &\int_{-h}^{\eta} \nabla \cdot (\mathbf{u}_h \circ \mathbf{u}) dz = \nabla_h \cdot \int_{-h}^{\eta} (\mathbf{u}_h \circ \mathbf{u}_h) dz \\ &- (\mathbf{u}_h \mid_{z=\eta})(\mathbf{u}_h \mid_{z=\eta} \cdot \nabla_h \eta) - (\mathbf{u}_h \mid_{z=-h})(\mathbf{u}_h \mid_{z=-h} \cdot \nabla_h h) \\ &+ \mathbf{u}_h w \mid_{z=\eta} - \mathbf{u}_h w \mid_{z=-h} . \end{aligned} \quad (6.81)$$

Substituting (6.78) and (6.81), Eq. (6.77) takes the following form:

$$\begin{aligned} &\frac{\partial H\bar{\mathbf{u}}_h}{\partial t} + \nabla_h \cdot \int_{-h}^{\eta} (\mathbf{u}_h \circ \mathbf{u}_h) dz + fH\mathbf{k} \times \bar{\mathbf{u}}_h \\ &- \left[\mathbf{u}_h \frac{\partial \eta}{\partial t} \right]_{z=\eta} - [\mathbf{u}_h(\mathbf{u}_h \cdot \nabla_h \eta)]_{z=\eta} - [\mathbf{u}_h(\mathbf{u}_h \cdot \nabla_h h)]_{z=-h} + [\mathbf{u}_h w]_{z=\eta} - [\mathbf{u}_h w]_{z=-h} \\ &= -\frac{H}{\rho} \nabla_h p^s - gH\nabla\eta + \frac{1}{\rho} (\tau^s - \tau^s) \end{aligned} \quad (6.82)$$

Note that a number of cancellations occur because of the kinematic boundary conditions (6.59a) and (6.59b). The second term left hand side can be further simplified by utilizing (6.69) to write

$$
\begin{aligned}
\int_{-h}^{\eta} (\mathbf{u}_h \circ \mathbf{u}_h) dz &= \int_{-h}^{\eta} (\bar{\mathbf{u}}_h + \mathbf{u}_h') \circ (\bar{\mathbf{u}}_h + \mathbf{u}_h') dz \\
&= \int_{-h}^{\eta} (\bar{\mathbf{u}}_h \circ \bar{\mathbf{u}}_h) dz + \int_{-h}^{\eta} (\bar{\mathbf{u}}_h \circ \mathbf{u}_h') dz \\
&+ \int_{h}^{\eta} (\mathbf{u}_h' \circ \bar{\mathbf{u}}_h) dz + \int_{-h}^{\eta} (\mathbf{u}_h' \circ \mathbf{u}_h') dz.
\end{aligned}
\tag{6.83}
$$

On the other hand, by virtue of (6.69) and (6.70)

$$
\begin{aligned}
\int_{-h}^{\eta} \mathbf{u}_h' dz &= \int_{-h}^{\eta} \mathbf{u}_h dz - \bar{\mathbf{u}}_h \int_{-h}^{\eta} dz \\
&= H\bar{\mathbf{u}}_h - H\bar{\mathbf{u}}_h = 0,
\end{aligned}
\tag{6.84}
$$

so that the second and third terms of (6.83) vanish, leaving

$$
\int_{-h}^{\eta} (\mathbf{u}_h \circ \mathbf{u}_h) dz = H\bar{\mathbf{u}}_h \circ \bar{\mathbf{u}}_h + \int_{-h}^{\eta} (\mathbf{u}_h' \circ \mathbf{u}_h') dz. \tag{6.85}
$$

Now, substituting (6.85) into (6.82) and defining a second order tensor

$$
\mathbf{F} = -\frac{1}{H} \int_{-h}^{\eta} (\mathbf{u}_h' \circ \mathbf{u}_h') dz, \tag{6.86}
$$

the momentum equations become:

$$
\begin{aligned}
&\frac{\partial H\bar{\mathbf{u}}_h}{\partial t} + \nabla_h \cdot (H\bar{\mathbf{u}}_h \circ \bar{\mathbf{u}}_h) + fH\mathbf{k} \times \bar{\mathbf{u}}_h \\
&= -gH\nabla\eta - \frac{H}{\rho}\nabla_h p^s + \frac{1}{\rho}\left(\boldsymbol{\tau}^s - \boldsymbol{\tau}^b\right) + \nabla_h \cdot H\mathbf{F}.
\end{aligned}
\tag{6.87}
$$

The first two terms can be expanded by making use of (1.28a) as

$$
\begin{aligned}
&\frac{\partial H\bar{\mathbf{u}}_h}{\partial t} + \nabla_h \cdot (\bar{\mathbf{u}}_h \circ H\bar{\mathbf{u}}_h) \\
&= H\frac{\partial \bar{\mathbf{u}}_h}{\partial t} + \bar{\mathbf{u}}_h \frac{\partial H}{\partial t} + \bar{\mathbf{u}}_h (\nabla_h \cdot H\bar{\mathbf{u}}_h) + (H\bar{\mathbf{u}}_h \cdot \nabla_h)\bar{\mathbf{u}}_h, \\
&= H\left\{\frac{\partial \bar{\mathbf{u}}_h}{\partial t} + \bar{\mathbf{u}}_h \cdot \nabla_h \bar{\mathbf{u}}_h\right\} + \bar{\mathbf{u}}_h \left\{\frac{\partial \eta}{\partial t} + \nabla_h \cdot \bar{\mathbf{u}}_h\right\}
\end{aligned}
\tag{6.88}
$$

the second term of which vanishes by virtue of (6.73). Therefore, using (6.88) in (6.87), the momentum equation becomes (dropping overbars and the subscript h hereafter):

$$\frac{\partial \mathbf{u}}{\partial t} + \mathbf{u} \cdot \nabla \mathbf{u} + f\mathbf{k} \times \mathbf{u} = -g\nabla\eta - \frac{1}{\rho}\nabla p^s + \frac{1}{\rho H}\left(\tau^s - \tau^b\right) + \frac{1}{H}\nabla \cdot H\mathbf{F}, \tag{6.89}$$

supplemented by (6.73), written in the same manner as

$$\frac{\partial \eta}{\partial t} + \nabla \cdot H\mathbf{u} = 0 \tag{6.90}$$

Equations (6.89) and (6.90) are called the *shallow water equations.*

In the momentum equation, the left hand side and the first term on the right hand side are familiar terms. The second term (left hand side) represents the pressure gradient arising due to variations of atmospheric pressure, and vanishes if it is uniformly distributed. The third term is the difference in surface and bottom shear stresses distributed per unit mass of the fluid column (divided by ρH). The bottom stresses can often be neglected and the remaining term represents forcing by wind stress on the sea surface. The last term is a weighed divergence of a tensor $\mathbf{F}$ with a form similar to that appearing in (3.8a). Therefore the tensor $\mathbf{F}$ defined in (6.86) is called *Reynold's stress tensor*. Since this term arises because of the vertical averaging of the $\mathbf{u}_h' \circ \mathbf{u}_h'$ tensor, and since it is expected that $\left|\mathbf{u}_h'\right| \ll |\bar{\mathbf{u}}_h|$, often it can be neglected.

6.2.6 Conservation Properties

Mass Conservation

Since $H(x, y, t) = h(x, y) + \eta(x, y, t)$, we can write continuity equation (6.90) as

$$\frac{\partial H}{\partial t} + \nabla \cdot H\mathbf{u} = 0 \tag{6.91}$$

or by regrouping the terms as

$$\frac{DH}{Dt} = \frac{\partial H}{\partial t} + \mathbf{u} \cdot \nabla H = -H\nabla \cdot \mathbf{u}. \tag{6.92}$$

Here, the horizontal divergence of the horizontal velocity represents the relative rate of change of the horizontal cross-sectional area of a material element:

$$\nabla \cdot \mathbf{u} = \frac{1}{A}\frac{dA}{dt} \tag{6.93}$$

so that (6.92) becomes

$$\frac{1}{H}\frac{DH}{Dt} + \frac{1}{A}\frac{dA}{dt} = \frac{D}{Dt} ln H + \frac{d}{dt} ln A = \frac{d}{dt} ln(HA) = 0 \tag{6.94}$$

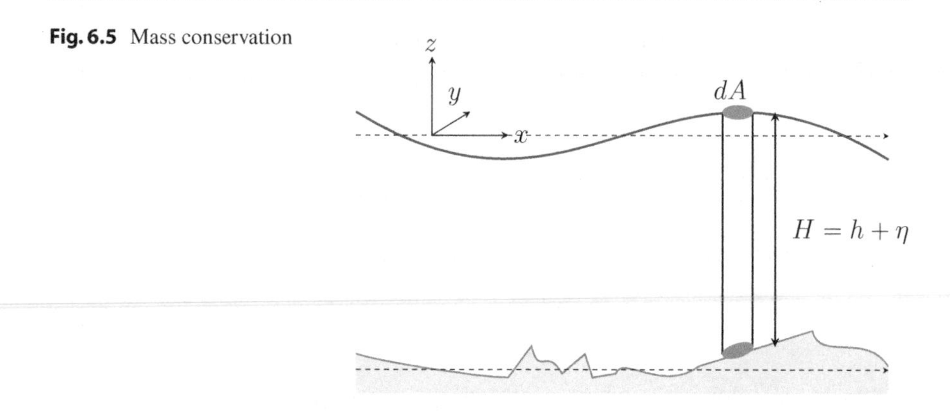

Fig. 6.5 Mass conservation

Equation (6.94) expresses the fact that the total volume HA of any fluid column is conserved. For an increase to occur in total depth, there must be a corresponding reduction in cross-sectional area as shown in Fig. 6.5 and vice versa.

We can also integrate the continuity equation (6.67) vertically to obtain vertical velocity, assuming that the horizontal velocity $\mathbf{u}$ is approximately independent of z:

$$w(x, y, z, t) = \int \frac{\partial w}{\partial z}\, dz \simeq -z(\nabla \cdot \mathbf{u}) + c(x, y, t) \tag{6.95}$$

where c is a constant of integration with respect to z. In the above equation $\nabla \cdot \mathbf{u}$ stands for $\nabla_h \cdot \mathbf{u}_h \simeq \nabla_h \cdot \bar{\mathbf{u}}_h$, but since subscripts have been dropped earlier, we choose this notation.

Equation (6.95) indicates that vertical velocity is approximately a linear function of z. In fact this is the only possibility to be able to satisfy kinematic boundary conditions (6.59a), (6.59b). Making use of (6.59b) the integration constant c is evaluated and

$$w = -(z + h)\nabla \cdot \mathbf{u} - \mathbf{u} \cdot \nabla h. \tag{6.96}$$

Substituting (6.92) and nothing that $h = h(x, y)$, we can write

$$w \equiv \frac{Dz}{Dt} = -\frac{z + h}{H}\frac{DH}{Dt} - \frac{Dh}{Dt} \tag{6.97}$$

or

$$\frac{D}{Dt}\left(\frac{z + h}{H}\right) = 0. \tag{6.98}$$

This result (6.98) shows that the vertical position of any material point measured relative to $z = -h$ and normalized by the total depth, i.e. $(z + h)/H$ is conserved following the motion as shown in Fig. 6.6.

Consider the flow contained by vertical fixed side walls on a boundary C enclosing a region A.

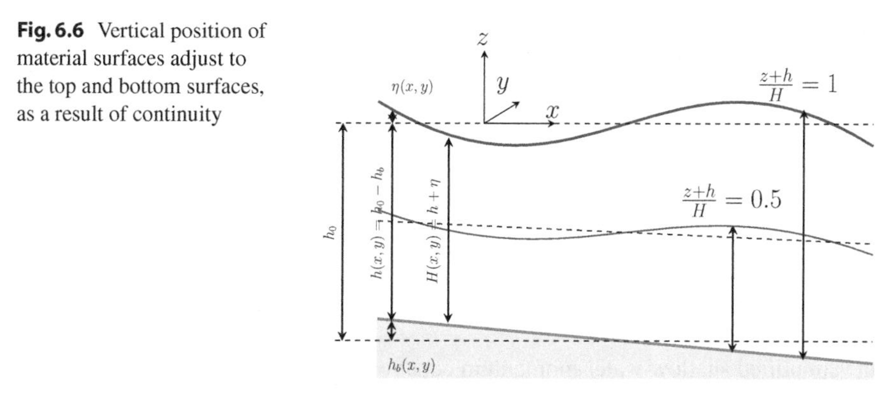

Fig. 6.6 Vertical position of material surfaces adjust to the top and bottom surfaces, as a result of continuity

At the side-walls there will be no normal fluxes,

$$\mathbf{u} \cdot \mathbf{n} = 0. \tag{6.99}$$

Integration of (6.90) across the area A yields

$$\int_A \frac{\partial \eta}{\partial t}\, dA = -\int_A \nabla \cdot H\mathbf{u}\, dA = -\oint_C H\mathbf{u} \cdot \mathbf{n}\, dl \tag{6.100}$$

through the use of the divergence theorem (1.29). For an open ocean, or a semi-enclosed basin, this yields

$$\frac{\partial}{\partial t} \int \eta\, dA = -\oint_C H\mathbf{u} \cdot \mathbf{n}\, dl, \tag{6.101a}$$

relating the mean sea-level to the volume flux through the open boundaries (i.e. a statement of mass conservation).

On the other hand, for a totally enclosed basin as in Fig. 6.7, (6.99) requires the right hand side of (6.100) to vanish, so that

$$\frac{\partial}{\partial t} \int \eta\, dA = 0. \tag{6.101b}$$

This equation states that the surface displacements integrated over the enclosed area should be constant at all times.

6.2.7 Vorticity Conservation

Neglecting the last three terms of (6.89) and making use of the vector identity (1.27e), namely

$$\mathbf{u} \cdot \nabla \mathbf{u} = \frac{1}{2} \nabla \mathbf{u} \cdot \mathbf{u} - \mathbf{u} \times \nabla \times \mathbf{u} \tag{6.102}$$

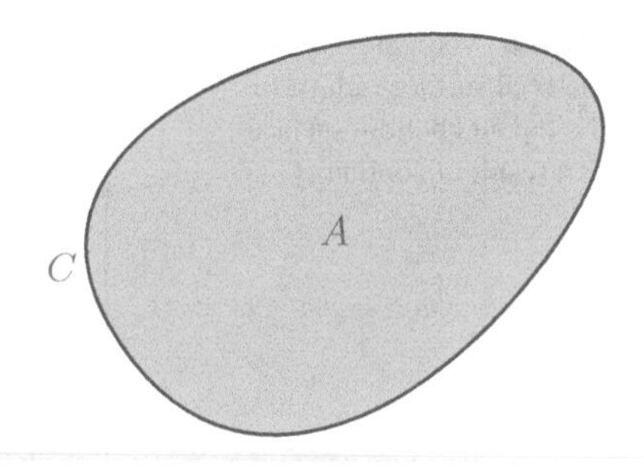

Fig. 6.7 A closed basin with surface area A bounded by a coast C

the simplified shallow water momentum equation can be re-written as

$$\frac{\partial \mathbf{u}}{\partial t} = -\nabla\left(g\eta + \frac{1}{2}\mathbf{u}\cdot\mathbf{u}\right) + \mathbf{u}\times(\nabla\times\mathbf{u} + f\mathbf{k}) \tag{6.103}$$

Here the "del" operator is one that is horizontal $\nabla = \nabla_h$, because the fields are only two-dimensional; and $\mathbf{u} = \mathbf{u}_h$ is the horizontal velocity, with subscripts dropped earlier. The term $\nabla\times\mathbf{u} = \nabla_h\times\mathbf{u}_h$ is the vertical component of the vorticity vector defined by

$$\mathbf{k}\cdot\omega = \mathbf{k}\cdot(\nabla_h\times\mathbf{u}_h) = \zeta \tag{6.104}$$

so that substituting (6.104) into (6.103) and taking curl yields

$$\frac{\partial \zeta\mathbf{k}}{\partial t} = \nabla\times(\mathbf{u}\times\mathbf{k}(\zeta + f)) \tag{6.105}$$

The second term of (6.103) vanishes upon taking the curl. Expanding the right hand side of (6.105) yields

$$\mathbf{k}\frac{\partial \zeta}{\partial t} = -\mathbf{u}\cdot\nabla\mathbf{k}(\zeta + f) - \mathbf{k}(\zeta + f)\nabla\cdot\mathbf{u} \tag{6.106}$$

or

$$\frac{\partial(\zeta + f)}{\partial t} + \mathbf{u}\cdot\nabla(\zeta + f) + (\zeta + f)\nabla\cdot\mathbf{u} \tag{6.107}$$

Substituting from (6.92) then gives

$$\frac{D(\zeta + f)}{Dt} - \frac{(\zeta + f)}{H}\frac{DH}{Dt} = 0 \tag{6.108}$$

or

$$\frac{D}{Dt}\left(\frac{\zeta + f}{H}\right) = 0 \tag{6.109}$$

Equation (6.109) states that the *potential vorticity* defined as $\frac{\zeta+f}{H}$ is conserved following any fluid column. ζ is often referred to as *relative vorticity*, and f as *planetary vorticity* which together make up the *absolute vorticity* $\zeta + f$. A fluid column will preserve its initial relative vorticity only if f and H are constant.

Integration of (6.103) along the boundary gives

$$\begin{aligned}\frac{\partial}{\partial t}\oint_C \mathbf{u}\cdot d\mathbf{r} &= -\oint_C \nabla\left(g\eta + \frac{1}{2}\mathbf{u}\cdot\mathbf{u}\right)\cdot d\mathbf{r} + \oint_C [\mathbf{u}\times(\zeta + f)\mathbf{k}]\cdot d\mathbf{r}\\ &= -\oint_C (\zeta + f)\mathbf{k}\cdot\mathbf{u}\times d\mathbf{r}, \\ &= -\oint_C (\zeta + f)\mathbf{u}\cdot\mathbf{n}dl\end{aligned} \tag{6.110}$$

yielding a relation between the circulation $\Gamma = \oint_C \mathbf{u}\cdot d\mathbf{r}$ for a closed path C, and the flux of vorticity through the boundaries.

Similarly, if Eq. (6.106) is integrated over the area A enclosed by the closed curve C, and if the divergence theorem (1.29) is used, we obtain

$$\frac{\partial}{\partial t}\int_A \zeta\, dA = -\int_A \nabla\cdot(\zeta + f)\mathbf{u}\, dA = -\oint_C (\zeta + f)\mathbf{u}\cdot\mathbf{n}\, dA. \tag{6.111}$$

Note that the *l.h.s.* of either Eq. (6.110) or (6.111) are equal by *Stokes' theorem* (3.71) and (3.72), written in the present context. The average vorticity in A is equal to the circulation of C, and changes only by fluxes of vorticity across the boundary C.

Consider now, a region enclosed by side-walls on boundary C, where $\mathbf{u}\cdot\mathbf{n} = 0$ (or $\mathbf{u}$ is parallel to $d\mathbf{r}$); in this case, the *r.h.s.* of (6.110) or (6.111) vanish:

$$\frac{\partial}{\partial t}\oint_C \mathbf{u}\cdot d\mathbf{r} = \frac{\partial}{\partial t}\int_A \zeta\, dA = 0. \tag{6.112}$$

Since there is no normal velocity across the solid boundary C the circulation around any rigid boundary, or equivalently, the area-averaged relative vorticity in an enclosed region A is conserved.

6.2.8 Energy Conservation

To derive the energy conservation equation, again we consider a region A enclosed by a boundary C. We first multiply (6.103) by $H\mathbf{u}$ to obtain

$$H\mathbf{u}\cdot\frac{\partial \mathbf{u}}{\partial t} + H\mathbf{u}\cdot\nabla\frac{1}{2}\mathbf{u}\cdot\mathbf{u} + H\mathbf{u}\cdot(\zeta + f)\mathbf{k}\times\mathbf{u} = -gH\mathbf{u}\cdot\nabla\eta \tag{6.113}$$

where, the last term on the *l.h.s* vanishes because $u\cdot\mathbf{k}\times\mathbf{u} \equiv 0$. We note, by making use of (6.92), that

$$H\mathbf{u}\cdot\frac{\partial \mathbf{u}}{\partial t} = \frac{\partial}{\partial t}\frac{1}{2}H\mathbf{u}\cdot\mathbf{u} - \frac{1}{2}\mathbf{u}\cdot\mathbf{u}\frac{\partial H}{\partial t} = \frac{\partial}{\partial t}\frac{1}{2}H\mathbf{u}\cdot\mathbf{u} + \frac{1}{2}\mathbf{u}\cdot\mathbf{u}\nabla\cdot H\mathbf{u} \tag{6.114}$$

then, by making repeated use of equation (6.92), (6.113) becomes

$$\begin{aligned}\frac{\partial}{\partial t}\left(\frac{1}{2}H\mathbf{u}\cdot\mathbf{u}\right)+\nabla\cdot\left[\left(\frac{1}{2}\mathbf{u}\cdot\mathbf{u}\right)H\mathbf{u}\right]&=-gH\mathbf{u}\cdot\nabla\eta\\&=-g\nabla\cdot\eta H\mathbf{u}+g\eta\nabla\cdot H\mathbf{u}\\&=-g\nabla\cdot\eta H\mathbf{u}-g\eta\frac{\partial H}{\partial t}\end{aligned}\quad(6.115)$$

Reorganizing (6.115),

$$\frac{\partial}{\partial t}\left(\frac{1}{2}H\mathbf{u}\cdot\mathbf{u}+\frac{1}{2}g\eta^2\right)=-\nabla\cdot\left[\left(\frac{1}{2}\mathbf{u}\cdot\mathbf{u}\right)H\mathbf{u}+g\eta H\mathbf{u}\right]\quad(6.116)$$

and, integrating over the domain A, using the divergence theorem (1.29) yields

$$\frac{\partial}{\partial t}\int_A\left(\frac{1}{2}H\mathbf{u}\cdot\mathbf{u}+\frac{1}{2}g\eta^2\right)dA=\oint_C\left(\frac{1}{2}\mathbf{u}\cdot\mathbf{u}+g\eta\right)H\mathbf{u}\cdot\mathbf{n}dl.\quad(6.117)$$

The individual terms on the *l.h.s* are defined as

$$\begin{aligned}KE&=\frac{1}{2}H\mathbf{u}\cdot\mathbf{u},\\PE&=\int_0^\eta gzdz=\frac{1}{2}g\eta^2,\end{aligned}\quad(6.118a, b)$$

representing the kinetic and potential energy per unit mass of the fluid column, so that (6.117) states the conservation of total mechanic energy. For open or semi-enclosed domains, the total energy in the region A changes by fluxes of kinetic and potential energy (the first and second terms on the *r.h.s* respectively), across the boundary C.

For an enclosed domain, with $\mathbf{u}\cdot\mathbf{n}=0$ on the solid boundary C, we have

$$\frac{\partial}{\partial t}\int_A(KE+PE)\,dA=0.\quad(6.119)$$

6.3 The f-Plane and the β-Plane Approximations

We already have assumed in Sect. 2.1 that the spherical geometry of earth can be reasonably approximated by a *tangent-plane* fitted to the region of interest. It should be noted however that the coriolis parameter

$$f=f(\phi)=2\Omega\sin\phi$$

is a function of the latitude angle ϕ. With respect to a fixed point on the earth (at latitude angle ϕ_0, where the tangent plane contacts the earth) the Coriolis parameter is expressed as

$$f = 2\Omega \sin(\phi_0 + \Delta\phi) = 2\Omega(\sin\phi_0 \cos\Delta\phi + \cos\phi_0 \sin\Delta\phi) \tag{6.120}$$

If the angle $\Delta\phi$ (which measures deviations from ϕ_0) is small, we can approximate

$$\cos\Delta\phi = 1 - \frac{(\Delta\phi)^2}{2!} + \frac{(\Delta\phi)^4}{4!} - \cdots \tag{6.121a}$$

$$\sin\Delta\phi = \Delta\phi - \frac{(\Delta\phi)^3}{3!} + \frac{(\Delta\phi)^5}{5!} - \cdots \tag{6.121b}$$

and, neglecting terms of $O(\Delta\phi)$ and smaller, (6.120) approximates to

$$f = f_0 \equiv 2\Omega \sin\phi_0 = \text{constant}. \tag{6.122}$$

This approximation for f is referred to as the *f-plane approximation*. Since $f = f_0$ is taken as constant, the effects of the latitudinal change in the coriolis parameter are not incorporated in the dynamics.

As the next level of approximation we can neglect terms of $O(\Delta\phi^2)$ and smaller, which yields

$$f = 2\Omega(\sin\phi_0 + \Delta\phi \cos\phi_0) \tag{6.123}$$

and since angle $\Delta\phi$ is small, it can be interchanged with

$$\Delta\phi = \frac{y}{r_0} \tag{6.124}$$

where y is the horizontal coordinate on the tangent plane pointing towards the north and r_0 is the earth's radius. Then Eq. (6.123) becomes

$$f = f_0 + \beta_0 y \tag{6.125}$$

where

$$\beta_0 = \frac{2\Omega \cos\phi_0}{r_0} = \frac{f_0 \cot\phi_0}{r_0} \tag{6.126}$$

and f_0 is given by (6.122). The variation of the Coriolis parameter f with latitude has been approximated by a linear function in (6.126), in order to incorporate this variation in the equations. This is called the β-plane approximation.

6.4 Simple Applications of the Potential Vorticity Conservation

Among the conservation laws derived in Sect. 2.2, the conservation of potential vorticity (6.109) is one of the most important and useful results in understanding the fundamental behavior of geophysical flows. We will consider several simple applications to emphasize the use of potential vorticity conservation. First we write (6.109) as

$$\frac{D}{Dt}\left(\frac{\zeta + f_0 + \beta y}{H}\right) = 0 \tag{6.127}$$

by making use of (6.125), i.e. including variations of the Coriolis parameter at the β-plane approximation level.

6.4.1 Geostrophic Flow

The geostrophic approximation (cf. Sect. 1.2.3) for shallow water equations (6.89) and (6.90) excluding the forcing terms of the momentum equation and unsteady terms of the continuity equation are:

$$f\mathbf{k} \times \mathbf{u} = -g\nabla\eta \tag{6.128a}$$

$$\nabla \cdot H\mathbf{u} = 0 \tag{6.128b}$$

Note that the above shallow water equations (6.128a), (6.128b) for geostrophic flow bounded in the vertical by a free surface and the bottom are somewhat different from the original geostrophic flow Eqs. (5.21) and (5.22) for an infinite domain as reviewed in Sect. 5.2.3. Although the only difference appears in the inclusion of the water depth H in the continuity equation (6.128b) as a result of vertical integration to obtain the shallow water equations, we can no longer claim that these equations are indeterminate. In fact, if we follow the same procedure as in Sect. 5.2.3 and substitute from (6.128a) into (6.128b), by making use of vector identities (1.27c) and (1.27b), we note

$$\begin{aligned}
\nabla \cdot H\mathbf{u} &= \left(\frac{g}{f}\right)\nabla \cdot (\mathbf{k} \times H\nabla\eta) \\
&= -\left(\frac{g}{f}\right)\mathbf{k} \cdot \nabla \times (H\nabla\eta) \\
&= -\left(\frac{g}{f}\right)\mathbf{k} \cdot \nabla H \times \nabla\eta \\
&= -\left(\frac{g}{f}\right)\mathbf{k} \cdot (\nabla h \times \nabla\eta + \nabla\eta \times \nabla\eta) \\
&= -\left(\frac{g}{f}\right)\mathbf{k} \cdot \nabla h \times \nabla\eta = 0,
\end{aligned}$$

which negates the redundancy of the shallow water version of geostrophic equations, with a new result showing

$$\nabla h \times \nabla \eta = 0 \tag{6.129}$$

The above result implies that the free surface elevation is aligned with depth variations, with their contours parallel to each other.

The conservation of vorticity can be re-derived for these equations. First by taking note of (6.128a) and expanding, we find

$$\begin{aligned} \nabla \times f\mathbf{k} \times \mathbf{u} &= \mathbf{u} \cdot \nabla f\mathbf{k} + f\mathbf{k}\nabla \cdot \mathbf{u} \\ &= -g\nabla \times \nabla \eta \equiv 0, \end{aligned}$$

resulting in

$$\mathbf{u} \cdot \nabla f + f\nabla \cdot \mathbf{u} = 0. \tag{6.130}$$

Then, utilizing (6.128b)

$$\nabla \cdot H\mathbf{u} = H\nabla \cdot \mathbf{u} + \mathbf{u} \cdot \nabla H = 0 \tag{6.131}$$

Equation (6.130) can be written as

$$\mathbf{u} \cdot \nabla f - \frac{f}{H}\mathbf{u} \cdot \nabla H = 0, \tag{6.132}$$

or, first dividing by H and collecting terms, this can in turn be written as

$$\mathbf{u} \cdot \nabla \left(\frac{f}{H}\right) = 0. \tag{6.133}$$

Now, since the motion is steady.This is equivalent to writing

$$\frac{D}{Dt}\left(\frac{f}{H}\right) = \frac{D}{Dt}\left(\frac{f_0 + \beta y}{H}\right) = 0 \tag{6.134}$$

Note that the relative vorticity ζ does not enter the conservation law (6.134). Since the inertial terms have been neglected in (6.128a). This does not actually imply that the vorticity is zero, since through (6.128a)

$$\zeta = \mathbf{k} \cdot \nabla \times \mathbf{u} = \mathbf{k} \cdot \nabla \times \left(\frac{g}{f}\mathbf{k} \times \nabla \eta\right) = \nabla \eta \cdot \nabla \frac{g}{f} + \frac{g}{f}\nabla \cdot \nabla \eta = \frac{g}{f^2}\nabla \eta \cdot \nabla f + \frac{g}{f}\nabla^2 \eta \tag{6.135}$$

and in the case that $f = f_0 = \text{constant}$, this reduces to

$$\zeta = \frac{g}{f_0}\nabla^2 \eta \tag{6.136}$$

which is consistent with the definition of pressure $(g\eta/f_0)$ as stream function in geostrophic flow (as shown in Sect. 5.2.3). In fact the equation of motion (6.128a)

$$\mathbf{u} = \frac{g}{f}\mathbf{k} \times \nabla\eta \tag{6.137}$$

implies that

$$\mathbf{u} \cdot \nabla\eta = \frac{g}{f}(\mathbf{k} \times \nabla\eta) \cdot (\nabla\eta) \equiv 0 \tag{6.138}$$

i.e. the horizontal velocity is everywhere parallel to the isolines of surface elevation $(\mathbf{u} \cdot \nabla\eta = 0)$, as indicated in Fig. 6.8.

The combined requirements of the above Eqs. (6.129) and (6.138) imply that the isolines of surface displacement η and depth h are parallel to each other.

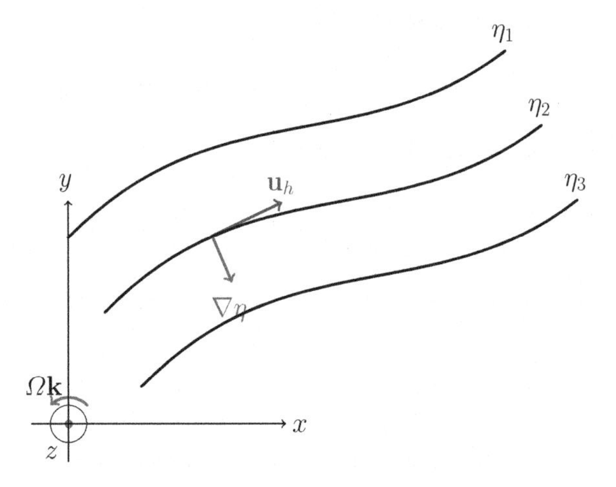

Fig. 6.8 Flow following the isolines of surface elevation

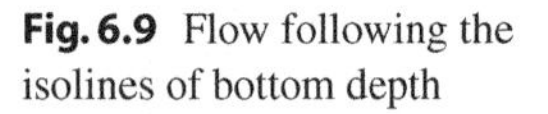

Fig. 6.9 Flow following the isolines of bottom depth

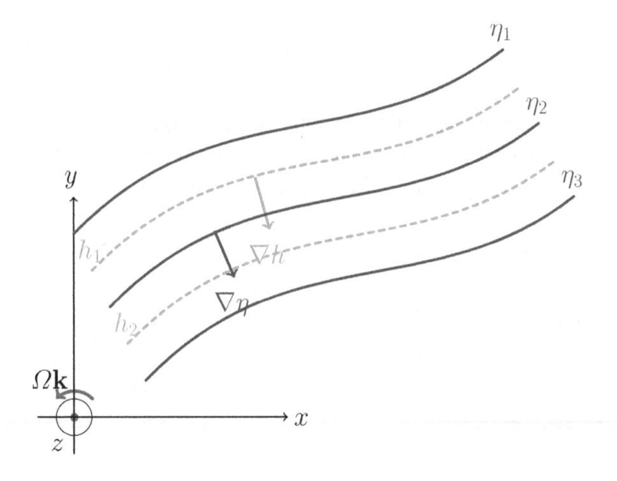

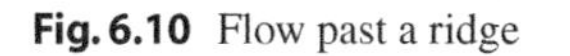

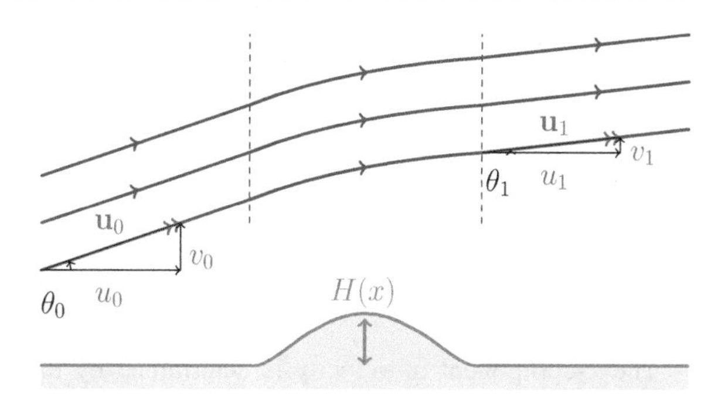

Fig. 6.10 Flow past a ridge

On the other hand, (6.134) implies that on an f-plane, ($\beta = 0,\ f = f_0$), any material element must move along isolines of H (total depth), since

$$\frac{D}{Dt}(f_0/H) = f_0\mathbf{u}\cdot\nabla(1/H) = 0$$

Approximately speaking, $H = h + \eta \simeq h$ ($\eta \ll h$), so that the above is equivalent to the requirement that fluid columns more along bottom contours, as in Fig. 6.9.

Multiplying (6.138) by f_0/H^2 and subtracting from (6.133) in the f-plane case ($f = f_0$) yields

$$\frac{f_0}{H^2}\mathbf{u}\cdot\nabla(H-\eta) = \frac{f_0}{H^2}\mathbf{u}\cdot\nabla h = 0, \tag{6.139}$$

which requires that velocity is parallel to bathymetric contours.

6.4.2 Flow over a Topographic Ridge (f-Plane)

As a second application consider the steady (but obviously not geostrophic) flow over a ridge shaped topographic barrier in Fig. 6.10.

Consider the flow to be independent of the y-coordinate (along the ridge), and let the incoming velocity be also independent of y, but assume it comes at an angle to the ridge with components (u_0, v_0) such that the initial vorticity $\zeta_0 = 0$.

Let the velocity components be (u_1, v_1) and the total depth be $H_1 = H_0$ after the ridge. Let the total depth on any point on the ridge be H, the velocity components be (u, v) and vorticity be ζ. Then potential vorticity conservation requires (for an f-plane)

$$\frac{f_0 + \zeta(x)}{H(x)} = \frac{f_0}{H_0} = \text{constant} \tag{6.140}$$

Since velocity components should be independent of y everywhere

$$\zeta(x) = \frac{dv(x)}{dx} \tag{6.141}$$

so that from (6.140)

$$\zeta = \frac{dv}{dx} = f_0 \left(\frac{H}{H_0} - 1 \right) \tag{6.142}$$

which is integrated to yield

$$v = v_0 + f_0 \int_{-\infty}^{x} \frac{H - H_0}{H_0} dx \tag{6.143}$$

The x-component of velocity is determined by the continuity equation (6.90), which is

$$\frac{\partial}{\partial x}(uH) = 0 \tag{6.144}$$

implying that

$$uH = u_0 H_0 \tag{6.145}$$

Then, on the downstream side of the topographic barrier and sufficiently far from it (6.143) and (6.145) give

$$v_1 = v_0 - \frac{f_0}{H_0} A \tag{6.146a}$$

$$u_1 = u_0 \frac{H_0}{H_1} = u_0 \tag{6.146b}$$

since $H_1 = H_0$, and where

$$A = \int_{-\infty}^{\infty} (H_0 - H)\, dx. \tag{6.146c}$$

Note that $H = h + \eta$, and if the surface displacement η is neglected, (6.146c) is

$$A \simeq \int_{-\infty}^{\infty} (h_0 - h(x))\, dx \tag{6.147}$$

which is approximately the cross-sectional area of the ridge.

The angle of incidence θ_0, and the angle of transmission θ_1, are given by

$$\tan\theta_0 = \frac{v_0}{u_0}, \tag{6.148a}$$

$$\tan\theta_1 = \frac{v_0 - \frac{fA}{h_0}}{u_0}, = \tan\theta_0 - S \tag{6.148b}$$

where

$$S = (fA/u_0 h_0). \tag{6.148c}$$

so that the angle through which the velocity vector turns upon passing the ridge ($\alpha = \theta_0 - \theta_1$) can be calculated from trigonometry:

$$\tan\alpha = \tan(\theta_0 - \theta_1) = \frac{\tan\theta_0 - \tan\theta_1}{1 + \tan\theta_0 \tan\theta_1} = \frac{u_0^2 S}{u_0^2 + v_0^2 - u_0 v_0 S} \tag{6.149a}$$

or

$$\tan\alpha = \frac{S}{1 + \tan^2\theta_0 - S\tan\theta_0}. \tag{6.149b}$$

Therefore, the flow will be deflected in a clockwise sense upon passing the ridge. Immediately over the ridge (6.142) can be approximated as

$$\zeta \simeq f_0 \left(\frac{h}{h_0} - 1 \right) \tag{6.150}$$

and since $h < h_0$, it is seen that a negative vorticity ($\zeta < 0$) is imparted on the fluid by the ridge:

Since negative (anticyclonic) vorticity is often associated with high pressure centers (as in the case of atmospheric highs), we expect a pressure excess (i.e., raised surface elevation) a top the ridge. Excess pressures are often observed on top of mountains in the atmosphere.

6.5 Topographic Effects

The governing *shallow water equations* (6.89) and (6.90) for a homogeneous layer of fluid in a rotating frame of reference have been derived in Sect. 6.2. We recall that the last term of (6.89) represented the lateral friction effects,

$$\mathbf{F} = -\frac{1}{H}\int_{-h}^{\eta} (\mathbf{u}_h' \circ \mathbf{u}_h') dz, \tag{6.151}$$

where $\underline{F}$ is the shear stress acting between neighboring fluid columns, such that $u_h'(x, y, z, t)$ is the difference between the three dimensional velocity and the vertically averaged horizontal velocity represented in Eqs. (6.89) and (6.90).

Since a priori information on the three dimensional velocity is typically nonexistent at this approximation level, the diffusive-dispersive term (6.152) is often parameterized, the simplest form of which is

$$\frac{1}{H}\nabla \cdot H\mathbf{F} \simeq \nu\nabla^2\mathbf{u}. \tag{6.152}$$

Making use of the above simple parameterization, the equations of motion (6.89) and (6.90) reduce to

$$\frac{D\mathbf{u}}{Dt} + f\mathbf{k} \times \mathbf{u} = -g\nabla\eta - \frac{1}{\rho}\nabla p^s + \frac{1}{\rho H}(\boldsymbol{\tau}^s - \boldsymbol{\tau}^b) + \nu\nabla^2\mathbf{u} \tag{6.153a}$$

$$\nabla \cdot H\mathbf{u} = 0 \tag{6.153b}$$

where $\mathbf{u}$ is the vertically averaged horizontal velocity, η is the free surface elevation, the external forcing elements p^s is the barometric pressure at the surface, $\boldsymbol{\tau}^s$ and $\boldsymbol{\tau}^b$ are the shear stresses applied at the surface and bottom.

To facilitate analysis, we define a *transport stream-function* ψ according to

$$H\mathbf{u} = \mathbf{k} \times \nabla\psi \tag{6.154}$$

which readily satisfies continuity equation (6.153b) by virtue of simple vector identities.

Next, we define *vorticity* in terms of the transport stream-function

$$\zeta = \mathbf{k} \cdot \nabla \times \mathbf{u} = \mathbf{k} \cdot \nabla \times \frac{1}{H}\mathbf{k} \times \nabla\psi = \nabla \cdot \frac{1}{H}\nabla\psi. \tag{6.155}$$

Taking curl of Eq. (6.153a) yields

$$\frac{\partial}{\partial t}\nabla \times \mathbf{u} + \nabla \times (\mathbf{u} \cdot \nabla\mathbf{u}) + f\nabla \times \mathbf{k} \times \mathbf{u} = \frac{1}{\rho}\nabla \times \frac{(\boldsymbol{\tau}^s - \boldsymbol{\tau}^b)}{H} + \nu\nabla^2\nabla \times \mathbf{u}, \tag{6.156}$$

where use have been made of certain vector identities. With further use of vector identities, we obtain

$$\nabla \times (\mathbf{u} \cdot \nabla\mathbf{u}) = \nabla \times \nabla\left(\frac{1}{2}\mathbf{u} \cdot \mathbf{u}\right) - \nabla \times (\mathbf{u} \times \nabla \times \mathbf{u}) = \nabla \times \zeta\mathbf{k} \times \mathbf{u}$$

with the first term vanishing by vector identity (1.27i), so that the vorticity equation (6.156) becomes

$$\frac{\partial}{\partial t}\zeta\mathbf{k} + \nabla \times [(\zeta + f)\mathbf{k} \times \mathbf{u}] = \frac{1}{\rho}\nabla \times \frac{(\boldsymbol{\tau}^s - \boldsymbol{\tau}^b)}{H} + \nu\mathbf{k}\nabla^2\zeta. \tag{6.157}$$

We note by vector identity (1.27d),

$$\nabla \times [(\zeta + f)\mathbf{k} \times \mathbf{u}] = (\mathbf{u} \cdot \nabla)(\zeta + f)\mathbf{k} - (\zeta + f)(\mathbf{k} \cdot \nabla)\mathbf{u} + \mathbf{k}(\zeta + f)\nabla \cdot \mathbf{u} - \mathbf{u}\nabla \cdot \mathbf{k}(\zeta + f)$$

where the second term drops out and by vector identities (1.27a), (1.27c) and (1.27j), the fourth term drops out

$$\nabla \cdot [\mathbf{k}(\zeta + f)] = \nabla \cdot \nabla \times \mathbf{u} + \nabla \cdot \mathbf{k}f = 0,$$

also re-writing the continuity equation,

$$\nabla \cdot \mathbf{u} = -\frac{1}{H}\mathbf{u} \cdot \nabla H = -\frac{1}{H}\frac{DH}{Dt}$$

then yields the second term of (6.157) to be written as

$$\nabla \times [(\zeta + f)\mathbf{k} \times \mathbf{u}] = \mathbf{u} \cdot \nabla(\zeta + f)\mathbf{k} - \mathbf{k}\frac{\zeta + f}{H}\frac{DH}{Dt},$$

and combining terms, we obtain

$$\frac{D}{Dt}\left(\frac{\zeta + f}{H}\right) = \frac{1}{\rho H}\mathbf{k} \cdot \nabla \times \frac{(\boldsymbol{\tau}^s - \boldsymbol{\tau}^b)}{H} + \nu \frac{1}{H}\nabla^2 \zeta \tag{6.158}$$

Note that in the vorticity equation the only external forcing is the surface wind stress, while the surface barometric pressure term existing in the original momentum equations has disappeared.

The bottom frictional stress τ^b can be parameterized in different ways. If we assume the well-stirred case, i.e. if the flow is turbulent, the bottom stress is related to the second power of velocity

$$\boldsymbol{\tau}^b = \rho C_d |\mathbf{u}| \mathbf{u} \tag{6.159a}$$

For shallow coastal flows the turbulent formulation is more appropriate.

However, this parameterization can be linearized over a suitable range of $\mathbf{u}$, assuming $r = C_d|\mathbf{u}|$, approximating r to be a constant, to give

$$\frac{\boldsymbol{\tau}^b}{\rho H} = r\frac{\mathbf{u}}{H}. \tag{6.159b}$$

With the linearized turbulent version chosen, (6.157) becomes

$$\frac{D}{Dt}\left(\frac{\zeta + f}{H}\right) = \frac{1}{\rho H}\mathbf{k} \cdot \nabla \times \frac{\boldsymbol{\tau}^s}{H} - r\frac{1}{H}\mathbf{k} \cdot \nabla \times \frac{\mathbf{u}}{H} + \nu \frac{1}{H}\nabla^2 \zeta. \tag{6.160}$$

By multiplying with H and expanding the first term we can also write the above equation as

$$\frac{\partial \zeta}{\partial t} + H\mathbf{u} \cdot \nabla\left(\frac{\zeta + f}{H}\right) = \frac{1}{\rho}\mathbf{k} \cdot \nabla \times \frac{\boldsymbol{\tau}^s}{H} - r\mathbf{k} \cdot \nabla \times \frac{\mathbf{u}}{H} + \nu \nabla^2 \zeta. \tag{6.161}$$

The bottom frictional term can alternatively be expressed as

$$\mathbf{k} \cdot \nabla \times \frac{\mathbf{u}}{H} = \mathbf{k} \cdot \nabla \times \left[\mathbf{k} \times \frac{1}{H^2}\nabla\psi\right] = \nabla \cdot \frac{1}{H^2}\nabla\psi. \tag{6.162}$$

Note that we can alternatively expand the bottom frictional term as

$$\nabla \cdot \frac{1}{H^2}\nabla\psi = \frac{1}{H}\nabla \cdot \frac{1}{H}\nabla\psi - \frac{1}{H^3}\nabla H \cdot \nabla\psi = \frac{1}{H^2}\nabla^2\psi - \frac{2}{H^3}\nabla H \cdot \nabla\psi.$$

Making use of equation (6.162), the governing equations are written as

$$\frac{\partial \zeta}{\partial t} + \mathbf{k} \times \nabla \psi \cdot \nabla \left(\frac{\zeta + f}{H} \right) = \frac{1}{\rho} \mathbf{k} \cdot \nabla \times \frac{\boldsymbol{\tau}^s}{H} - r \nabla \cdot \frac{1}{H^2} \nabla \psi + \nu \nabla^2 \zeta \quad (6.163a)$$

$$\nabla \cdot \frac{1}{H} \nabla \psi = \zeta \quad (6.163b)$$

to constitute closed form equations for stream-function ψ and vorticity ζ, presenting different forms of the vorticity equation.

We can develop non-dimensional Forms of the vorticity equation by defining the scales

$$x \sim L_0, \quad t \sim f_0^{-1}, \quad H \sim H_0, \quad |\mathbf{u}| \sim u_0, \quad |\boldsymbol{\tau}^s| \sim \tau_0,$$

which implies

$$\psi \sim u_0 H_0 L_0, \quad \zeta \sim u_0 / L_0.$$

Substituting these scales results in the non-dimensional equations

$$\frac{\partial \zeta}{\partial t} + \mathbf{k} \times \nabla \psi \cdot \nabla \left(\mathbf{R} \zeta + 1 \right) = \lambda \mathbf{k} \cdot \nabla \times \frac{\boldsymbol{\tau}^s}{H} - \mu \nabla \cdot \frac{1}{H^2} \nabla \psi + \mathbf{E} \nabla^2 \zeta \quad (6.164a)$$

$$\nabla \cdot \frac{1}{H} \nabla \psi = \zeta \quad (6.164b)$$

where

$$\mathbf{R} = \frac{u_0}{f_0 L_0}, \quad \mathbf{E} = \frac{\nu}{f_0 L_0^2}$$

are the *Rossby* and *Ekman* numbers, and

$$\mu = \frac{r}{f_0 H_0}, \quad \lambda = \frac{\tau_0}{\rho f_0 u_0 H_0}$$

are non-dimensional numbers measuring the relative importance of bottom friction and wind stress terms.

The values for parameters for typical shelf flows can be taken as $u_0 = 0.1\,\mathrm{m/s}$, $f_0 = 10^{-4}\mathrm{s}^{-1}$, $L_0 = 100\,\mathrm{km}$, $H_0 = 100\,\mathrm{m}$, $r = 10^{-3}$, $\nu = 100\,\mathrm{m^2/s}$, the typical magnitude estimates for the non-dimensional parameters are $\mathbf{R} \simeq 1$, $\mathbf{E} = 10^{-4}$, $\mu \simeq 0.1$ and $\lambda = 100$.

6.5.1 Solutions for Uniform Shelf

The solutions for a uniform shelf topography, e.g. $H = H(x)$ can be obtained for a steady ($\partial/\partial t = 0$), linearized ($\mathbf{R} \to 0$) inviscid ($\nu = 0$) flow as follows. With these approximations, we obtain from (6.163a), (6.163b)

$$-\frac{f_0}{H^2}\mathbf{k} \times \nabla\psi \cdot \nabla H = -r\nabla \cdot \frac{1}{H^2}\nabla\psi, \tag{6.165}$$

and since $H = H(x)$, this is reduced to

$$-\frac{f_0}{H^2}\psi_y H_x = -r\left[\left(\frac{1}{H^2}\psi_x\right)_x + \left(\frac{1}{H^2}\psi_y\right)_y\right]. \tag{6.166}$$

After rearranging, this becomes

$$-\alpha(x)\psi_y + \gamma(x)\psi_x = \frac{1}{2}\nabla^2\psi \tag{6.167a}$$

where

$$\alpha(x) = \frac{f_0}{2r}H_x \quad \text{and} \quad \gamma(x) = \frac{H_x}{H}. \tag{6.167c, d}$$

Let us consider a channel confined between $x = x_c$ and $x = x_l$, and assume that the flow (e.g. the streamfunction) is specified at $y = 0$, $\psi(x, 0) = \psi_0(x)$. We want to obtain solutions for all y subject to this initial condition. At the side boundaries, the boundary conditions are $\psi(x_c, y) = \psi_c$ and $\psi(x_l, y) = \psi_l$, where ψ_c and psi_l are constants.

We can try separation of variables as a method of solution,

$$\psi(x, y) = F(x)G(y)$$

so that equation (6.167a) takes the form

$$-\alpha(x)FG_y + \gamma(x)F_xG = \frac{1}{2}\left(F_{xx} + G_{yy}\right), \tag{6.168a}$$

or

$$-\frac{1}{2}\frac{F_{xx}}{F} + \gamma(x)\frac{F_x}{F} = \frac{1}{2}\frac{G_{yy}}{G} + \alpha(x)\frac{G_y}{G}. \tag{6.168b}$$

Since we have a non-constant coefficient $\alpha(x)$ on the right hand side, separation of variables can not be achieved in general. However, for a specific form of the depth profile, i.e. $H(x) = H_c + a(x - x_c)$, $\alpha = af_0/2r$ becomes a constant and $\gamma(x) = a/H(x)$, so that we can write

$$-\frac{1}{2}\frac{F_{xx}}{F} + \gamma(x)\frac{F_x}{F} = \frac{1}{2}\frac{G_{yy}}{G} + \alpha\frac{G_y}{G} = \frac{1}{2}\lambda^2, \tag{6.169}$$

where λ is a separation constant. For this specific case, the separated equations are:

$$F_{xx} - 2\gamma(x)F_x + \lambda^2 F = 0, \tag{6.170a}$$

$$G_{yy} + 2\alpha G_y - \lambda^2 G = 0. \tag{6.170b}$$

The roots of the characteristic equation for the second equation are

$$m_+ = -\alpha + \sqrt{\alpha^2 + \lambda^2} \quad \text{and} \quad m_- = -\alpha - \sqrt{\alpha^2 + \lambda^2}.$$

It is seen that the character of the solution for $y > 0$ and $y < 0$ differ greatly. For an initial value $G(0) = \psi_0$ specified at cross-section $y = 0$ of the shelf, the decaying (evanescent) solutions m_+ for $y > 0$ and m_- for $y < 0$ must be chosen. It is clear that the solution for $y > 0$ decays at a much higher rate than the solution for $y < 0$.

For the more general case of an arbitrary $H = H(x)$, either analytical or numerical solutions must be sought, since separation of variables can not be applied.

Note that after some distance from the initial conditions at $y = 0$ either in the positive or negative y domain, the solution approaches an asymptotic state. In essence this asymptotic state represents the adjustment of the flow to the depth contours, and vanishing of the dependence on y. Under these conditions, the equation is much simplified to yield

$$\left(\frac{1}{H^2}\psi_x\right)_x = 0. \tag{6.171}$$

Integrating the above equation twice yields the general solution

$$\psi(x) = c\int_{x_c}^{x} H^2(x)\,dx + \psi_c. \tag{6.172}$$

The velocity along the shelf is calculated as

$$v = \frac{1}{H}\psi_x = \frac{1}{H}(cH^2) = cH(x) \tag{6.173}$$

and consequently the total flux of volume carried through the shelf is

$$Q = \int_{x_c}^{x_l} Hv\,dx = c\int_{x_c}^{x_l} H^2(x)\,dx = \psi_l - \psi_c, \tag{6.174}$$

which allows the calculation of the constant c to give

$$c = Q\Big/\int_{x_c}^{x_l} H^2(x)\,dx. \tag{6.175}$$

The vorticity of the flow can also be calculated:

$$\zeta = \frac{\partial}{\partial x}\left(\frac{1}{H}\frac{\partial \psi}{\partial x}\right) = \frac{\partial}{\partial x} cH(x). \tag{6.176}$$

The solution for the linear depth profile

$$H(x) = H_c + S(x - x_c), \quad \text{where} \quad S = \frac{H_l - H_c}{x_l - x_c}, \quad H_l = H(x_l), \quad H_c = H(x_c) \tag{6.177}$$

is obtained as

$$\psi(x) - \psi_c = c \int_{x_c}^{x} [H_c + S(x - x_c)]^2 \, dx = \frac{c}{3S}\{[H_c + S(x - x_c)]^3 - H_c^3.\} \tag{6.178}$$

The constant c is evaluated as follows:

$$\psi(x_l) - \psi_c = \psi_l - \psi_c = Q = \frac{c}{3S}(H_l^3 - H_c^3). \tag{6.179}$$

6.5.2 Nonlinear Inviscid Solutions

Under steady, inviscid conditions, the nonlinear vorticity equation reduces to

$$-\frac{f_0}{H^2}\mathbf{k} \times \nabla\psi \cdot \nabla\left(\frac{\zeta + f}{H}\right) \tag{6.180}$$

which is simply the *Jacobian*

$$J(\psi, q) = 0 \tag{6.181}$$

where q is the potential vorticity defined as

$$q = \frac{\zeta + f}{H} = \frac{\nabla\frac{1}{H}\nabla\psi + f}{H}. \tag{6.182}$$

The above Eq. (6.181) has a special solution $q = K(\psi)$ which makes the Jacobian vanish and therefore renders the original equation linear, i.e.

$$Hq = \nabla\frac{1}{H}\nabla\psi + f = H\ K(\psi) \tag{6.183}$$

so that

$$\nabla^2\psi - \frac{H_x}{H}\psi_x + \frac{H_y}{H}\psi_y + fH = H^2 K(\psi). \tag{6.184}$$

The solution to the above equation is not unique and depends on the choice of $K(\psi)$ which has to satisfy the above equation as well as the initial and boundary conditions. The approaching flow would set an initial condition that would be sufficient to determine $K(\psi)$. The uniform flow adjusted to the uniform depth variations $H = H(x)$ at a continental shelf has to satisfy

$$\psi_{xx} - \frac{H_x}{H}\psi_x + fH = H^2 K(\psi). \tag{6.185}$$

6.5.3 Flow over a Depth Discontinuity (β−Plane)

We consider a flow approaching a depth discontinuity in the zonal direction as shown in Fig. 6.11.

We consider the flow on a β−plane, and assume that the y−direction is aligned towards the north (northern hemisphere). We assume that the surface displacement η is small compared to the total depth ($\eta \ll H_0$). In fact, in order to be able to solve the problem we use the *rigid lid approximation* and take $\eta = 0$. Therefore, the depths $H_0,\ H_1$ on two sides are replaced by $h_0,\ h_1$ respectively.

By virtue of the continuity equation (6.181) we can define a stream function ψ by $H\mathbf{u} = \mathbf{k} \times \nabla\psi$. On the other hand, vorticity is expressed by $\zeta = \nabla H^{-1}\nabla\psi$ as in the last section.

Led by the last section, the solution to the steady state problem must be obtained from (6.180) or (6.181), which is in non-linear form. As reviewed in the last section, a special solution to the non-linear equation can be obtained that makes the Jacobian vanish in the case that the potential vorticity is a function of the stream-function $q = K(\psi)$

$$q = \frac{\zeta + f}{H} = \frac{\nabla H^{-1}\nabla\psi + f_0 + \beta y}{H} = K(\psi). \tag{6.186}$$

The function $K(\psi)$ is determined by the upstream conditions, requiring potential vorticity conservation

$$\frac{\zeta + f_0 + \beta y}{h_1} = \frac{h_1^{-1}\nabla^2\psi + f_0 + \beta y}{h_1} = K(\psi) = \frac{f_0 + \beta y_0}{h_0} \tag{6.187}$$

Fig. 6.11 Flow approaching a depth discontinuity

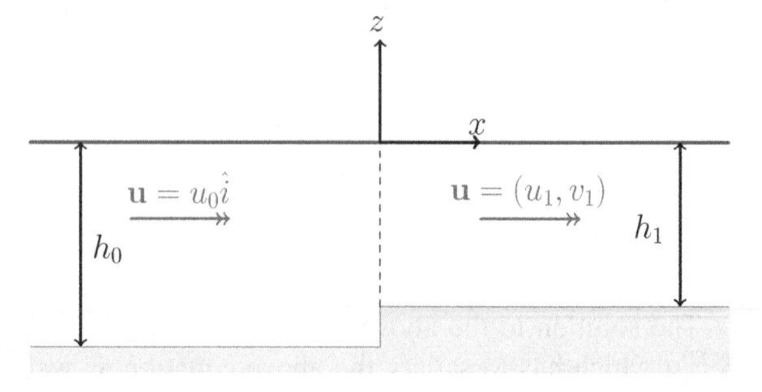

where the approaching flow has no vorticity ($\zeta = 0$ for $x < 0$), and y_0 represents the y-position of any approaching fluid particle. The velocity (for $x > 0$) is no longer uniform in the y-direction by virtue of (6.186). Horizontal fluxes satisfy the continuity equation (6.181) at the depth discontinuity, and the stream function immediately near $x = 0$, approaching from $x > 0$, is given by

$$h_0\, u_0 = h_1\, u_1 = -\frac{\partial \psi}{\partial y} \text{ at } x = 0 \tag{6.188a}$$

so that

$$\psi = -h_1\, u_1\, y = -h_0\, u_0\, y_0 \text{ at } x = 0. \tag{6.188b}$$

Now, from (6.188b) we can express

$$y_0 = -\frac{1}{u_0 h_0}\psi$$

to evaluate

$$K(\psi) = \frac{f_0 - \frac{\beta}{u_0 h_0}\psi}{h_0}$$

and therefore substituting into (6.187) we have

$$\frac{1}{h_1}(\nabla^2 \psi + p^2 \psi) = G(y) \tag{6.189}$$

where $G(y)$ defines the right hand side of the above equation and

$$r \equiv \frac{h_1}{h_0}, \quad G(y) \equiv -(1 - r) f_0 - \beta y, \quad p^2 = \frac{\beta}{u_0} r^2. \tag{6.190a, c}$$

A combination of homogeneous (ψ_h) and particular (ψ_h) solutions for Eq. (6.189) can be written as

$$\frac{1}{h_1}\psi = \frac{1}{h_1}(\psi_h + \psi_p) = \frac{u_0}{\beta r^2}\{E(y)F(x) + G(y)\} \tag{6.91}$$

Substituting this solution into (6.195), amounting to the *separation of variables* technique, we observe that this yields an ordinary differential equation

$$F_{xx} + p^2 F = 0 \tag{6.192a}$$

where $F_{xx} = \frac{d^2 F}{dx^2}$ is the second differential of $F(x)$ in short.

One of the boundary conditions is obtained by requiring (6.188a) at $x = 0$ and making use of (6.190)

$$h_1\, u_1 = -\frac{\partial \psi}{\partial y}|_{x=0} = -h_1 \frac{u_0}{\beta r^2}\{E_y F(0) + G_y\} = u_0\, h_0$$

where subscripts y denote differentiation with respect to y. Substituting for $G(y)$ from (6.189) we find

$$E_y F(0) = \beta(1 - r) \tag{6.192b}$$

Similarly we also require that the y-component of velocity should vanish at $x = 0$

$$h_1\, v_1|_{x=0} = \frac{\partial \psi}{\partial x}|_{x=0} = h_1 \frac{u_0}{\beta r^2} E F_x(0) = h_0 v_0 = 0 \tag{6.192c}$$

since there is no latitudinal component of the approach velocity.

Finally, we also should ensure that the transport stream-function is continuous at the step,

$$\frac{1}{h_1}\psi = \frac{u_0}{\beta r^2}\{E(y)F(0) + G(y)\} = -u_1 y = -u_0 \left(\frac{h_0}{h_1}\right) y = -u_0 r y$$

which by substituting from (6.190) amounts to

$$E(y)F(0) = (1 - r)\beta y + (1 - r) f_0 \tag{6.192d}$$

Understanding that $E(y)$ is at most a linear function

$$E(y) = ay + b$$

and letting $F(0) = 1$ and $E(y)$ to absorb any remaining constants, it is apparent that

$$E(y) = (1 - r)(f_0 + \beta y) \tag{6.193d}$$

Noting that Eqs. (6.198a–d) are essentially equivalent to the initial value problem

$$\begin{aligned} &F_{xx} + \frac{\beta}{u_0} r^2 F = 0 \\ &F(0) = 1 \\ &F_x(0) = 0 \end{aligned} \tag{6.194a-c}$$

the solution is obtained by integration of (6.194a) with boundary conditions (6.194b, c)

$$F(x) = \cos\left(r\sqrt{\frac{\beta}{u_0}}x\right), \tag{6.195}$$

and by substitution of (6.193c) and (6.195) into (6.190) the solution becomes

$$\frac{1}{h_1}\psi = \frac{u_0}{\beta}\frac{1 - r}{r^2}\left\{(f_0 + \beta y)\cos\left(r\sqrt{\frac{\beta}{u_0}}x\right) - \frac{\beta}{1 - r}y - f_0\right\}$$

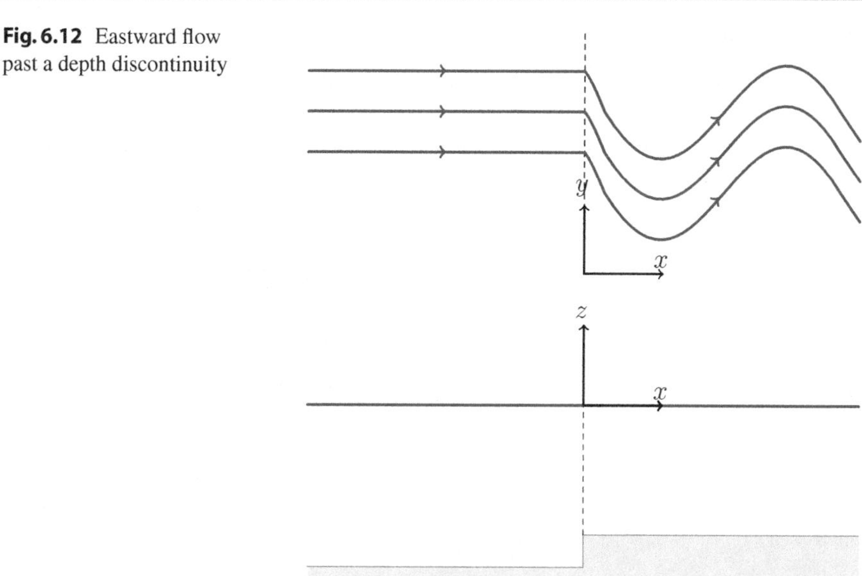

Fig. 6.12 Eastward flow past a depth discontinuity

and noting that

$$\Delta h \equiv h_0 - h_1, \quad r \equiv \frac{h_1}{h_0}, \quad 1 - r = \frac{\Delta h}{h_0}, \quad \frac{1-r}{r^2} = \frac{h_0 \Delta h}{h_1^2},$$

the full solution is obtained as follows

$$\psi = \frac{u_0}{\beta}\frac{h_0}{h_1}\Delta h \left\{ (f_0 + \beta y)\cos\left(\frac{h_1}{h_0}\sqrt{\frac{\beta}{u_0}}x\right) - \frac{\beta h_0}{\Delta h}y - f_0 \right\}. \tag{6.196}$$

The solution is sketched in Fig. 6.12. For real valued p, (i.e., if $u_0 > 0$ in the northern hemisphere) the solution is oscillatory in $x > 0$, with a wave length of $2\pi/p = 2\pi\sqrt{u_0/\beta}$. These waves, resulting from restoring forces in a β–plane, are called *Rossby waves*. Note that the amplitude of the oscillatory part of (6.198b) increases with latitude y. Note also that if we were to calculate an average position of a streamline for $y > 0$, by averaging the solution over one wavelength, we observe that the streamline is displaced to the south by a net distance $\Delta y = -\frac{f_0}{\beta}\frac{\Delta h}{h_1}$.

At the step, the jet is deflected in anti-cyclonic sense to the south by vorticity conservation, and as it moves south, gains vorticity by compensation of the decrease in planetary vorticity (β effect), and eventually has to turn back north when the initially negative fluid vorticity becomes positive. If the depth increased rather than decreased across the step, i.e. $\Delta h < 0$, then the jet would deflect north by an equal amount, gain positive vorticity which would decrease by compensation against increasing

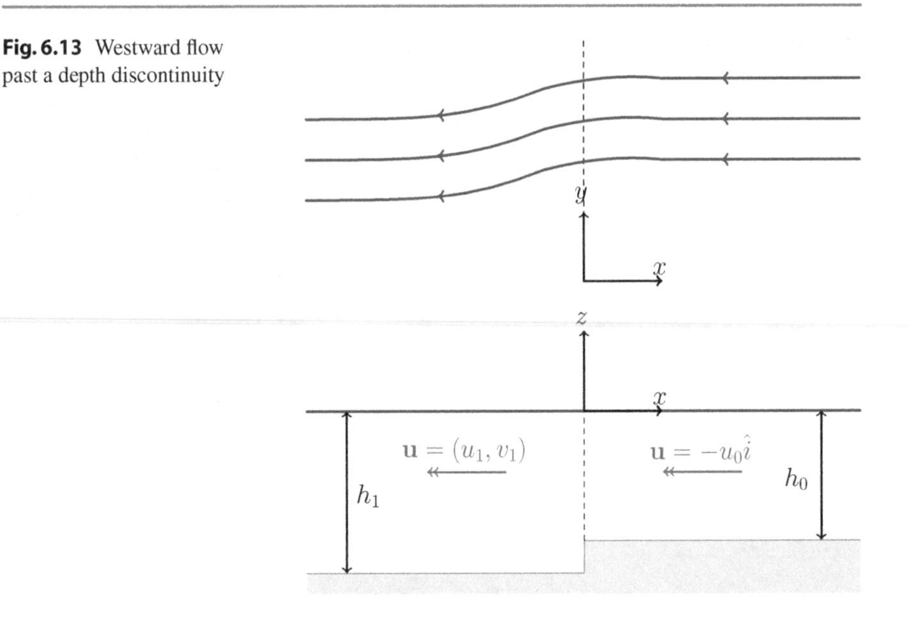

Fig. 6.13 Westward flow past a depth discontinuity

planetary vorticity, and again form a wave motion with mean position of streamlines displaced to the north.

Note that if $p^2 < 0$ (if $u_1 < 0$ or if $u_0 < 0$, as in a westward flow shown in Fig. 6.13, the solution would be exponentially decaying after the step, and since the first derivatives would have to be matched at the step, the curvature of the solution would have to continue before the step, and therefore the approaching flow would feel the step beforehand, as shown below.

The nature of the westward and eastward flows approaching a step are vastly different; which points us to the basic asymmetry in geophysical flows. All of these new features are direct results of the *β-effect.*

Application: Jet Stream and Planetary Waves

Rossby waves are easier observed in the atmosphere than the ocean. This is because the typical horizontal scales are often too large for ocean basins, but they can fit global dimensions by circumnavigating the globe. Often 3–5 wave-lengths fit the polar circumference, with variable amplitudes.

In the past there have been efforts to correlate phase of the periodic pattern with topography of land masses and land-ocean global distribution. A typical configuration of the wave pattern in the northern hemisphere often shows phase-locking with the north American continent. Often the initial phase of the Rossby wave pattern seems to match the western coast of America, especially coming on to the continent from Pacific Ocean and facing the Cascades and Rocky Mountain ranges, with some similarity to the model of step-wise change in topography reviewed in the present section.

The Rossby wave typically generates a trough immediately to the east of the western edge of the continent as in Fig. 6.14, with subsequent undulations as the flow

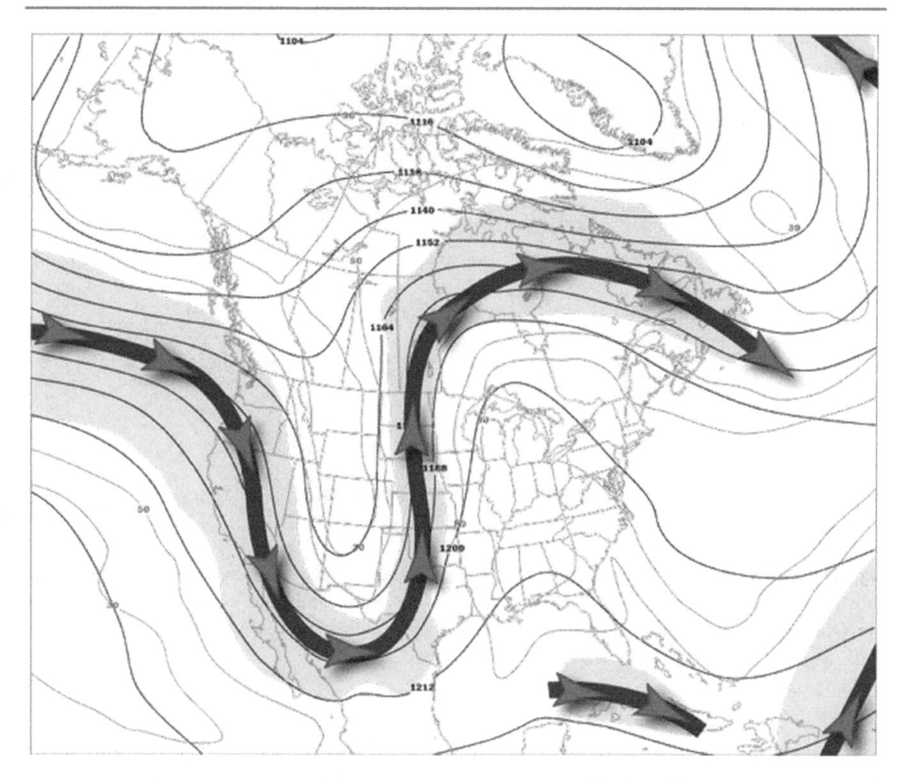

Fig. 6.14 An upper atmospheric jet-stream at 300 mb pressure level, showing geopotential thickness isolines and the region where upper winds exceed a speed of 130 km/h (blue shading). Credits: USA National Weather Service, educational web pages on the Jet Stream structure supported by NOAA / National Weather Service of the USA (https://www.weather.gov/jetstream/300mb)

proceeds to the Atlantic and European regions. During certain occasions, extreme deepening of the north American trough brings extreme weather events influencing continental climate. A case in the winter of 2017–2018 illustrated in Fig. 6.15 has developed a very sharp warm-and-cold contrast between the western and eastern regions of north America.

6.6 Oscillatory Motions

6.6.1 Planetary (Rossby) Waves

We have seen in the last section that Rossby waves are generated by a uniform flow impinging on a north-south aligned depth discontinuity. These waves are a direct result of the variation of the coriolis parameters with latitude.

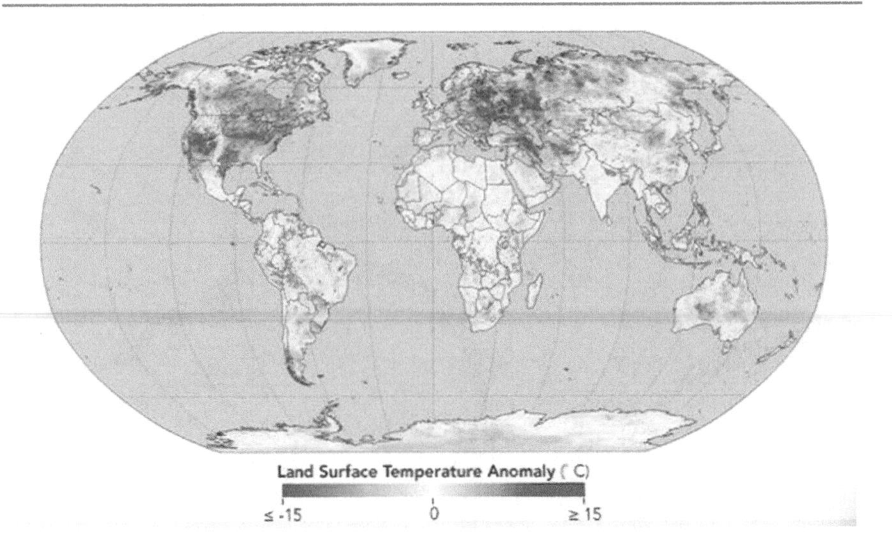

Fig. 6.15 Anomaly comparing average winter land surface temperature (26/12/2017–02/01/2018) with the same period of 2001–2010. Red areas show warm and blue cold anomalies. Credits: NASA Earth Observatory (https://earthobservatory.nasa.gov/images/91517/its-coldand-hotin-north-america) supported by NASA Earth Observatory

It is easy to show, by setting $\beta = 0$ in the solutions of Sect. 6.5.3, that the wave motion ceases to exist. Since only linear variations of f on a β-plane were considered to the lowest order, and since the f-variation actually occurs on a planetary scale, these waves are alternatively referred to as *planetary waves*. These are not the only the waves that can be sustained in rotating flow, but planetary waves are those waves directly resulting from β-effect, i.e. the latitudinal variation of the Coriolis parameter.

The case of constant total depth, $H = H_0$, is the simplest configuration to show the existence of planetary waves. We shall also assume that the total depth can be approximated by the still-water depth h_0, i.e., $H_0 = h_0 + \eta \simeq h_0$ since $\eta \ll h$ (the rigid-lid approximation). With appropriate definitions of a stream-function and vorticity

$$\mathbf{u} = \mathbf{k} \times \nabla\psi \tag{6.197}$$

$$\zeta = \nabla^2\psi, \tag{6.198}$$

we express potential vorticity conservation via Eq. (6.127):

$$\begin{aligned}\frac{D}{Dt}\left(\frac{f_0+\beta y+\zeta}{H_0}\right) &\simeq \frac{1}{h_0}\frac{D}{Dt}(\beta y+\nabla^2\psi)\\ &= \frac{1}{h_0}\left\{\frac{\partial}{\partial t}\nabla^2\psi+\mathbf{u}\cdot\nabla(\beta y+\nabla^2\psi)\right\}\\ &= \frac{1}{h_0}\left\{\frac{\partial}{\partial t}\nabla^2\psi+(\mathbf{k}\times\nabla\psi)\cdot\nabla(\beta y+\nabla^2\psi)\right\}\\ &= \frac{1}{h_0}\left\{\frac{\partial}{\partial t}\nabla^2\psi+\beta\frac{\partial\psi}{\partial x}+\mathbf{k}\times\nabla\psi\cdot\nabla(\nabla^2\psi)\right\}=0\end{aligned} \tag{6.199}$$

Here, we may introduce the definition of a *Jacobian*:

$$J(A,B)=\frac{\partial A}{\partial x}\frac{\partial B}{\partial y}-\frac{\partial A}{\partial y}\frac{\partial B}{\partial x}=(\mathbf{k}\times\nabla A)\cdot\nabla B=\mathbf{k}\cdot\nabla A\times\nabla B. \tag{6.200}$$

Setting $A=\psi,\ B=\nabla^2\psi$ in (6.200) and comparing with (6.199), the vorticity conservation equation becomes:

$$\frac{\partial}{\partial t}\nabla^2\psi+\beta\frac{\partial\psi}{\partial x}+J(\psi,\nabla^2\psi)=0. \tag{6.201}$$

Note that this is a nonlinear equation since the Jacobian (6.200) is nonlinear by definition. If this equation is somehow linearized (assuming the Jacobian term is much smaller than the other terms), it would read

$$\frac{\partial}{\partial t}\nabla^2\psi+\beta\frac{\partial\psi}{\partial x}=0 \tag{6.202}$$

Next we investigate the existence of a *plane-wave solution* of the form

$$\psi=\psi_0 e^{i(kx+ly-\omega t)}=\psi_0 e^{i\theta(x,y,t)}. \tag{6.203}$$

where $k,\ l$ represent *wave number* in the $x,\ y$ directions respectively, and ω represents the *angular frequency* of the motion. $\theta=(kx+ly-\omega t)$ is called the *phase* of the wave motion; when $\theta=$constant it follows that $\psi=$constant. We sketch the wave motion at a fixed time in Fig. 6.16, as follows:

Now, in substituting (6.203) into (6.202) we first note that

$$\nabla^2\psi=-\psi_0(k^2+l^2)e^{i(kx+ly-\omega t)}=(k^2+l^2)\psi$$

and therefore the Jacobian term would identically vanish, although we have not made an a priori assumption of linearity. It can also be verified to be true by virtue of an

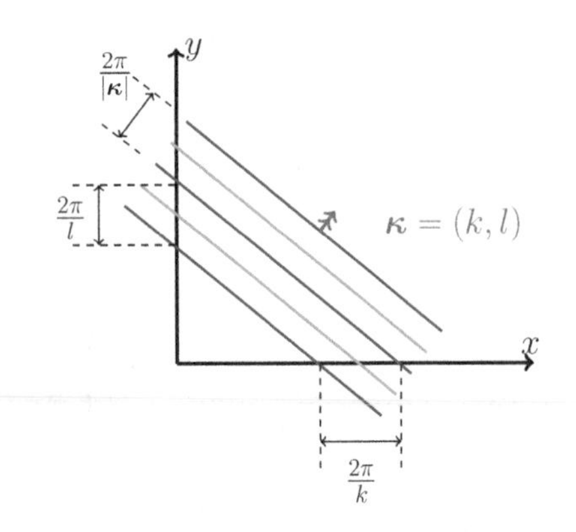

Fig. 6.16 Plane waves

important property of the Jacobian: that if A and B are *linearly dependent*, i.e., if $B = f(A)$, then (6.200) yields

$$J(A, B) = J(A, f(A)) = \frac{\partial A}{\partial x}\frac{\partial f}{\partial A}\frac{\partial A}{\partial y} - \frac{\partial A}{\partial y}\frac{\partial f}{\partial A}\frac{\partial A}{\partial x} \equiv 0 \tag{6.204}$$

Therefore, plane wave solutions of the form (6.203) do not contribute to the nonlinear terms, and the nonlinear equation (6.201) is equivalent to (6.202) in this special case.

Substituting (6.203) into (6.202) yields

$$-i\omega(-k^2 - l^2) + \beta i k = 0$$

or

$$\omega = -\frac{\beta k}{k^2 + l^2} \tag{6.205}$$

which is the *dispersion relation* for planetary waves. The *phase speed* in the x and y directions are respectively calculated as

$$C_x = \frac{\omega}{k} = -\frac{\beta}{k^2 + l^2} \tag{6.206a}$$

$$C_y = \frac{\omega}{l} = -\frac{\beta}{k^2 + l^2}\frac{k}{l} \tag{6.206b}$$

Note that $C_x < 0$ for all possible values of $k,\ l$, the wave propagation is always in the negative x-direction, i.e. *towards the west*. We also note that these waves can only exist if $\beta \neq 0$, i.e., they arise as a result of the β-effect.

We can also define a *wave number vector* $\boldsymbol{\kappa}$ with components $\boldsymbol{\kappa} = (k,\ l)$. The phase speed in the direction of the wave-number vector (or in the direction of phase propagation) is

$$C = \frac{\omega}{|\boldsymbol{\kappa}|} = \frac{\omega}{\sqrt{k^2 + l^2}} = -\frac{\beta k}{(k^2 + l^2)^{3/2}} = -\beta\frac{k}{|\boldsymbol{\kappa}|^3} \tag{6.207}$$

Note that the phase-speed does not satisfy vector decomposition (!) since the resultant of (6.206a, b) does not satisfy (6.207). This means that the direction of wave propagation is not normal to the *wave fronts*. It is also seen that the phase speed is maximum for westward propagating waves ($|\boldsymbol{\kappa}| = k$ and zero for waves propagating in the north-south direction, i.e. wave patterns with north-south orientation do not propagate.

Note that, although the nonlinear terms vanish for any single plane-wave solution, the superposition of a number of plane-waves does not necessarily constitute a solution for the nonlinear equation (6.201).

The *group velocity* is the velocity of propagation of wave energy (i.e., of wave packets), with components defined as

$$C_{gx} = \frac{\partial \omega}{\partial k}, \quad C_{gy} = \frac{\partial \omega}{\partial l} \tag{6.208a, b}$$

These can be evaluated from (6.205) to be:

$$C_{gx} = \frac{\beta(k^2 - l^2)}{k^2 + l^2} = \frac{\beta(k^4 - l^4)}{k^2 + l^2} \tag{6.209a}$$

$$C_{gy} = -\frac{2\beta kl}{(k^2 + l^2)^2} \tag{6.209b}$$

It can be observed that although the phase velocity in the x-direction is always negative (westward phase propagation), the group velocity can be positive ($k > l$) or negative ($k < l$), i.e. while the wave is propagating west it can transfer energy in either of the horizontal directions. It is useful to construct a *group velocity vector* by using (6.208a,b):

$$\mathbf{C}_g = \nabla_{\kappa}\omega = \mathbf{i}\frac{\partial \omega}{\partial k} + \mathbf{j}\frac{\partial \omega}{\partial l} \tag{6.210}$$

where ∇_{κ} denotes gradient in the direction of the wavenumber vector $\boldsymbol{\kappa}$.

Through definitions (6.207), the above can be modified to read as

$$\mathbf{C}_g = \nabla_{\kappa}|\boldsymbol{\kappa}|C = |\boldsymbol{\kappa}|\nabla_{\kappa}C + C\nabla_{\kappa}|\boldsymbol{\kappa}| = |\boldsymbol{\kappa}|\nabla_{\kappa}C + 2\frac{\boldsymbol{\kappa}}{|\boldsymbol{\kappa}|}C, \tag{6.211}$$

where $\mathbf{C}$ denotes phase speed (vector) in the direction of the wave number $\boldsymbol{\kappa}$. Since C depends on $\boldsymbol{\kappa}$ by virtue of (6.172), $\mathbf{C}_g$ and $\mathbf{C}$ differ in magnitude and direction. Such waves are called dispersive waves. [Nondispersive waves are those for which phase speed is independent of wave number, or for which group and phase velocities are equal].

Note that a Rossby wave is that it is a *transverse wave*. By making the rigid-lid assumption $\eta = 0$, and considering a constant depth case, the continuity equation (6.90) becomes

$$\nabla \cdot \mathbf{u} = 0. \tag{6.212}$$

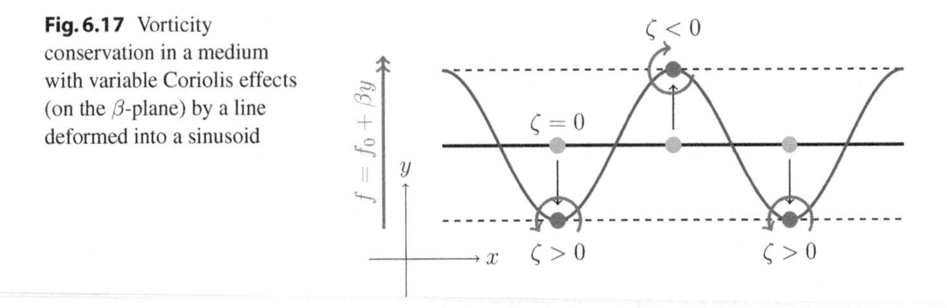

Fig. 6.17 Vorticity conservation in a medium with variable Coriolis effects (on the β-plane) by a line deformed into a sinusoid

By substituting the plane-wave solution

$$\mathbf{u} = \mathbf{u}_0 e^{i(kx+ly-\omega t)} = \mathbf{u}_0 e^{i(\boldsymbol{\kappa}\cdot\mathbf{x}-\omega t)} \tag{6.213}$$

into (6.177), it can be shown that

$$\boldsymbol{\kappa} \cdot \mathbf{u}_0 = 0, \tag{6.214}$$

i.e., the fluid velocity $\mathbf{u}$ is always perpendicular to the wave propagation direction $\boldsymbol{\kappa}$. This is a very important result, since it explains why the nonlinear terms $\mathbf{u} \cdot \nabla\mathbf{u}$ do not make any contribution for this type of wave, although we have not specifically made the linearity assumption.

In order to interpret the physical reason for the westward propagation of Rossby waves, consider the vorticity equation (6.167), which alternatively could be written as

$$\frac{\partial \zeta}{\partial t} + \beta v = 0 \tag{6.215}$$

According to this equation, a fluid particle with positive velocity in the y-direction ($v > 0$) will gain negative vorticity, and a particle with $v < 0$ will gain positive vorticity. Therefore, if we consider an initially straight line of fluid particles having zero initial vorticity, and displace this line to give it a sinusoidal form, the resulting vorticity distribution could be sketched as in Fig. 6.17.

The train of vortices generated by the displacements will have a self induced velocity field carrying the pattern towards west. By virtue of (6.206a, b), the solutions would be stationary for an observer moving west with speed C_x. Similarly, if the waves were superposed on an easterly current exactly opposing C_x, the resulting pattern would be stationary. This is the case for the step problem discussed earlier in Sect. 6.5.3, where the wave adjusts its structure (i.e., wavelength) to a stationary pattern in order match the boundary conditions at the step.

In the more general case of superposition of an easterly flowing current and planetary waves, whether the waves would appear to be propagating to the east or west depends on whether or not the speed of the current overcomes the westerly phase speed of the waves. For instance in the tropics, the mean flow in the atmosphere is in a westerly direction ("easterlies") so that atmospheric systems which typically have higher phase speeds most often travel west. In the mid-latitudes there are strong "westerlies" (i.e., easterly flowing mean currents) which often overcome the phase speed of planetary waves, and therefore mid-latitude weather systems are often observed moving east.

6.6.2 Planetary Waves with a Mean Current

We now look into what would happen if the Rossby waves with a velocity field $\mathbf{u}'$ is superposed on some mean current with constant horizontal velocity $\mathbf{U} = (U, V)$, by letting

$$\mathbf{u} = \mathbf{U} + \mathbf{u}' = \mathbf{U} + \mathbf{k} \times \nabla\psi$$

whereupon Eq. (6.199) takes the form

$$\frac{\partial}{\partial t}\nabla^2\psi + [\mathbf{U} + \mathbf{k} \times \nabla\psi] \cdot \nabla(\beta y + \nabla^2\psi).$$

Expanding terms we have

$$\frac{\partial}{\partial t}\zeta + \mathbf{U} \cdot \nabla\zeta + \beta V + \beta\frac{\partial\psi}{\partial x} + J(\psi, \zeta) = 0.$$

where $\zeta = \nabla \times \mathbf{u} = \nabla^2\psi$. Alternatively, since the Jacobian, i.e. the last term, is zero, we write

$$\frac{\partial}{\partial t}\zeta + U\frac{\partial\zeta}{\partial x} + V\frac{\partial\zeta}{\partial y} + \beta V + \beta\frac{\partial\psi}{\partial x} = 0 \tag{6.216}$$

Imposing the wave solution (6.203) in (6.206), we obtain

$$(-k^2 - l^2)(-i\omega + ikU + ilV) + \beta(V + ik) = 0$$

or

$$\omega = kU + lV - \frac{\beta(-iV + k)}{(k^2 + l^2)} \tag{6.217}$$

as the dispersion relationship in this case.

We note the additional terms related to the mean current setting the difference of this equation compared to (6.205). Firstly, the frequency of the wave motion can no longer be a real number since the equation is complex. In fact, there will be real and imaginary parts of the frequency if $V \neq 0$, which means that in the presence of a latitudinal component of velocity, the wave motion will have a decaying or growing oscillations. Therefore we cannot accept this for a plane-wave solution and must set $V = 0$. Furthermore, we have

$$C_x = \frac{\omega}{k} = U - \frac{\beta}{(k^2 + l^2)} \tag{6.218a}$$

and

$$C_{gx} = \frac{\partial \omega}{\partial k} = U + \frac{\beta(k^2 - l^2)}{(k^2 + l^2)^2} \tag{6.218b}$$

respectively for the phase speed and group speed of these waves in the x-direction. Although Rossby waves (represented by the second term) originally would have directional preference of traveling to the west, in this case the waves are swept back by the positive velocity U towards the east. This behavior typical, for instance, of atmospheric Rossy waves, with wave-like mid-latitude weather patterns traveling predominantly east under the influence of the atmospheric *jet-stream* influencing the same latitudes.

For the case of propagation only in the longitudinal direction, setting $l = 0$, we observe that the wave pattern may become stationary if the wave-number k and hence the wave-length $\lambda_c = 2\pi/k_c$ would satisfy the critical value

$$k_c = \sqrt{\beta/U}.$$

For typical values of $U = 20\,\text{km/s}$ and $\beta \simeq 2 \times 10^{-11}$ the wave-length of the stationary wave can be estimated as $\lambda_c = 2\pi\sqrt{U/\beta} \simeq 6000\,\text{km}$, which seems to be the same order of length scale one would observe in meteorological cases. Note that for wavelengths $\lambda < \lambda_c$ the wave pattern will propagate east (typical for most observed atmospheric cases), while for $\lambda > \lambda_c$ it will propagate west.

Note that although the phase speed (6.118a) may indicate mainly eastward propagation with possible westerly propagating cases for very long wavelengths $\lambda > \lambda_c$, the group velocity (6.218b) is even more exclusively of eastern orientation, for only $k > l$, which is often the case in the atmosphere as planetary waves travel around the globe following latitudes. At exactly the critical wave-length $k_c = \sqrt{\beta/U}$, for $l = 0$, we note that the group speed becomes $C_{gx} = 2U$ according to (6.218b), at double the current speed towards the east.

The above considerations appear relevant in connection to Sect. 6.5.3, where we have seen the creation of a Rossby standing wave past a depth discontinuity, exactly with the wave-number $k_c = (h_1/h_2)\sqrt{\beta/u_0}$ accounting for the additional vorticity induced by the depth change. In that case, the standing Rossby wave carries the signal east from the discontinuity although there is no phase propagation.

6.6.3 Small Amplitude Motions with a Free Surface

In the foregoing sections, we have used the rigid-lid assumption ($\eta = 0$). To see the effects of surface displacement, we reconsider the shallow water equations (6.89) and (6.90) simplified by neglecting the forcing terms on the right hand side of (6.89). Furthermore, small amplitude motions will be considered by neglecting the nonlinear terms. For example, we assume

$$\frac{\partial \mathbf{u}}{\partial t} \gg \mathbf{u} \cdot \nabla \mathbf{u}$$

in Eq. (6.89), and in the second term of (6.90) expressed as

$$\nabla \cdot H\mathbf{u} = \nabla \cdot \eta\mathbf{u} + \nabla \cdot h\mathbf{u},$$

assuming that the nonlinear first term on the right hand side is much smaller than the second term ($\eta \ll h$). Then Eqs. (6.89) and (6.90) are simplified to give

$$\frac{\partial \mathbf{u}}{\partial t} + f\mathbf{k} \times \mathbf{u} = -g\nabla\eta, \tag{6.219a}$$

$$\frac{\partial \eta}{\partial t} + \nabla \cdot h\mathbf{u} = 0. \tag{6.219b}$$

These are linear equations from which the unknowns $\mathbf{u}$ and η can be solved. We also assume that $f = f_0 =$ constant. In order to eliminate one of the unknowns, say $\mathbf{u}$, from the equations, we can multiply (6.219a) by h, and take first the divergence, and then, the curl of this equation, yielding

$$\frac{\partial}{\partial t}\nabla \cdot h\mathbf{u} - f\mathbf{k} \cdot \nabla \times h\mathbf{u} = -g\nabla \cdot h\nabla\eta \tag{6.220a}$$

$$\frac{\partial}{\partial t}\nabla \times h\mathbf{u} + f\mathbf{k}\nabla \cdot h\mathbf{u} = -g\nabla h \times \nabla\eta \tag{6.220b}$$

Note that use has been made of the vector identities (1.27b, c, d). Then, $\nabla \times h\mathbf{u}$ is eliminated from (6.219a), (6.219b), yielding

$$\left(\frac{\partial^2}{\partial t^2} + f^2\right)\nabla \cdot h\mathbf{u} = -g\left\{\frac{\partial}{\partial t}\nabla \cdot h\nabla\eta + f\mathbf{k} \cdot \nabla h \times \nabla\eta\right\}. \tag{6.221}$$

Utilization of (6.219b) eliminates $\nabla \cdot h\mathbf{u}$ from the above equations:

$$\frac{\partial}{\partial t}\left\{\left(\frac{\partial^2}{\partial t^2} + f^2\right)\eta - g\nabla \cdot h\nabla\eta\right\} - gfJ(h, \eta) = 0, \tag{6.222}$$

where the Jacobian $J(h, \eta)$ is defined through (6.200). The last equation is essentially a wave equation for η and is much similar to the wave equation

$$\frac{\partial^2 \eta}{\partial t^2} = C_0^2\nabla^2\eta \tag{6.223}$$

(which occurs only if $f = 0$, $h = h_0$=constant), where

$$C_0^2 = gh_0, \tag{6.224}$$

C_0 being the phase speed of wave solutions that can be obtained for the simple case (6.223). Once η is obtained as solution of the Eq. (6.222), the velocity field can be obtained from (6.219a), or following some manipulation, from the following equation:

$$\left(\frac{\partial^2}{\partial t^2} + f^2\right)\mathbf{u} = -g\left(\frac{\partial}{\partial t}\nabla\eta - f\mathbf{k} \times \nabla\eta\right). \tag{6.225}$$

6.6.4 Plane Waves for Constant Depth

Consider free oscillations in the case of a constant depth $h = h_0$. Further, assume plane wave solutions for (6.222), of the form

$$\eta = \Re\{\eta_0 e^{i(kx+ly-\omega t)}\} = \Re\{\eta_0 e^{i(\boldsymbol{\kappa}\cdot\mathbf{x}-\omega t)}\}. \tag{6.226}$$

Substitution of (6.226) into (6.222) gives

$$i\omega\eta_0[f^2 - \omega^2 + gh_0(k^2 + l^2)] = 0 \tag{6.227}$$

which is a dispersion relation for these waves. Letting $\kappa = |\boldsymbol{\kappa}|$ be the magnitude of the wave number vector $\boldsymbol{\kappa} = (k, l)$, the above equation is simplified to

$$\omega = \pm\sqrt{f^2 + C_0^2\kappa^2} \tag{6.228}$$

where C_0 is given by (6.224). Since the wave frequency ω depends on κ, the waves are *dispersive*. For each value of the wave number κ there are two waves with opposite phase speeds of

$$C = \frac{\omega}{\kappa} = \pm\sqrt{C_0^2 + \frac{f^2}{\kappa^2}} \tag{6.229}$$

In the non-rotating case of $f = 0$, $C = C_0 = \pm\sqrt{gh_0}$, and the waves propagate with the classical shallow water wave speed, in a *non-dispersive* mode. Therefore, we can see that rotation introduces dispersion, as well as increasing the wave speed. Also by virtue of (6.228), it is clear that all possible wave frequencies $|\omega|$ must exceed f, corresponding to *super-inertial* frequencies ($|\omega| > f$). The velocity field can be obtained by substituting

$$\mathbf{u} = \Re\{\mathbf{u}_0 e^{i(\boldsymbol{\kappa}\cdot\mathbf{x}-\omega t)}\} \tag{6.230}$$

and (6.226) into (6.225), yielding

$$(f^2 - \omega^2)\mathbf{u}_0 = -g(\omega\boldsymbol{\kappa} - if\mathbf{k} \times \boldsymbol{\kappa})\eta_0. \tag{6.231}$$

The right hand side has two terms: one parallel and the other perpendicular to the $\boldsymbol{\kappa}$ vector. For $\omega \gg \kappa$ (*gravity waves*) the above equation reduces to

$$\mathbf{u}_0 \simeq \frac{g}{\omega}\eta_0\boldsymbol{\kappa} \tag{6.232}$$

and the particle velocity is parallel to the wavenumber vector, i.e. the motion is a *longitudinal wave*. In the general case, Eq. (6.231) indicates that the velocity vector makes an angle with the propagation direction. This more general case of motion is called an *inertia-gravity wave*, characterized by a mixture of longitudinal and transverse modes. The vector $\mathbf{u}_0$ can be decomposed into two components $\mathbf{u}_0 = (u_{0_\parallel},\ v_{0_\perp})$ such that (using 6.231),

$$u_{0_\parallel} = \mathbf{u}_0 \cdot \frac{\boldsymbol{\kappa}}{|\boldsymbol{\kappa}|} = -\frac{g\omega|\boldsymbol{\kappa}|}{f^2-\omega^2}\eta_0, \tag{6.233a}$$

$$u_{0_\perp} = \mathbf{u}_0 \cdot \frac{\mathbf{k}\times\boldsymbol{\kappa}}{|\boldsymbol{\kappa}|} = i\frac{gf|\boldsymbol{\kappa}|}{f^2-\omega^2}\eta_0. \tag{6.233b}$$

Substituting from (6.228), and using $\omega/\kappa = C_0 = \sqrt{gh_0}$, one obtains

$$u_{0_\parallel} = \frac{g\omega\kappa\eta_0}{C_0^2\kappa^2} = -\frac{C_0}{h_0}\eta_0 \tag{6.234a}$$

$$u_{0_\perp} = \frac{gf\kappa\eta_0}{C_0^2\kappa^2} = i\frac{C_0}{\eta_0}\frac{f}{\omega}\eta_0 \tag{6.234b}$$

Note that, in general, η_0, $u_{0_\parallel}$, $u_{0_\perp}$ are complex numbers. Velocity components can be calculated from (6.230), by similarly decomposing the velocity vector as $\mathbf{u} = (u_\parallel, u_\perp)$,

$$\begin{aligned} u_\parallel &= \Re\left\{-\frac{C_0}{h_0}\eta_0 e^{i(\boldsymbol{\kappa}\cdot\mathbf{x}-\omega t)}\right\} \\ &= -\frac{C_0}{h_0}|\eta_0|\cos(kx+ly-\omega t+\phi) \end{aligned} \tag{6.235a}$$

$$\begin{aligned} u_\perp &= \Re\left\{i\frac{C_0}{h_0}\frac{f}{\omega}\eta_0 e^{i(\boldsymbol{\kappa}\cdot\mathbf{x}-\omega t)}\right\} \\ &= -\frac{C_0}{h_0}\frac{f}{\omega}|\eta_0|\sin(kx+ly-\omega t+\phi) \end{aligned} \tag{6.235b}$$

where ϕ is the phase of η_0, such that $\eta_0 = |\eta_0|e^{i\phi}$. It can thus be shown, by combining (6.235a), (6.235b), that

$$u_\parallel{}^2 + u_\perp{}^2\left(\frac{\omega}{f}\right)^2 = \left(\frac{C_0|\eta_0|}{h_0}\right)^2 \tag{6.236}$$

which is the equation for an ellipse elongated in the propagation direction as shown in Fig. 6.18.

Note that by virtue of (6.235a) and (6.235b) the current vector is rotating in the clockwise direction, because of the clockwise deflection of the Coriolis effect. For $\omega \gg f$ (*gravity waves*), the ellipse becomes narrower and becomes aligned in the direction of propagation ($u_{0_\perp}$ becomes vanishingly small by virtue of (6.234b). On the other hand, as $\omega \to f$ (*inertial motion*), both components of velocity become equal and the ellipse becomes a circle. In the general case of *inertia-gravity waves* ($\omega > f$) the current ellipse is aligned in the propagation direction. While it is not

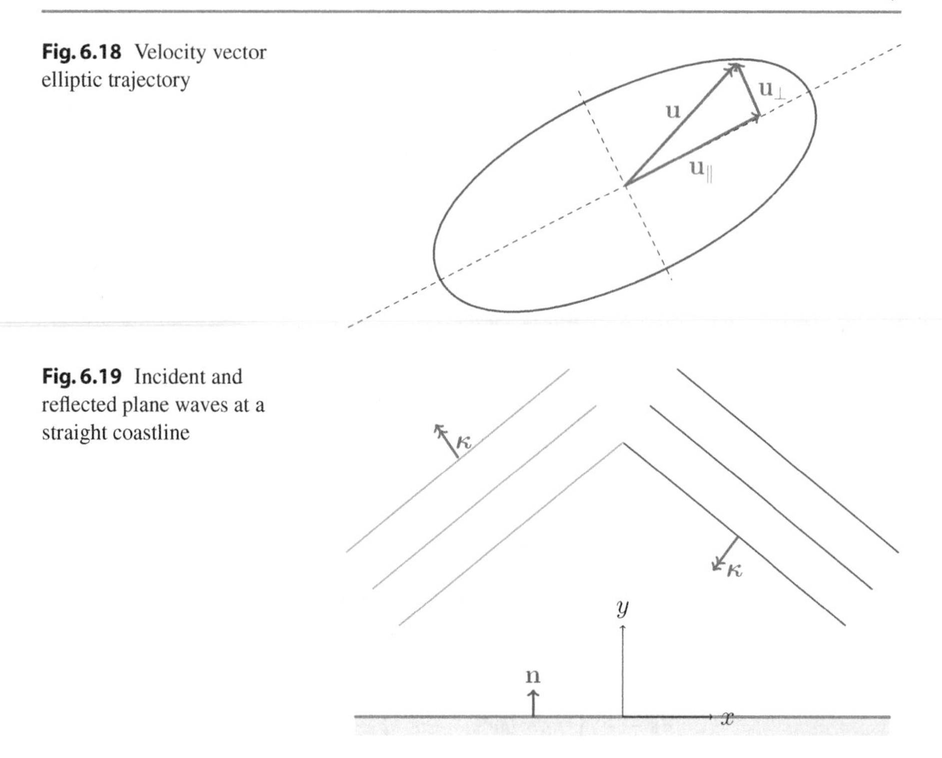

Fig. 6.18 Velocity vector elliptic trajectory

Fig. 6.19 Incident and reflected plane waves at a straight coastline

directly applicable here, it is worth noting that for the case $\omega < f$, the ellipse is oriented perpendicular to the propagation, and only in the limit $\omega \to 0$ (*geostrophic motion*) the current becomes perpendicular to the isolines of surface elevation (isobars). Therefore, it can be seen that, in general, inertia-gravity waves are far from being geostrophic. Certainly they are another class of waves much different from Rossby waves.

6.6.5 Poincaré and Kelvin Waves

Consider now a coast aligned with the x-axis as in as in Fig. 6.19.

Let the depth be constant, $h = h_0$. Then Eq. (6.222) reduces to

$$\frac{\partial}{\partial t}\left\{\left(\frac{\partial^2}{\partial t^2} + f^2\right)\eta - C_0^2\nabla^2\eta\right\} = 0, \tag{6.237}$$

where $C_0^2 = gh_0$ as in (6.224). The boundary condition at the coast is

$$\mathbf{u}\cdot\mathbf{n} = \mathbf{u}\cdot\mathbf{i} = 0 \tag{6.238}$$

i.e., the normal velocity at the coast must vanish. This boundary condition can be conveniently expressed, using (6.225) as

$$\left(\frac{\partial}{\partial t}\nabla\eta - f\mathbf{k}\times\nabla\eta\right)\cdot\mathbf{i} = -\frac{1}{g}\left(\frac{\partial^2}{\partial t} + f^2\right)\mathbf{u}\cdot\mathbf{n} = 0 \text{ on } y = 0 \tag{6.239}$$

or as:

$$\frac{\partial^2\eta}{\partial t\partial y} - f\frac{\partial\eta}{\partial x} = 0 \text{ on } y = 0. \tag{6.240}$$

Another boundary condition is needed at the open boundary, which will be the requirement that

$$\eta \to \text{ bounded, as } y \to \infty. \tag{6.241}$$

We will consider solutions which are periodic along the coast and with respect to time:

$$\eta = \Re\{N(y)e^{i(kx-\omega t)}\} \tag{6.242}$$

which, upon substitution into (6.237), (6.240) and (6.241), gives the set

$$\frac{d^2N}{dy^2} + \left\{\frac{\omega^2 - f^2}{C_0^2} - k^2\right\}N = 0 \tag{6.243}$$

$$\frac{dN}{dy} + f\frac{k}{\omega}N = 0 \text{ on } y = 0 \tag{6.244}$$

$$N \to \text{ bounded, as } y \to \infty. \tag{6.245}$$

For simplicity, let

$$l^2 = \frac{\omega^2 - f^2}{C_0^2} - k^2, \tag{6.246}$$

which is the same expression as the dispersion relation (6.227) or (6.128).

Now the solution has different character for different values of l, as shown in the following cases.

Solution I—Poincaré Waves ($\mathbf{l^2 \geq 0}$)

For $l^2 \geq 0$ (l is a real number), the solution to (6.243) is

$$N(y) = Ae^{ily} + Be^{-ily} \tag{6.247}$$

upon substituting in boundary condition (6.244), this gives

$$\left(il + f\frac{k}{\omega}\right)A + \left(-il + f\frac{k}{\omega}\right)B = 0. \tag{6.248}$$

The boundary condition (6.245) is readily satisfied by the form of the solution. The solution (6.242) is then:

$$\eta = \Re\{Ae^{i(kx+ly-\omega t)} + Be^{i(kx-ly-\omega t)}\} \tag{6.249}$$

This solution, then, represents an incident wave with amplitude B and a reflected wave with amplitude A, whose ratio is given by (6.248)

$$\frac{A}{B} = \frac{is-1}{is+1} = -\frac{(is-1)^2}{1+s^2} = -\frac{1-s^2-2is}{1+s^2} \tag{6.250}$$

where

$$s = \frac{l}{k}\frac{\omega}{f}. \tag{6.251}$$

Substituting

$$B = |B|e^{i\delta} \tag{6.252}$$

yields

$$\eta = |B|\Re\left\{e^{i(kx-ly-\omega t+\delta)} - \frac{1-s^2-2is}{1+s^2}e^{i(kx+ly-\omega t+\delta)}\right\} \tag{6.253}$$

as the solution.

The superposed waves will have the pattern shown in Figs. 6.20 and 6.21. Note that the coefficient of the reflected wave can be written as

$$\begin{aligned}
-\frac{1-s^2-2is}{1+s^2} &= -\frac{\sqrt{(1-s^2)^2+(2s)^2}}{1+s^2}e^{-i\ \tan^{-1}\left(\frac{2s}{1-s^2}\right)} \\
&= -\frac{\sqrt{1+s^4-2s^2+4s^2}}{1+s^2}e^{-i\ \tan^{-1}\left(\frac{2s}{1-s^2}\right)} \\
&= -1e^{-i\ \tan^{-1}\left(\frac{2s}{1-s^2}\right)} = e^{i\theta}
\end{aligned} \tag{6.254}$$

with θ replacing

$$\theta = \pi - \tan^{-1}\left(\frac{2s}{1-s^2}\right),$$

so that (6.253) becomes

$$\eta = |B|\{\cos(kx-ly-\omega t+\delta) + \cos(kx+ly-\omega t+\delta+\theta)\} \tag{6.255}$$

i.e. the only change in the reflected wave is that it merely suffers a phase shift.

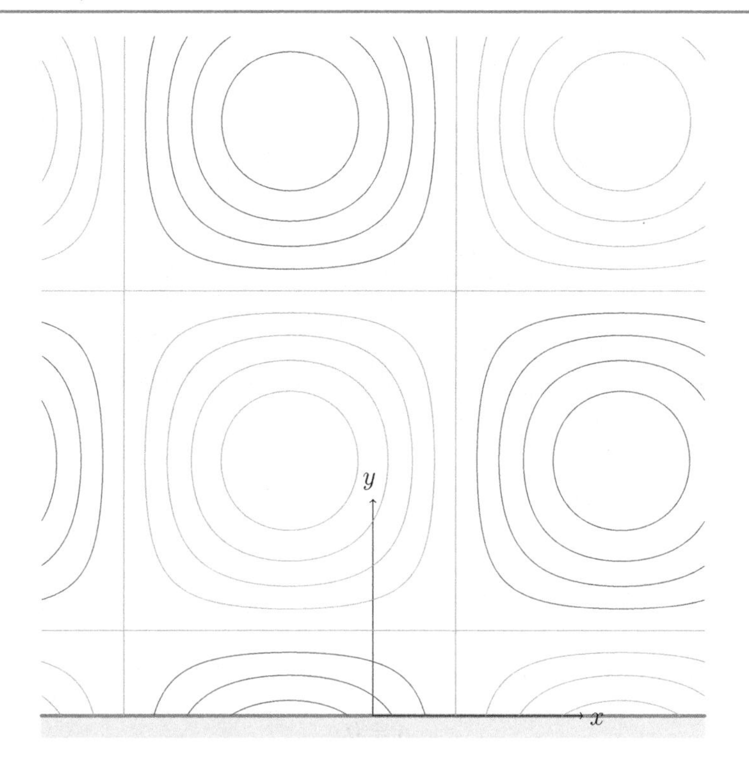

Fig. 6.20 Wave pattern resulting from the reflection of an incident Poincaré wave at the coast

These waves are called *Poincaré waves*. A special case occurs when $l = 0$, (i.e. waves propagating along the coast) in which

$$\omega^2 = \omega_0^2 = f^2 + C_0^2 k^2 \tag{6.256}$$

gives the minimum frequency ω_0 possible for Poincaré waves. By virtue of (6.246), Poincaré waves with $l \neq 0$ always have

$$\omega > \omega_0. \tag{6.257}$$

Note also that for all possible cases, $\omega > f$ (period of motion is less than the inertial period).

Solution II—Kelvin Wave ($\mathbf{l^2 < 0}$)

$$l^2 = -\alpha^2 < 0 \tag{6.258}$$

For $l^2 < 0$ (l is an imaginary number), we can interchange l with α, such that $\alpha^2 = -l^2 > 0$ (α is real). The solution to (6.243) is expressed as

$$N = Ae^{+\alpha y} + Be^{-\alpha y}. \tag{6.259}$$

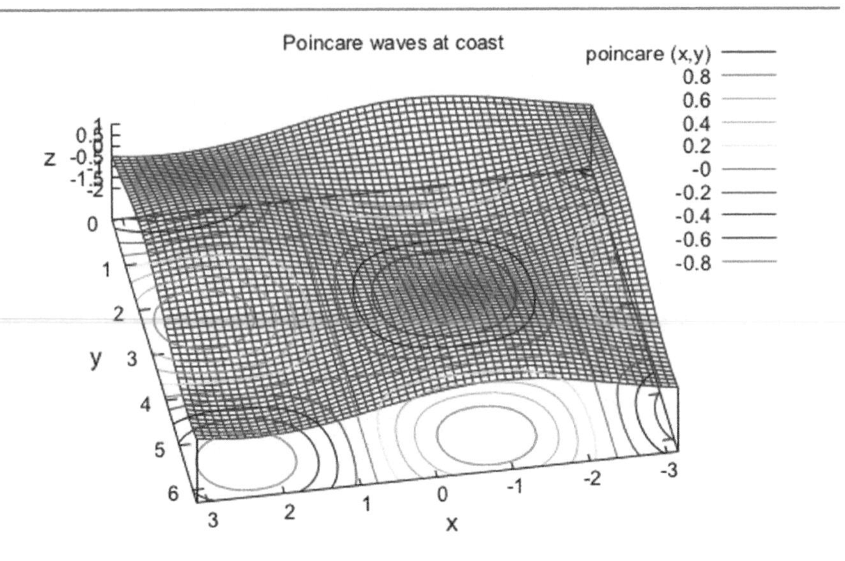

Fig. 6.21 3D view of Poincaré waves at the coast

Since the first term can not satisfy the boundary condition (6.245), we must have

$$A = 0. \tag{6.260}$$

The boundary condition (6.244) requires

$$\left(-\alpha + f\frac{k}{\omega}\right) B = 0 \tag{6.261}$$

and using (6.246) and (6.258) in (6.261),

$$\alpha^2 = k^2 - \frac{\omega^2 - f^2}{C_0^2} = \frac{f^2 k^2}{\omega^2}, \tag{6.262}$$

and this reduces to

$$\frac{\omega^2 - f^2}{C_0^2} = \left(1 - \frac{f^2}{\omega^2}\right) k^2 = \frac{\omega^2 - f^2}{\omega^2} k^2,$$

or simply

$$\left(\frac{\omega}{k}\right)^2 = C_0^2 \tag{6.263}$$

This equation has both positive and negative roots $\omega/k = \pm C_0$. On the other hand, since $\alpha > 0$, Eq. (6.261) requires that only a positive root can be accepted:

$$\frac{\omega}{k} = +C_0 = \sqrt{gh_0}. \tag{6.264}$$

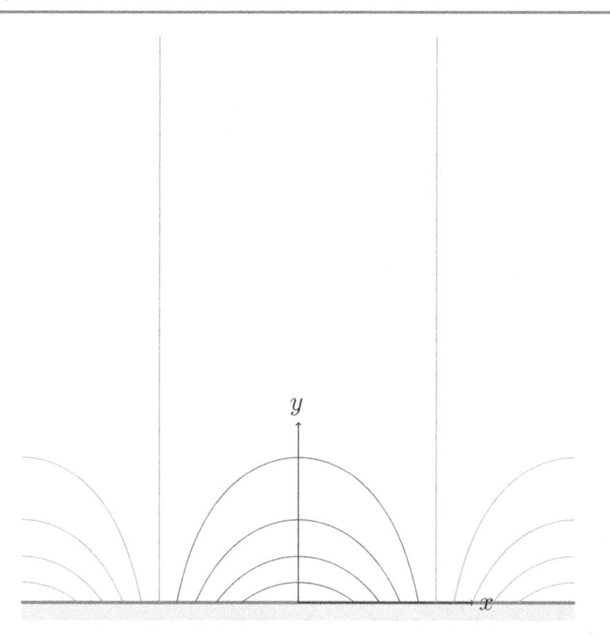

Fig. 6.22 Kelvin wave at the coast

The choice of only one of the roots imposes a preferential direction to the motion. Furthermore, this simple result in itself is remarkable, yielding a *non-dispersive wave solution*, propagating with the classical shallow water phase speed C_0, (for a non-rotating fluid), despite the fact that the wave motion owes its existence to rotation. The decay parameter alpha is calculated from (6.262) and (6.264) as

$$\alpha = \frac{f}{C_0} = \frac{f}{\sqrt{gh_0}}. \tag{6.265}$$

Substituting (6.265) and (6.264), and letting $B = \eta_0$ (a real number) without loss of generality, the solution (6.242) takes the following form:

$$\begin{aligned} \eta &= \eta_0 \Re\{e^{-\frac{f}{C_0}y} e^{ik(x - C_0 t)}\} \\ &= \eta_0 e^{-\frac{f}{C_0}y} \cos\left[k(x - C_0 t)\right] \end{aligned} \tag{6.266}$$

The wave motion is sketched in Figas in Figs. 6.22 and 6.23.

Since $C_0 > 0$, (6.266) indicates that the *Kelvin wave* always travels in the positive x-direction for increasing t, i.e. *it takes the coast to its right.* The wave motion has a maximum amplitude of η_0 at the coast and decays offshore within an e-folding distance of

$$R = \frac{C_0}{f} = \frac{\sqrt{gh_0}}{f}, \tag{6.267}$$

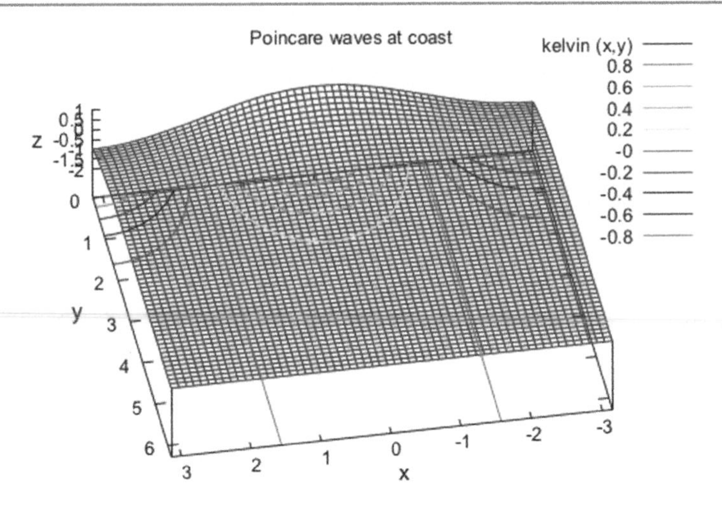

Fig. 6.23 3D view of Kelvin wave at the coast

which is the *Rossby radius of deformation*. An estimate, with $h_0 = 1000\,\text{m}$, $f = 10^{-4}$ rad/s, for the ocean gives $R \simeq 1000\,\text{km}$; i.e. the wave amplitude decays within several thousands of kilometers from the coast (!). This solution does not seem to be of physical significance in the ocean, but its analogue in the atmosphere is feasible. In the ocean, a similar pattern of motion, the internal Kelvin wave, occurs only in a stratified fluid. The velocity $\mathbf{u}$ is obtained from (6.187), by assuming a form of

$$\mathbf{u} = \Re\{\mathbf{u}_0 e^{-\frac{f}{C_0}y} e^{ik(x-C_0t)}\} \tag{6.268}$$

and by substituting this in (6.225):

$$\begin{aligned}(f^2 - k^2C_0^2)\mathbf{u}_0 &= -g\left\{-ikC_0\left(ik\mathbf{i} - \frac{f}{C_0}\mathbf{j}\right) + f\left(-\frac{f}{C_0}\mathbf{i} - ik\mathbf{j}\right)\right\}\eta_0 \\ &= -g\left\{\left(k^2C_0 - \frac{f^2}{C_0}\right)\mathbf{i} + i(fk - fk)\mathbf{j}\right\}\eta_0 \\ &= g\frac{\eta_0}{C_0}(f^2 - k^2C_0^2)\mathbf{i}.\end{aligned} \tag{6.269}$$

This result shows that the velocity field has only a u component. We have

$$u_0 = \mathbf{u}_0 \cdot \mathbf{i} = \frac{g\eta_0}{C_0} = \frac{\eta_0}{h_0}C_0, \quad v_0 = \mathbf{u}_0 \cdot \mathbf{j} = 0, \tag{6.270a, b}$$

and by (6.266) and (6.268), we obtain for the velocity components

$$u = \mathbf{u} \cdot \mathbf{i} = \frac{\eta_0}{h_0}C_0 e^{-\frac{f}{C_0}y} \cos k(x - C_0t), \quad v = \mathbf{u} \cdot \mathbf{j} = 0. \tag{6.271a, b}$$

Since the v-velocity vanishes everywhere, the corresponding terms in the shallow water equations (6.219a), (6.219b) vanish, to yield

$$\frac{\partial u}{\partial t} = -g\frac{\partial \eta}{\partial x} \tag{6.272a}$$

$$fu = -g\frac{\partial \eta}{\partial y} \tag{6.272b}$$

$$\frac{\partial \eta}{\partial t} + h_0\frac{\partial u}{\partial x} = 0 \tag{6.272c}$$

The equations with the remaining terms indicate an interesting balance. While the alongshore acceleration is driven by the alongshore pressure gradient, the velocity is in geostrophic balance with the cross-shore pressure gradient. Elimination of u from (6.272a), (6.272c) gives

$$\frac{\partial^2 \eta}{\partial t^2} = C_0^2\frac{\partial^2 \eta}{\partial x^2} \tag{6.273}$$

i.e., a classic wave equation for the propagation (with speed C_0) in the x-direction. Of the two possible wave solutions with phase speeds $\pm C_0$, only the positive valued solution can satisfy the boundary conditions.

For both of the above solutions, i.e. the Poincaré and Kelvin waves, a dispersion diagram can be constructed as in as in Fig. 6.24.

The Poincaré waves occupy the shaded regions in which all frequencies that satisfy the dispersion relation (6.246) are allowable. For the Kelvin wave, there is only one wave number satisfying the dispersion relation (6.264) at each frequency.

6.6.6 Topographic Rossby Waves

One Dimensional Depth Variations

Let us now consider the effects of depth variation. The governing Eq. (6.222) can be written as

$$\frac{\partial}{\partial t}\left\{\left(\frac{\partial_2}{\partial t^2} + f^2\right)\eta - g\left(\frac{\partial}{\partial x}h\frac{\partial \eta}{\partial x} + \frac{\partial}{\partial y}h\frac{\partial \eta}{\partial y}\right)\right\} - fg\left(\frac{\partial h}{\partial x}\frac{\partial \eta}{\partial y} - \frac{\partial h}{\partial y}\frac{\partial \eta}{\partial x}\right) = 0. \tag{6.274}$$

Without loss of generality, consider depth variations in the y-direction, $h = h(y)$, as in as in Fig. 6.25. Then (6.274) reduces to

$$\frac{\partial}{\partial t}\left\{\left(\frac{\partial^2}{\partial t^2} + f^2\right)\eta - g\left(h\frac{\partial^2 \eta}{\partial x^2} + \frac{\partial}{\partial y}h\frac{\partial \eta}{\partial y}\right)\right\} + fg\frac{\partial h}{\partial y}\frac{\partial \eta}{\partial x} = 0. \tag{6.275}$$

The depth variations induce a restoring effect, as seen through the potential vorticity conservation (6.109), in a much similar way to the β-effect. Motion across

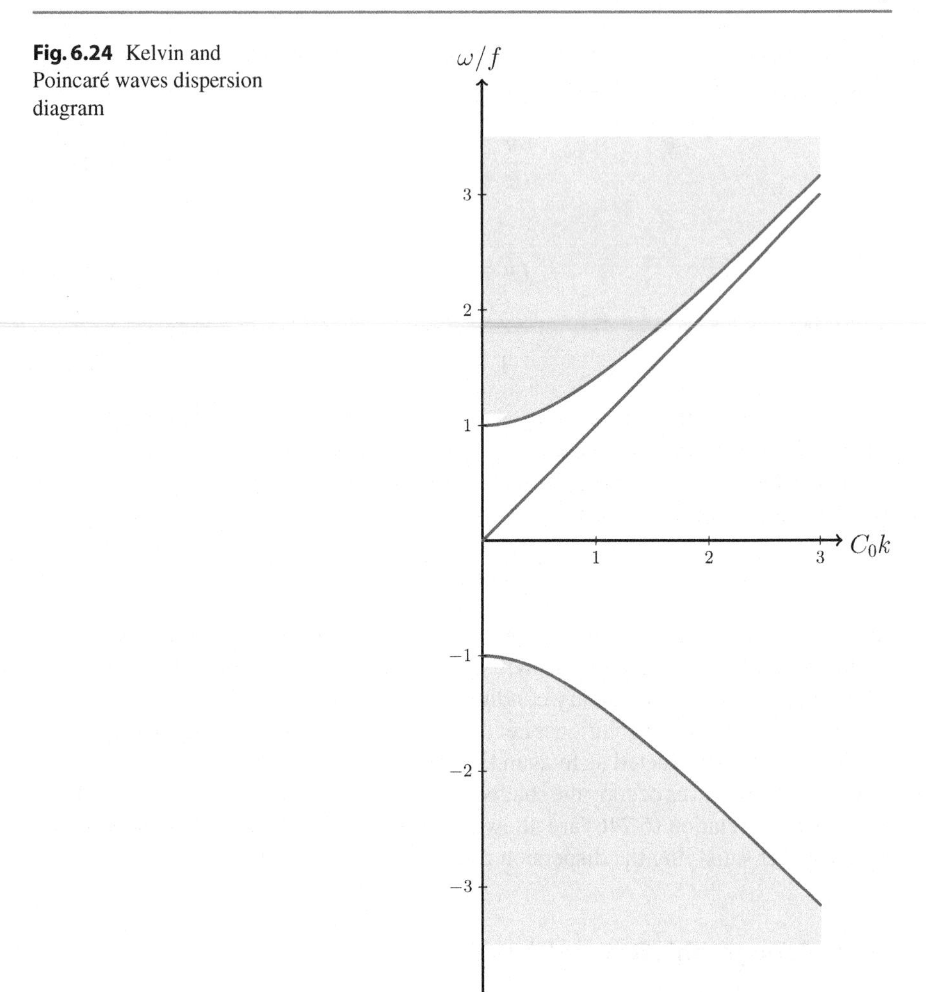

Fig. 6.24 Kelvin and Poincaré waves dispersion diagram

isobaths will cause stretching or extension of fluid columns and thus generate vorticity.

Therefore we may expect topographic Rossby waves of the form

$$\eta = \Re\{N(y)e^{i(kx-\omega t)}\} \tag{6.276}$$

which reduces (6.275) to

$$-i\omega\left\{(f^2-\omega^2)N - g\left(-k^2hN + h\frac{d^2N}{dy^2} + \frac{dh}{dy}\frac{dN}{dy}\right)\right\} + fg\left(-ikN\frac{dh}{dy}\right) = 0$$

or

$$N'' + \left(\frac{h'}{h}\right)N' - \left(\frac{f^2-\omega^2}{gh} + k^2 - f\frac{k}{\omega}\frac{h'}{h}\right)N = 0, \tag{6.277}$$

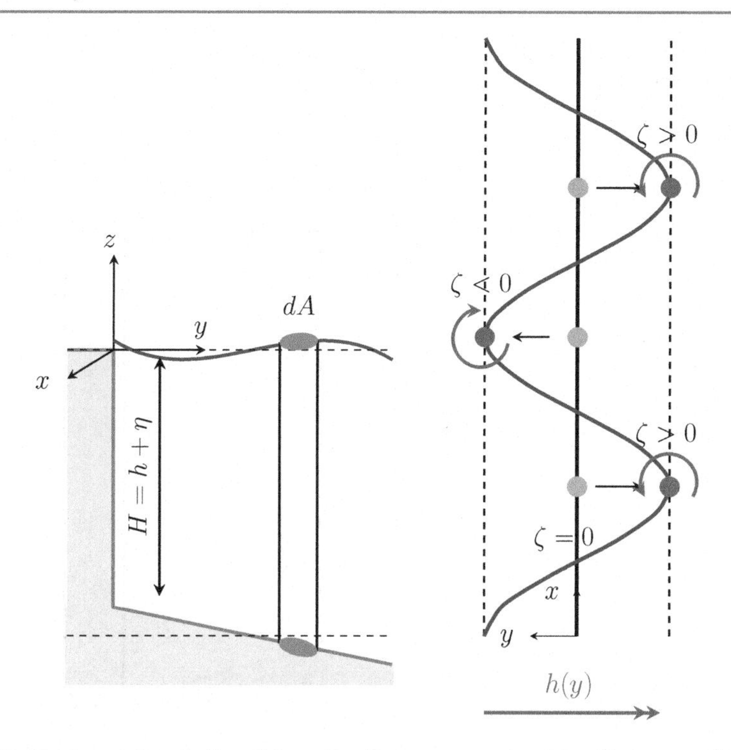

Fig. 6.25 Depth variations in the offshore direction near a coast and its effects on vorticity of a displaced fluid parcels

where primes denote differentiation with respect to y. We can further denote

$$F(y) = \frac{1}{h}\frac{dh}{dy} = \frac{h'}{h} \tag{6.278}$$

and substitute

$$N(y) = M(y)e^{-\frac{1}{2}\int F(y)dy} \tag{6.279}$$

to transform the variables into a new form:

$$\begin{aligned} N &= Me^{-\frac{1}{2}\int Fdy} \\ N' &= \left(M' - \frac{1}{2}FM\right)e^{-\frac{1}{2}\int Fdy} \\ N'' &= \left[M'' - FM' + \left(\frac{1}{4}F^2 - \frac{1}{2}F'\right)M\right]e^{-\frac{1}{2}\int Fdy}. \end{aligned} \tag{6.280a-c}$$

Then, using these substitutions in (6.277) reduces it to

$$M'' - \left[\frac{f^2-\omega^2}{gh} + k^2 - \frac{f}{\omega}\frac{h'}{h}k + \frac{1}{4}\left(\frac{h'}{h}\right)^2 + \frac{1}{2}\left(\frac{h'}{h}\right)'\right]M = 0. \tag{6.281}$$

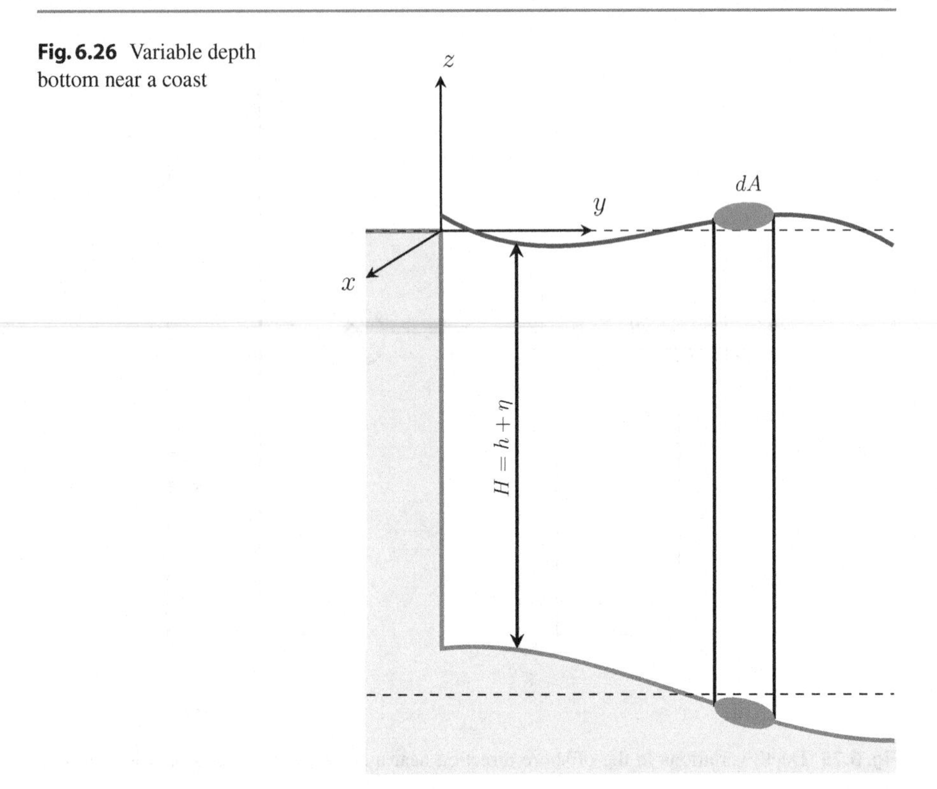

Fig. 6.26 Variable depth bottom near a coast

Consider the flow on a sloping bottom adjacent to a coast in as in Fig. 6.26.

At the coast ($y = 0$), we must require the normal velocity to vanish $\mathbf{u} \cdot \mathbf{i} = 0$, so that boundary condition (6.244) applies. With the substitutions (6.270a,b), (6.244) reads

$$M' + \left(\frac{f}{\omega}k - \frac{1}{2}\frac{h'}{h}\right) M = 0 \ \text{ on } y = 0. \tag{6.282}$$

Far away from the coast, the motion must vanish, so that boundary condition (6.245) applies. But if N is bounded as $y \to \infty$, we have by virtue of (6.279),

$$M \to \text{ bounded, as } y \to \infty. \tag{6.283}$$

A solution can then be obtained for (6.281), with boundary conditions (6.282) and (6.283):

$$\begin{aligned} N &= M(y)e^{-\frac{1}{2}\int \frac{1}{h}\frac{dh}{dy}dy} \\ &= M(y)e^{-\frac{1}{2}\int \frac{d\ln h}{dy}dy} \\ &= M(y)[e^{\ln h}]^{-\frac{1}{2}} \\ &= h^{-\frac{1}{2}}(y)M(y), \end{aligned} \tag{6.284}$$

so that the assumed solution (6.276) can be written as

$$\eta = \Re\{h^{-\frac{1}{2}}(y)M(y)e^{i(kx-\omega t)}\} \tag{6.285}$$

where $M(y)$ is determined by solving (6.281) with boundary conditions of the type (6.282) and (6.283) imposed.

Note that (6.281) is a differential equation with non-constant coefficients. Furthermore Eq. (6.281) with boundary conditions of type (6.282) and (6.283) constitute an eigenvalue problem, which must yield eigenvalues which relate ω to k. However the terms in square brackets of (6.243) are extremely complex. For example, assuming an eigenvalue is determined, these terms are cubic in frequency ω and quadratic in wave number k. Therefore analytic solutions can be obtained only in simple cases where a certain simple topography is assumed and with further possible simplifications.

A further simplification can often be made in (6.281) with respect to the first term in the brackets. In fact, let L be the horizontal y-scale of the motion, and let h_0 be a depth scale; then comparing the first and second terms of (6.281), we have

$$\frac{\left(\frac{f^2-\omega^2}{gh}M\right)}{\left(\frac{d^2M}{dy^2}\right)} = \frac{O\left(\frac{f^2M}{gh_0}\right)}{O\left(\frac{M}{L^2}\right)} = O\left(\frac{f^2L^2}{gh_0}\right) = O(\delta) \tag{6.286}$$

where

$$\delta \equiv \frac{f^2L^2}{gh_0} \tag{6.287}$$

is defined as the *divergence parameter*. Note that this parameter is equal to the ratio

$$\delta = \left(\frac{L}{R}\right)^2 \tag{6.288}$$

where R is the Rossby radius of deformation defined in (6.267). The scale L can be thought of as the horizontal scale in which depth variations occur (i.e., the shelf width).

For typical values of $f = 10^4$ rad/s, $L = 10$ km, $g = 10\,\text{m/s}^2$, $h_0 = 100$ m; δ is calculated to be $\delta \simeq 10^{-3} \ll 1$, and therefore the second term of (6.281) is much smaller than the first. With this approximation ($\delta \ll 1$) in place, (6.281) reads

$$M'' - \left[k^2 - f\frac{k}{\omega}\left(\frac{h'}{h}\right) + \frac{1}{4}\frac{h'^2}{h} + \frac{1}{2}\left(\frac{h'}{h}\right)'\right]M = 0. \tag{6.289}$$

The approximation for the horizontal scale being much less than the Rossby radius ($\delta \ll 1$) is in fact equivalent to the rigid-lid approximation, and amounts to neglecting the $\frac{\partial \eta}{\partial t}$ term in the continuity equation (6.119b). This can simply be verified by making the appropriate substitution in Eq. (6.221), i.e. $\nabla \cdot h\mathbf{u} = 0$, and similarly causes the first term of (6.281) to vanish.

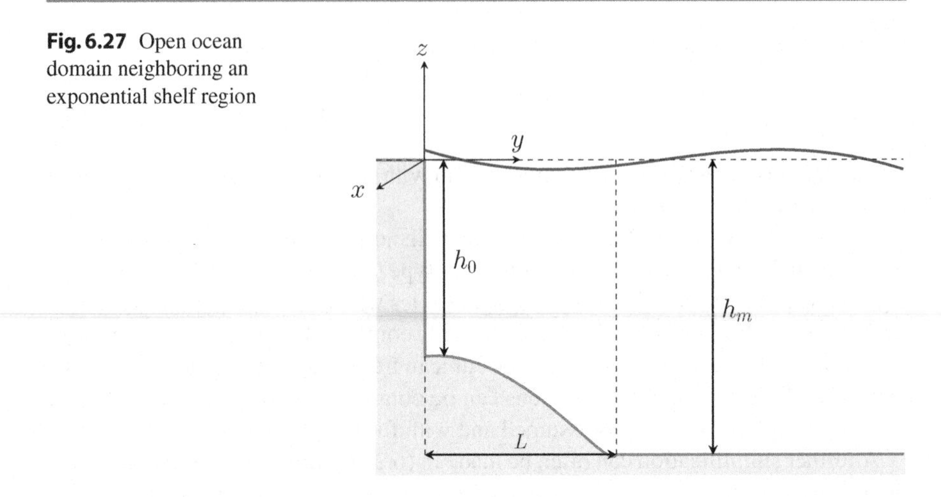

Fig. 6.27 Open ocean domain neighboring an exponential shelf region

Shelf Waves—Exponential Shelf

As an example of simple case with an analytical solution, consider the coast adjoined by the following shelf topography

$$h(y) = \begin{cases} h_0 e^{2by}, & 0 < y < l \\ h_0 e^{2bL} = h_m, & L < y < \infty, \end{cases} \tag{6.290}$$

visualised in Fig. 6.27.

Region 1 and 2 are the continental shelf and deep ocean regions respectively. In region 1 (the shelf), (6.240) is a constant

$$F(y) = \frac{h'}{h} = \frac{2bh_0 e^{2by}}{h_0 e^{2by}} = 2b \tag{6.291}$$

and in the deep ocean region

$$F(y) = 0, \tag{6.292}$$

so that the corresponding Eq. (6.243) become:
Region 1:

$$M_1'' - \left[\frac{f^2 - w^2}{gh_0} e^{-2by} + k^2 - \frac{f}{\omega} 2bk + b^2 \right] M_1 = 0 \tag{6.293a}$$

Region 2:

$$M_2'' - \left[\frac{f^2 - w^2}{gh_m} + k^2 \right] M_2 = 0 \tag{6.293b}$$

The second term of (6.293a), which gives rise to non-constant coefficients, can be neglected with the rigid-lid approximation, $\delta \ll 1$, yielding:

Region 1:

$$M_1'' - \left[k^2 - \frac{f}{\omega}2bk + b^2\right] M_1 = 0 \tag{6.294a}$$

Region 2:

$$M_2'' - k^2 M_2 = 0 \tag{6.294b}$$

We define

$$\gamma^2 = k^2 - \frac{f}{\omega}2bk + b^2 \tag{6.295}$$

to replace the constant coefficient in the first equation.

The boundary condition at the coast is obtained by substituting from (6.253),

$$M_1' + \left[f\frac{k}{\omega} - b\right] M_1 = 0 \text{ at } y = 0 \tag{6.296a}$$

We will be interested in waves that are trapped in the shelf region, so that far from the coast we not only want the solution be bounded, but also to vanish:

$$M_2 \to 0 \text{ as } y \to \infty \tag{6.296b}$$

Finally we need jump conditions at $x = L$ where the shelf joins the deep ocean. For this we require that η and $\mathbf{u} \cdot \mathbf{i}$ be continuous at the junction, yielding:

$$M_1 = M_2 \text{ at } y = L \tag{6.296c}$$

$$M_1' + \left(\frac{f}{\omega}k - b\right) M_1 = M_2' + \left(\frac{f}{\omega}k\right) M_2 \text{ at } y = L \tag{6.296d}$$

The solution to (6.294a), (6.294b) can be written as

$$M_1 = Ae^{\gamma(y-L)} + Be^{-\gamma(y-L)} \tag{6.297a}$$

$$M_2 = Ce^{k(y-L)} + De^{-k(y-L)} \tag{6.297b}$$

Boundary condition (6.296b) then requires that

$$C = 0,$$

while boundary conditions (6.258a, c, d) require:

$$(\gamma + \frac{f}{\omega}k - b)e^{-\gamma L} A + (-\gamma + \frac{f}{\omega}k - b)e^{\gamma L} B = 0 \tag{6.298a}$$

$$A + B = D \tag{6.298b}$$

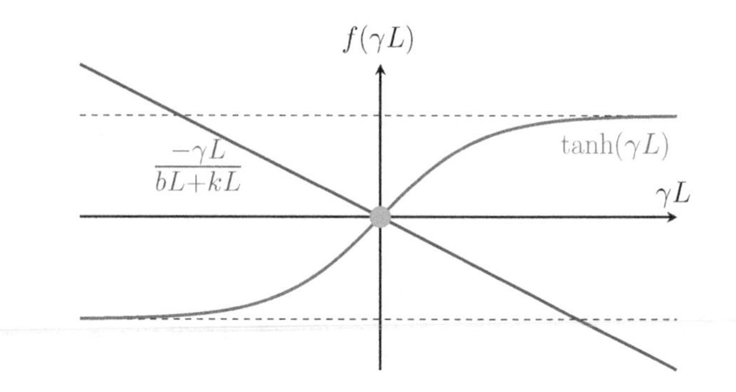

Fig. 6.28 Solution for the transcendental equation γ

$$\left(\gamma + \frac{f}{\omega}k - b\right) A + \left(-\gamma + \frac{f}{\omega}k - b\right) B = (-k + sk)D \tag{6.298c}$$

and eliminating D from (2.298b, c) we have

$$(\gamma + k - b)\, A + (-\gamma + k - b)\, B = 0 \tag{6.299}$$

For non trivial values of A and B, (6.298a) and (6.299) then yield

$$\left(\gamma + \frac{f}{\omega}k - b\right)(-\gamma + k - b)e^{-\gamma L} - \left(-\gamma + \frac{f}{\omega}k - b\right)(\gamma + k - b)e^{\gamma L} = 0, \tag{6.300}$$

or, rearranging terms, and utilizing (6.295)

$$\tanh \gamma L = -\frac{\gamma}{b+k}. \tag{6.301}$$

This is a transcendental equation for which the roots γ_n must be obtained. Graphically the right and left hand sides can be sketched as in Fig. 6.28.

The only possible root is $\gamma = 0$ which is trivial. In (6.295) we have imperatively assumed that $\gamma^2 > 0$ (γ is real), but we now find that non-trivial solutions are not possible. However, if we let $\gamma^2 < 0$ then (γ is imaginary),

$$\gamma^2 = -\mu^2 < 0 \tag{6.302}$$

($\mu^2 > 0,\ \mu$ is real), then (6.295) becomes

$$\mu^2 = \frac{f}{\omega}2bk - b^2 - k^2 \tag{6.303}$$

and (6.301) becomes (with $\gamma = i\mu$),

$$\tanh i\mu L = i \tan \mu L = -i\frac{\mu}{b+k}. \tag{6.304}$$

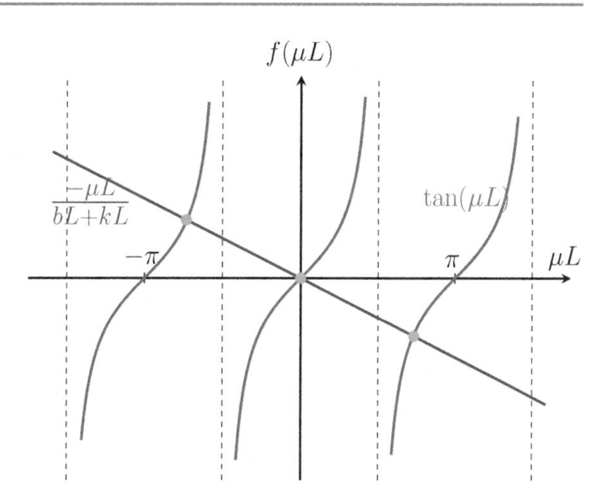

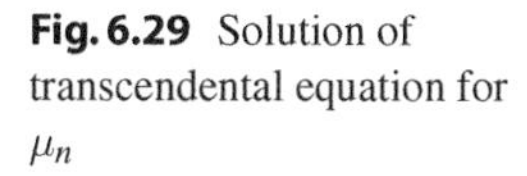
Fig. 6.29 Solution of transcendental equation for μ_n

This equation possesses an infinite number of discrete roots $\mu_n = \mu_1, \mu_2, \dots (n > 0)$ represented graphically in Fig. 6.29 and by virtue of (6.303) an infinite number of discrete frequencies

$$\frac{\omega}{f} = \frac{2bk}{b^2 + k^2 + \mu_n^2}, \tag{6.305}$$

where μ_n are the roots of (6.304) for $n > 0$. The solutions (6.297a), (6.297b) become

$$M_1 = A_n e^{i\mu_n(y-L)i} + B_n e^{-i\mu_n(y-L)} \tag{6.306a}$$

$$M_2 = D_n e^{-k(y-L)} \tag{6.306b}$$

and the relations between the complex-valued coefficients A_n, B_n and D_n are given by (6.298b) and (6.299). By making use of (6.304),

$$\begin{aligned}
\frac{B_n}{A_n} &= -\frac{k - b + i\mu_n}{k - b - i\mu} \\
&= -\frac{(k-b) - (k+b)\tanh i\mu_n L}{(k-b) + (k+b)\tanh i\mu_n L} \\
&= -\frac{k(1 - \tanh i\mu_n L) - b(1 + \tanh i\mu_n L)}{k(1 + \tanh i\mu_n L) - b(1 - \tanh i\mu_n L)} \\
&= -\frac{ke^{-i\mu_n L} - be^{i\mu_n L}}{ke^{i\mu_n L} - be^{-i\mu_n L}}
\end{aligned} \tag{6.307a}$$

$$\begin{aligned}\frac{D_n}{A_n} &= 1 + \frac{B_n}{A_n} = -\frac{2i\mu_n}{k-b-i\mu_n}\\ &= \frac{2(k+b)\tanh i\mu_n L}{(k-b)+(k+b)\tanh i\mu_n L}\\ &= \frac{(k+b)(e^{i\mu_n L}-e^{-i\mu_n L})}{ke^{i\mu_n L}-be^{-i\mu_n L}}\\ &= \frac{(k+b)2i\sin\mu_n L}{ke^{i\mu_n L}-be^{i\mu_n L}}.\end{aligned} \tag{6.307b}$$

Letting

$$A_n = \frac{1}{2i}\bar{A}_n(ke^{i\mu_n L}-be^{-i\mu_n L}) \tag{6.308}$$

The solutions (6.306a), (6.306b) become

$$\begin{aligned}M_1 &= \frac{\bar{A}_n}{2i}\left\{[ke^{i\mu_n L}-be^{-i\mu_n L}]e^{i\mu_n(y-L)}-[ke^{-i\mu_n L}-be^{i\mu_n L}]e^{i\mu_n(y-L)}\right\}\\ &= \frac{\bar{A}_n}{2i}\left\{k[e^{i\mu_n y}-e^{-i\mu_n y}]-b[e^{i\mu_n(y-2L)}-e^{-i\mu_n(y-2L)}]\right\}\\ &= \bar{A}_n\{k\sin\mu_n y-b\sin\mu_n(y-2L)\}\end{aligned} \tag{6.309}$$

and

$$M_2 = \bar{A}_n(k+b)\sin\mu_n L e^{-k(y-L)}. \tag{6.310}$$

Substituting

$$\eta_0 = b(\sin 2\mu_n L)h_0^{-1/2}\bar{A}_n \tag{6.311}$$

and utilizing (6.285), (6.290), (6.309) and (6.310), the full solution can be written as

$$\eta = \begin{cases}\frac{\eta_0}{b\sin 2\mu_n L}[k\sin\mu_n y-b\sin\mu_n(y-2b)]e^{-by}\cos(kx-\omega_n t), & 0< y<L,\\ \frac{\eta_0\sin\mu_n L}{b\sin 2\mu_n L}(k+b)e^{-bL}e^{-k(y-L)}\cos(kx-w_n t), & L< y<\infty\end{cases} \tag{6.312}$$

where η_0 is the amplitude of the wave at the coast. The amplitude of these waves with offshore distance can be sketched as follows.

The wave motion is thus trapped in the continental shelf region. Note that by virtue of (6.267), the frequency ω_n is positive for any positive k, so that

$$\cos(kx-\omega_n t)$$

always represents a wave moving in the positive x-direction.

The waves preferentially take the coast to their right (similar to the case for Kelvin waves).

The dispersion relation (6.305) yields an infinite number of frequencies ω_n corresponding to the roots μ_n. Since μ_n are ordered in an increasing sequence for $n = 1, 2, \ldots$, higher modes have smaller frequencies, i.e. for

$$\mu_1 < \mu_2 < \mu_3 \ldots$$

the frequencies are ordered as:

$$\omega_1 > \omega_2 > \omega_3 \ldots$$

Note that (6.267) has a maximum when

$$k^2 = k_{\max}^2 = b^2 + \mu_n^2, \tag{6.313}$$

which corresponds to a maximum frequency of

$$\left(\frac{\omega_n}{f}\right)_{\max} = \frac{b}{\sqrt{b^2 + \mu_n^2}} = \frac{b}{k_{\max}} \tag{6.314}$$

for each mode. Note that for any possible mode, $\omega_{n,\max} < f$ (since $\mu_n^2 > 0$), i.e. possible frequencies are always smaller than the *inertial frequency*, or the Coriolis parameter f.

Furthermore, note that for a fixed frequency ω_n of any single mode, (6.305) yields two possible wavenumbers, k_n^+ and k_n^-, since (6.305) can be written as

$$k^2 - \left(2b\frac{f}{\omega_n}\right)k + b^2 + \mu_n^2 = 0 \tag{6.315}$$

and it follows that

$$\begin{aligned} k_n^+,\ k_n^- &= b\left(\frac{f}{\omega_n}\right) \pm \sqrt{\left(b\frac{f}{\omega_n}\right)^2 - (b^2 + \mu_n^2)} \\ &= b\left(\frac{f}{\omega_n}\right)\left[1 \pm \sqrt{1 - \frac{b^2 + \mu_n^2)}{(bf/\omega_n)^2}}\right] \\ &= \left(\frac{\omega_{n,\max}}{\omega_n}\right)k_{\max}\left\{1 \pm \sqrt{1 - \left(\frac{\omega_n}{\omega_{n,\max}}\right)^2}\right\} \end{aligned} \tag{6.316}$$

where (6.313) and (6.314) have been utilized.

In the above analytical solution displayed in Fig. 6.30, we have excluded the Poincare and Kelvin waves because we have used the rigid-lid approximation. However, it is often found that the dispersion characteristics of these waves are only slightly modified by the presence of bottom topography. The general form of the dispersion diagram is sketched in Fig. 6.31.

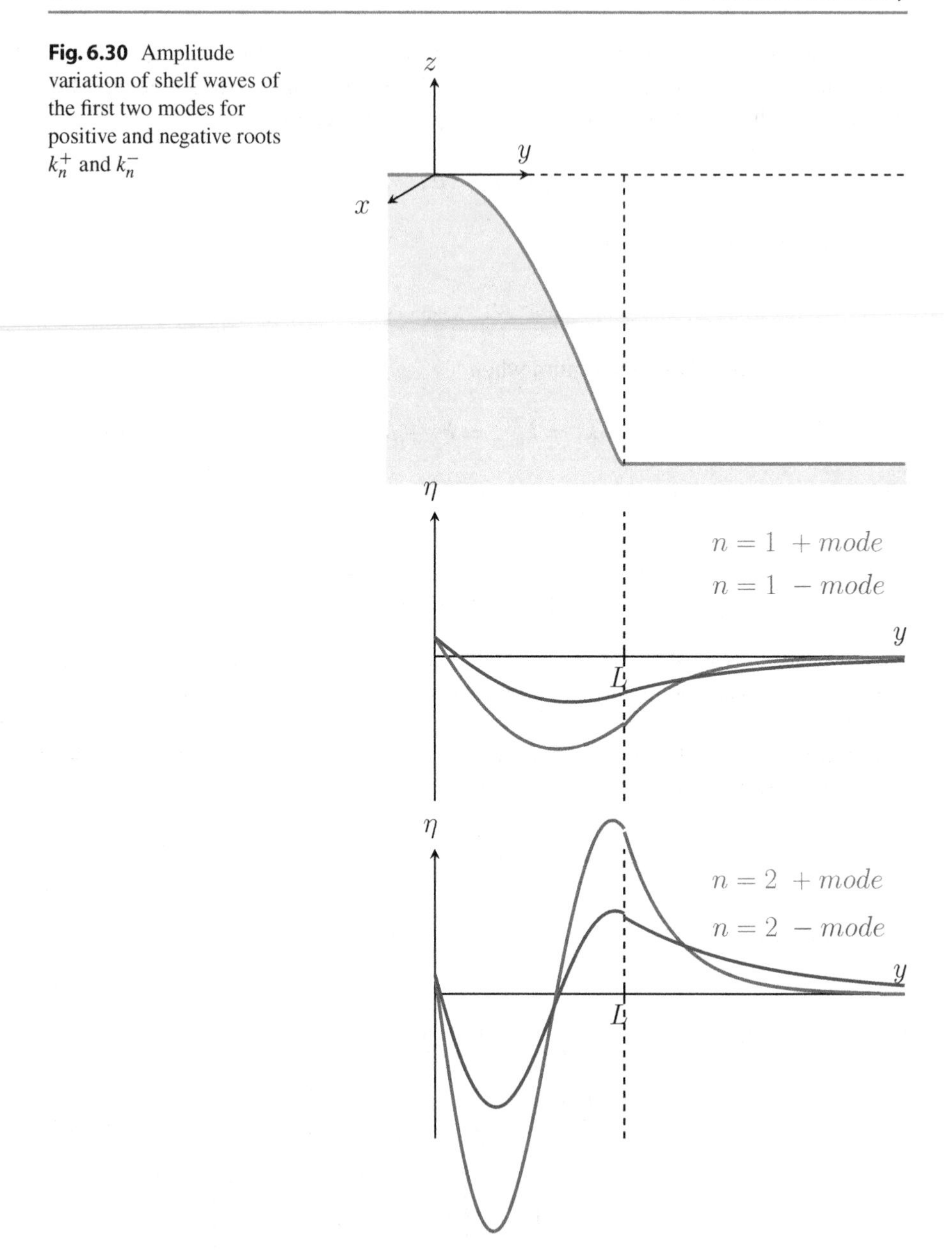

Fig. 6.30 Amplitude variation of shelf waves of the first two modes for positive and negative roots k_n^+ and k_n^-

The phase speed of shelf waves is always in the positive x-direction since these waves always propagate with the coast on the right hand side,

$$C_x = \frac{\omega}{k} = \frac{2b}{(b^2 + k^2 + \mu_n^2)} > 0 \tag{6.317}$$

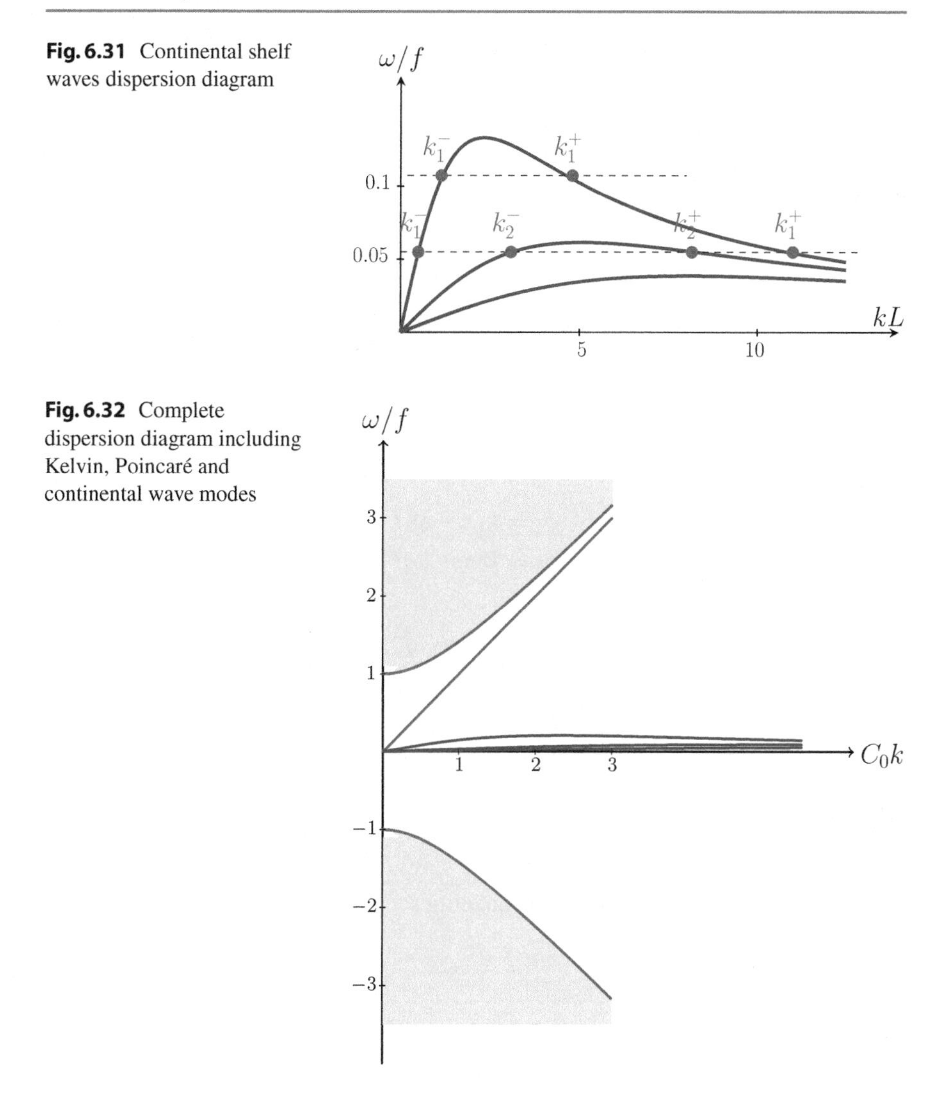

Fig. 6.31 Continental shelf waves dispersion diagram

Fig. 6.32 Complete dispersion diagram including Kelvin, Poincaré and continental wave modes

A calculation of the group velocity can be made as follows:

$$C_{gx} = \frac{\partial \omega}{\partial k} = 2bf \frac{b^2 + \mu_n^2 - k^2}{(b^2 + \mu_n^2 + k^2)^2} \tag{6.318}$$

Now, note that $C_{gx} < 0$ if $k > k_{max}$. Therefore it can be verified that for a fixed frequency ω_n of any mode the wave with wavenumber k^- carries energy in the same direction as phase propagation, while the wave with wavenumber k^+ carries energy in the opposite direction. Also note that at $k = k_{\max}$ the group velocity vanishes, implying that no energy can be transmitted by such waves.

The combined dispersion diagram including Poincaré, Kelvin and topographic Rossby waves can be sketched as in Fig. 6.32.

Exercises

Exercise 1

Consider a wave motion described by a superposition of two waves at different frequency and wavenumber:

$$\eta = \cos(k_1 x - \omega_1 t) + \cos(k_2 x - \omega_2 t)$$

where

$$\phi_1 = k_1 x - \omega_1 t$$
$$\phi_2 = k_2 x - \omega_2 t$$

and

$$\omega_1 = \omega - \Delta\omega/2$$
$$\omega_2 = \omega + \Delta\omega/2$$
$$k_1 = k - \Delta k/2$$
$$k_2 = k + \Delta k/2$$

such that

$$\Delta\omega \ll \omega \quad \text{and} \quad \Delta k \ll k$$

Plot this function, and interpret the quantities

$$c_p = \omega/k$$

and

$$c_g = \Delta\omega/\Delta k.$$

Exercise 2

Consider a fluid in semi-infinite horizontal domain of constant depth h_0 bounded by an oscillating side-wall consisting of a wave-maker as shown in Fig. 6.33. The wave-maker oscillates with displacement x_0 of small amplitude a_0

$$x_0 = \Re\{a_0 e^{i\omega t}\}.$$

Consider only small-amplitude motion in x, uniform in y (i.e. $\frac{\partial}{\partial y} = 0$) and consider a linear regime for the motions generated.

(i) Obtain solutions for frequencies $\omega > f$,
(ii) obtain solutions for frequencies $\omega < f$,

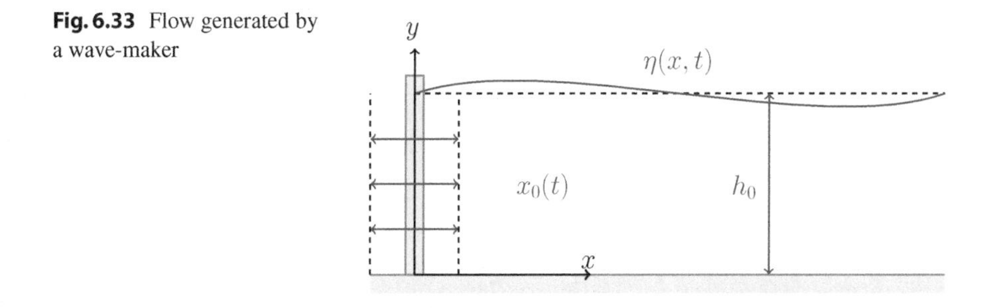

Fig. 6.33 Flow generated by a wave-maker

Investigate particle velocities u, v and sketch velocity as a function of time and distance from the side-wall wave-maker, for each of the above cases. Compare the frequency regimes with respect to longitudinal or transverse wave motions.

(iii) What happens when $\omega \to f$ and when $\omega \to 0$? How is the length scale of the waves related to the Rossby radius of deformation?

Exercise 3

Consider a depth discontinuity in a fluid of infinite horizontal extent rotating about the vertical axis, with depth sharply changing from h_1 to h_2, at $x = 0$

$$h = \begin{cases} h_1 & \text{for } x < 0 \\ h_2 & \text{for } x > 0 \end{cases}$$

(i) If plane-waves were to exist in these domains, what would be the specific conditions in frequency and wave-number space?

How would an incident plane-wave pass from one half-space to the other? Are there any constraints on frequency and wave-number? Is total reflection possible? In which parameter range would this occur?

(ii) Determine and apply the matching conditions required at the depth discontinuity. Calculate transmission and reflection coefficients for an incident plane wave of arbitrary frequency and directional wave-number. Define and derive expressions for the angles of incidence, reflection and transmission, with accompanying physical interpretation.

(iii) In which parameter range could evanescent modes occur in either half-space? What would be their shape and propagation characteristics?

Exercise 4

An estuary/lagoon basin with surface area A_b is connected to the open ocean through a channel of length ℓ, uniform width w and depth h, as shown in Fig. 6.34.

For the channel, the equations of motion are:

$$\frac{\partial u}{\partial t} = -g\frac{\partial \eta}{\partial x} - \nu u,$$

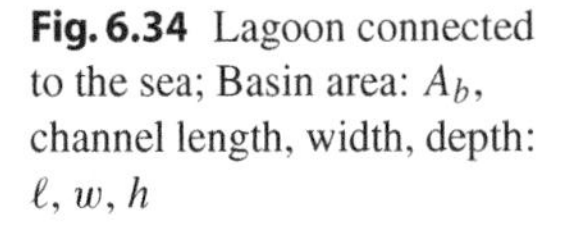

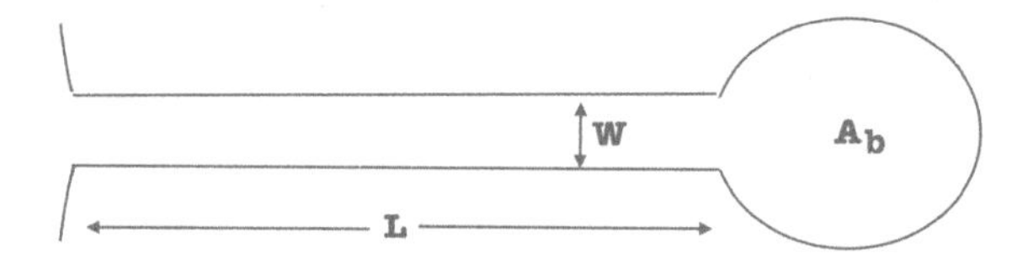

Fig. 6.34 Lagoon connected to the sea; Basin area: A_b, channel length, width, depth: ℓ, w, h

and

$$\frac{\partial \eta}{\partial t} + h\frac{\partial u}{\partial x} = 0,$$

which are the linearized momentum and continuity equations with a linear friction term $-\nu u$ added.

Storage in the basin is a function of the volume flux at the junction of channel, specified by the continuity equation

$$uhw = A_b \frac{\partial \eta_b}{\partial t} \ at \ x = \ell$$

where the basin is assumed to have a uniform sea level response, η_b.

At the ocean side there is tidal forcing specified as

$$\eta = a_0 e^{-i\omega t}.$$

With the above equations determine the solution for the tide in the channel and the basin.

`(i) case 1–large basin` If the basin is too large the sea-level response is so small that we can neglect tidal in influence the in the basin. Describe the solution in this case, taking $\eta_b = 0$.

`(ii) case 2–small basin`

η_b not negligible

Describe the solution, when the basin size A_b is small, so that the ocean forcing reaches the basin producing tidal oscillations $\eta_b(t)$ in the basin. Describe the motion in this case.

`Exercise 5`

Consider the shallow-water model with wind-stress forcing at the surface, approximated with the following linear equations of motion:

$$\frac{\partial \mathbf{u}}{\partial t} + f\mathbf{k} \times \mathbf{u} = -g\nabla\eta - \frac{1}{\rho}\nabla p^s + \frac{1}{\rho h}\boldsymbol{\tau}^s$$

$$\frac{\partial \eta}{\partial t} + \nabla \cdot h\mathbf{u} = 0$$

where $\boldsymbol{\tau}$ is the wind-stress and $h = h(x, y)$ is the still water depth. By manipulating these equations, obtain equation that can be solved for η and $\mathbf{u}$ fields following a similar development as in Sect. 6.6.3, including atmospheric forcing as additional effects. Describe the contribution of each term in the governing equations.

(i) Considering time-periodic solutions subject to periodic forcing, reduce the above governing equations to obtain closed forms for surface elevation η and velocity $\mathbf{u}$ with forcing terms due to barometric pressure and wind-stress.

(ii) Next, consider the time dependent flow adjacent to a coast with constant depth h_0, driven by a periodic wind-stress that is spatially uniform. Let the coast be aligned with the x-axis and assume the flow to be independent of the alongshore direction x, depending only on the offshore coordinate y, with amplitudes $N(y)$ and $\mathbf{U}(y)$, while the wind-stress amplitude is constant. How are the equations further reduced under these conditions? Where is the forcing applied? What would be the boundary conditions at the coast and for the solution to remain bounded as $y- > \infty$?

(iii) What is the solution to equations developed in part (ii)? How do the characteristics change with the frequency ω and direction of wind-stress? What happens as $\omega \rightarrow f$ and $\omega \rightarrow 0$?

(iv) Consider the case with a periodic, propagating wind-stress in the alongshore direction x:

$$\tau(x,t) = \mathbf{T}' e^{-ikx+\omega t}$$

where $\mathbf{T}$ is constant. What would be the solution in this case? How do the characteristics change with the frequency ω and direction of wind-stress?

What happens as $\omega \rightarrow f$ and $\omega \rightarrow 0$? What is the difference between solutions propagating in opposite directions along the coast, $k > 0$ and $k < 0$?

Quasigeostrophic Theory 7

7.1 An Overview and Derivation of Quasi-geostrophic Equations

In Chap. 2 we derived the shallow water equations, and studied possible solutions to these equations. In general terms, we have always seeked to represent the complex geophysical flows by simplified equations which could be used to better understand the possible motions. We have done this in two ways: (i) in Sect. 2.4, we have made the rigid lid approximation and utilized the potential vorticity conservation to investigate low frequency motions such as Rossby waves, (ii) in Sect. 2.5, we have allowed surface displacement, but in order to simplify the equations, we have neglected nonlinear terms, upon which we discovered new types of motions which were mainly of high frequency.

In this section, we will develop the approximate theory for the first type of motions considered above. We have seen in Sect. 2.4.1 that the motion becomes geostrophic in the steady limit. We have also shown that the low frequency motion (Rossby waves) carry many features of the geostrophic case, such as the particle motions being transverse to the surface elevation gradient and feeling topographic steering effects. The motion is than essentially close to being geostrophic (which we have shown to be a degenerate case), and hence called *quasi-geostrophic*, because of the set of simplifications consistently approximating the primitive equations.

Let us re-write the shallow water Eqs. (6.89) and (6.90) excluding the barometric pressure and frictional effects:

$$\frac{\partial \mathbf{u}}{\partial t} + \mathbf{u} \cdot \nabla \mathbf{u} + f\mathbf{k} \times \mathbf{u} = -g\nabla\eta \tag{7.1}$$

$$\frac{\partial \eta}{\partial t} + \nabla \cdot H\mathbf{u} = 0 \tag{7.2}$$

© Springer Nature Switzerland AG 2020

E. Özsoy, *Geophysical Fluid Dynamics I*, Springer Textbooks in Earth Sciences, Geography and Environment, https://doi.org/10.1007/978-3-030-16973-2_7

where

$$H = \eta + h \tag{7.3}$$

is the total depth and

$$f = f_0 + \beta y \tag{7.4}$$

is the Coriolis parameter, as it was approximated earlier in Sect. 2.3.

We now want to focus our attention on motions with time scales much larger than the (internal period) f_0^{-1}. Therefore we want to scale the equations accordingly, choosing the following scales:

$$\begin{gathered} \mathbf{x} \sim L,\ t \sim T \\ H \sim H_0,\ \beta \sim \tfrac{U}{L^2},\ f \sim f_0 \\ \eta \sim \tfrac{fUL}{g},\ \mathbf{u} \sim U \end{gathered} \tag{7.5}$$

upon which the non-dimensional forms of the Eqs. (7.1) and (7.2) become

$$\epsilon_T \frac{\partial \mathbf{u}}{\partial t} + \epsilon(\mathbf{u} \cdot \nabla \mathbf{u} + \beta y \mathbf{k} \times \mathbf{u}) + \mathbf{k} \times \mathbf{u} = -\nabla \eta \tag{7.6}$$

$$\delta \epsilon_T \frac{\partial \eta}{\partial t} + \nabla \cdot H\mathbf{u} = 0 \tag{7.7}$$

where

$$\epsilon = \frac{U}{f_0 L} \tag{7.8a}$$

$$\epsilon_T = \frac{1}{f_0 T} \tag{7.8b}$$

$$\delta = \frac{f_0^2 L^2}{g H_0} \tag{7.8c}$$

The first of these nondimensional numbers, ϵ, is the Rossby number, the second one ϵ_T measures the ratio of the inertial time scale f_0^{-1} to the time scale T of the motion. For the type of motion considered, we insist that the rotational effects are important, $\epsilon \ll 1$, and that the time scale of the motion is much grater than the inertial time scale, i.e. the motion is of low frequency so that $\epsilon_T \ll 1$. While both of these numbers are small, their ratio

$$\frac{\epsilon}{\epsilon_T} = \frac{UT}{L} \tag{7.9a}$$

determines the relative importance of the nonlinear terms in Eqs. (7.6) and (7.7). When the ratio (7.9a) is small the equations can be linearized. Since we want to keep the nonlinearity, we must also insist that the ratio is unity, or

$$\epsilon_T = \epsilon \tag{7.9b}$$

Finally the third parameter δ in (7.8c) is called the *divergence parameter*. We have seen earlier that (cf.Eq. 2.250)

$$\delta = \left(\frac{L}{R}\right)^2 \tag{7.10}$$

where R is the Rossby Radius of deformation. Its order of magnitude depends on the application. For example, in the context of shelf waves, we have seen that $\delta \ll 1$, and such motions are called *quasi-nondivergent* if corresponding terms have a small contribution, or *nondivergent* if these terms are completely neglected. The nondivergent case is also known as the rigid-lid approximation. Here, we will assume that

$$\delta = O(1) \tag{7.11}$$

and keep the corresponding terms. Therefore, the only small parameter in the equations is the Rossby number, ϵ. We can now make a perturbation analysis in terms of the small parameter ϵ.

Before carrying out this analysis, let us first consider (7.3) and make further approximations about the depth variations. Let us assume a mean depth H_0 (which was used as a scale earlier) and call the bottom topographic deviations from this mean depth h_b(x,y). Equation (7.3) becomes

$$H = \eta + h = \eta + H_0 - h_b, \tag{7.12a}$$

and dividing by H_0,

$$\frac{H}{H_0} = \frac{H_0 + \eta - h_b}{H_0} = 1 + \frac{\eta}{H_0} - \frac{h_b}{H_0}.$$

Then, utilizing the scales (7.5), the nondimensional form of (7.12a) becomes

$$H = 1 + \epsilon\delta\eta - \gamma, \tag{7.12b}$$

where since $\epsilon\delta = fUL/gH_0$ and γ is defined as

$$\gamma = \frac{h_b}{H_0} \tag{7.13}$$

Upon substitution of (7.12b), Eq. (7.7) becomes

$$\delta\epsilon\frac{\partial\eta}{\partial t} + H\nabla\cdot\mathbf{u} + \mathbf{u}\cdot\nabla H = \delta\epsilon\frac{\partial\eta}{\partial t} + (1+\epsilon\delta\eta-\gamma)\nabla\cdot\mathbf{u} + \mathbf{u}\cdot\nabla(\epsilon\delta\eta-\gamma) \tag{7.14}$$

A formal perturbation expansion of the unknown variables in the small parameter ϵ proceeds as

$$\mathbf{u}(\mathbf{x},t,\epsilon) = \mathbf{u}^{(0)}(\mathbf{x},t) + \epsilon(\mathbf{u}^{(1)}(\mathbf{x},t) + \epsilon^2\mathbf{u}^{(2)}(\mathbf{x},t) + \cdots \tag{7.15a}$$

$$(\mathbf{x},t,\epsilon) = \eta^{(0)}(\mathbf{x},t) + \epsilon\eta^{(1)}(\mathbf{x},t) + \epsilon^2\eta^{(2)}(\mathbf{x},t) + \cdots \tag{7.15b}$$

and substituting the above expressions in Eqs. (7.6) and (7.14), and collecting terms with respect to the powers of ϵ yields:

$$\left[\mathbf{k}\times\mathbf{u}^{(0)} + \nabla\eta^{(0)}\right] + \epsilon\left[\frac{\partial\mathbf{u}^{(0)}}{\partial t} + \mathbf{u}^{(0)}\cdot\nabla\mathbf{u}^{(0)} + \beta y\mathbf{k}\times\mathbf{u}^{(0)} + \mathbf{k}\times\mathbf{u}^{(1)} + \nabla\eta^{(1)}\right] + \epsilon^2[\ldots] + \cdots = 0 \tag{7.16a}$$

$$\left[(1-\gamma)\nabla\cdot\mathbf{u}^{(0)} - \mathbf{u}^{(0)}\cdot\nabla\zeta\right] + \epsilon\left[\delta\left(\frac{\partial\eta^{(0)}}{\partial t} + \eta^{(0)}\nabla\cdot\mathbf{u}^{(0)} + \mathbf{u}^{(0)}\cdot\nabla\eta^{(0)}\right) + \nabla\cdot\mathbf{u}^{(1)}\right] + \epsilon^2[\ldots] + \cdots = 0 \tag{7.16b}$$

The coefficients for each term in powers of ϵ must vanish, since the equations must be valid for arbitrary values of ϵ and for all x, y, t. Therefore, we get separate equations setting each of the concerned brackets in Eqs. (7.16a, 7.16b) to zero. For example, the first order equations, for terms of $O(\epsilon^0) = O(1)$ are:

$O(\epsilon^0)$:

$$\mathbf{k}\times\mathbf{u}^{(0)} = -\nabla\eta^{(0)} \tag{7.17a}$$

$$(1-\gamma)\nabla\cdot\mathbf{u}^{(0)} - \mathbf{u}^{(0)}\cdot\nabla\gamma = 0 \tag{7.17b}$$

We are mainly interested in the first order flow, since ϵ is small. However, if we only keep the $O(\epsilon^0)$ terms, Eqs. (7.17a) and (7.17b) would be the same as the geostrophic Eqs. (5.21 and 5.22), if we were to drop the bottom topography term γ. In fact, using vector identities and taking the curl of (7.17a) yields

$$\nabla\times\mathbf{k}\times\mathbf{u}^{(0)} = -\nabla\times\nabla\eta^{(0)} = 0$$

or

$$\nabla\cdot\mathbf{u}^{(0)} = 0. \tag{7.18}$$

Substituting into (7.17b) gives

$$\mathbf{u}^{(0)}\cdot\nabla\gamma = 0. \tag{7.19a}$$

In dimensional terms, this is

$$\mathbf{u}^{(0)} \cdot \nabla h_b = 0 \tag{7.19b}$$

i.e. the first order (pure geostrophic) motion must follow the isobaths. Since H_0 is constant, (7.19b) can also be written in the same form as Eq. (6.139):

$$\mathbf{u}^{(0)} \cdot \nabla(H_0 - h_b) = \mathbf{u}^{(0)} \cdot \nabla h = -u^{(0)} \cdot \nabla h_b = 0. \tag{7.19c}$$

The requirement for the flow to follow the isobaths is a very strong constraint, and would fail when $H = H_0 =$ constant. At this point, we have another choice: perhaps we should limit the effects of bottom topography, by requiring that the bottom variations are of the same order as the Rossby number, i.e.

$$\gamma = \frac{h_b(x, y)}{H_0} = \epsilon \eta_b(x, y) = O(\epsilon) \ll 1 \tag{7.20a}$$

such that

$$h_b = O(\epsilon H_0) \ll H_0. \tag{7.20b}$$

With this choice of orders, it is assumed that the bottom topography variations (deviations from the mean depth) are much smaller than the mean depth.

With approximation (7.20) imposed, the terms proportional to γ in (7.16b) are carried over from order one terms to order ϵ terms, and the first order equations are reduced to

$O(\epsilon^0)$:

$$\mathbf{k} \times \mathbf{u}^{(0)} = -\nabla \eta^{(0)} \tag{7.21a}$$

$$\nabla \cdot \mathbf{u}^{(0)} = 0 \tag{7.21b}$$

i.e. the geostrophic equations. As we have seen earlier, the geostrophic equations are degenerate (linearly dependent), since manipulation of (7.21a) directly yields (7.18), making one of the two equations redundant. It is evident that the description of the first order flow will only be possible by obtaining corrections from the second order. Substituting (7.20) in (7.16a, 7.16b) yields the second order equations

O(ϵ^1):

$$\mathbf{k} \times \mathbf{u}^{(1)} + \nabla \eta^{(1)} + \frac{\partial \mathbf{u}^{(0)}}{\partial t} + \mathbf{u}^{(0)} \cdot \nabla \mathbf{u}^{(0)} + \beta y \mathbf{k} \times \mathbf{u}^{(0)} = 0 \tag{7.22a}$$

$$\nabla \cdot \mathbf{u}^{(1)} + \delta \frac{\partial \eta^{(0)}}{\partial t} + (\delta \eta^{(0)} - \eta_b) \nabla \cdot \mathbf{u}^{(0)} + \mathbf{u}^{(0)} \nabla (\delta \eta^{(0)} - \eta_b) = 0 \tag{7.22b}$$

Note that the third term of (7.22b) vanishes by virtue of (7.21b). We then take the curl of (7.22a), first using vector identities to write

$$\begin{aligned}\mathbf{u}^{(0)} \cdot \nabla \mathbf{u}^{(0)} &= \frac{1}{2}\nabla(\mathbf{u}^{(0)} \cdot \mathbf{u}^{(0)}) + (\nabla \times \mathbf{u}^{(0)}) \times \mathbf{u}^{(0)} \\ &= \frac{1}{2}\nabla(\mathbf{u}^{(0)} \cdot \mathbf{u}^{(0)}) + \zeta^{(0)}\mathbf{k} \times \mathbf{u}^{(0)}\end{aligned}$$

and

$$\begin{aligned}\nabla \times (\mathbf{u}^{(0)} \cdot \nabla \mathbf{u}^{(0)}) &= \frac{1}{2}\nabla \times \nabla(\mathbf{u}^{(0)} \cdot \mathbf{u}^{(0)}) + \nabla \times (\zeta^{(0)}\mathbf{k} \times \mathbf{u}^{(0)}) \\ &= \mathbf{u}^{(0)} \cdot \nabla\zeta^{(0)}\mathbf{k} + \zeta^{(0)}\mathbf{k}\nabla \cdot \mathbf{u}^{(0)} \\ &= \mathbf{k}\mathbf{u}^{(0)} \cdot \nabla\zeta^{(0)}\end{aligned}$$

by virtue of (7.21b). Note that some terms have disappeared by vector identities, and the definition of vorticity (6.104) have been used in the above, where

$$\omega^{(0)} = \zeta^{(0)}\mathbf{k} = \nabla \times \mathbf{u}^{(0)}.$$

Similarly,

$$\begin{aligned}\nabla \times \beta y\mathbf{k} \times \mathbf{u}^{(0)} &= \mathbf{u}^{(0)} \cdot \nabla\beta y\mathbf{k} + \beta y\mathbf{k}\nabla \cdot \mathbf{u}^{(0)} \\ &= \mathbf{k}\mathbf{u}^{(0)} \cdot \nabla\beta y \\ &= \mathbf{k}\beta v^{(0)}\end{aligned}$$

and

$$\nabla \times \mathbf{k} \times \mathbf{u}^{(1)} = \mathbf{k}\nabla \cdot \mathbf{u}^{(1)}.$$

With the above substitution, the curl of Eq. (7.22b) becomes

$$\nabla \cdot \mathbf{u}^{(1)} + \frac{\partial\zeta^{(0)}}{\partial t} + \mathbf{u}^{(0)} \cdot \nabla\zeta^{(0)} + \beta v^{(0)} = 0 \tag{7.23}$$

and eliminating $\nabla u^{(1)}$ between (7.23) and (7.22b) then yields:

$$\frac{\partial\zeta^{(0)}}{\partial t} + \mathbf{u}^{(0)} \cdot \nabla\zeta^{(0)} + \beta v^{(0)} - \delta\frac{\partial\eta^{(0)}}{\partial t} - \mathbf{u}^{(0)} \cdot \nabla(\delta\eta^{(0)} - \eta_b) = 0, \tag{7.24}$$

or,

$$\frac{D}{Dt}\left(\zeta^{(0)} + \eta_b - \delta\eta^{(0)}\right) + \beta v^{(0)} = 0 \tag{7.25}$$

or using $\beta v^{(0)} = \mathbf{u}^{(0)} \cdot \nabla\beta y$,

$$\frac{D}{Dt}\left(\zeta^{(0)} + \eta_b - \delta\eta^{(0)} + \beta y\right) = 0. \tag{7.26}$$

We essentially have derived the approximate form of the potential vorticity equation, i.e. the *quasi-geostrophic vorticity equation*.

In fact, the same equation can be obtained through the exact form of the nondimensional form of the potential vorticity Eq. (6.127)

$$\frac{D}{Dt}\left(\frac{\epsilon\zeta + 1 + \epsilon\beta y}{H}\right) = 0 \tag{7.27}$$

To see this, we first expand

$$\begin{aligned} \zeta = \mathbf{k}\cdot\times\mathbf{u} &= \mathbf{k}\cdot\times(\mathbf{u}^{(0)} + \epsilon\mathbf{u}^{(1)} + \cdots) \\ &= \zeta^{(0)} + \epsilon\zeta^{(1)} + \cdots \end{aligned} \tag{7.28}$$

and using (7.12b) and (7.20a), write

$$\frac{1}{H} = \frac{1}{1 + \epsilon(\delta\eta - \eta_b)} = \frac{1}{1 + \epsilon(\delta\eta^{(0)} - \eta_b) + \epsilon^2\delta\eta^{(1)} + \cdots} \tag{7.29}$$

Since $\epsilon \ll 1$, (7.29) can be approximated as

$$\frac{1}{H} = 1 - \epsilon(\delta\eta^{(0)} - \eta_b) + \epsilon^2(\ldots) + \cdots \tag{7.30}$$

So that (7.27) can equivalently be approximated as

$$\frac{D}{Dt}\{1 + \epsilon[\beta y + \zeta^{(0)} + \delta\eta^{(0)} - \eta_b] + \epsilon^2[\ldots] + \cdots\} = 0 \tag{7.31}$$

which then, to first order yields (7.26).

The finishing touch to Eq. (7.25) or (7.26) comes from the definition or vorticity. Reorganizing (7.21a) and taking curl:

$$\mathbf{u}^{(0)} = \mathbf{k}\times\nabla\eta^{(0)} \tag{7.32}$$

that one obtains:

$$\begin{aligned} \zeta^{(0)} = \mathbf{k}\cdot\nabla\times\mathbf{u}^{(0)} &= \mathbf{k}\cdot\nabla\times\mathbf{k}\times\nabla\eta^{(0)} \\ &= \nabla\cdot\nabla\eta^{(0)} = \nabla^2\eta^{(0)} \end{aligned} \tag{7.33}$$

Alternatively, a stream function

$$\psi = \eta^{(0)} \tag{7.34}$$

can be defined since (7.32) is in the form of the definition of this stream function which readily satisfies continuity Eq. (7.21). Therefore, substituting (7.34) and

$$\zeta^{(0)} = \nabla^2\psi \tag{7.35}$$

in (7.26) gives

$$\frac{D}{Dt}\left(\nabla^2\psi - \delta\psi + \beta y + \eta_b\right) = 0 \tag{7.36a}$$

We also note:

$$\begin{aligned}\frac{D}{Dt} &= \frac{\partial}{\partial t} + \mathbf{u}^{(0)} \cdot \nabla \\ &= \frac{\partial}{\partial t} + (\mathbf{k} \times \nabla\psi) \cdot \nabla\end{aligned}$$

so that (7.36) can be written as

$$\left[\frac{\partial}{\partial t} + \frac{\partial\psi}{\partial x}\frac{\partial}{\partial y} - \frac{\partial\psi}{\partial y}\frac{\partial}{\partial x}\right]\left[\nabla^2\psi - \delta\psi + \beta y + \eta_b\right] = 0 \tag{7.36b}$$

or alternatively

$$\frac{\partial}{\partial t}[\nabla^2\psi - \delta\psi] + J[\psi, (\nabla^2\psi - \delta\psi + \beta y + \eta_b)] = 0 \tag{7.36c}$$

and equivalently as

$$\frac{\partial}{\partial t}\left[\nabla^2\psi - \delta\psi\right] + \beta\frac{\partial\psi}{\partial x} + J[\psi, (\nabla^2\psi - \delta\psi + \eta_b)] = 0 \tag{7.36d}$$

Note that when $\delta = 0$ (rigid-lid) and $\eta_b = 0$ (constant depth) the above equation reduces to (6.166), derived earlier in Sect. 6.5.3.

There are several points to note in the quasi-geostrophic theory:

(*i*) While the flow is in geostrophic balance by virtue of Eqs. (7.21), the actual dynamics is governed by a simple Eq. (7.36) in the stream function $\psi = \eta^{(0)}$ which equivalently represents the dynamic pressure. This equation essentially incorporates the corrections to geostrophy by the surface elevation, β-effect, small bathymetric influences, and nonlinearity.

(*ii*) Once the stream function (or pressure) is determined from (7.36) then the velocity field can be obtained geostrophically from (7.32).

(*iii*) We have assumed earlier that the depth variations, are small as compared to the total depth, $(h_b/H_0) = \epsilon\eta_b = O(\epsilon)$, where $\eta_b = O(1)$. If we relax this assumption by insisting that $(h_b/H_0) = O(1)$, then $\eta_b = O(1/\epsilon)$ becomes large as compared to the other terms in (7.36a) so that

$$\frac{D\eta_b}{Dt} = \mathbf{u}^{(0)} \cdot \nabla\eta_b = 0 \tag{7.37}$$

which is the same as the constraint (7.19) obtained without making the assumption of small depth variations. Therefore, that possibility is readily contained in the present theory.

(*iv*) We have noted that Eq. (7.26) is a statement for conservation of *quasi-geostrophic potential vorticity* defined as

$$\Pi_g = \zeta^{(0)} + \eta_b - \delta\eta^{(0)} + \beta y \tag{7.38a}$$

in the one dimensional variables. In dimensional variables this is equivalent to the following:

$$\Pi_g = \zeta^{(0)} + \left(\frac{f_0}{H_0}\right)(h_b - \eta^{(0)}) + \beta y. \tag{7.38b}$$

The above definitions of quasi-geostrophic potential vorticity differs from the exact definition by a fixed constant which is of no relevance, and is the first order approximate form of the latter. Note that relative vorticity $\zeta^{(0)}$ and planetary vorticity βy contribute to the *quasi-geostrophic potential vorticity* as well as the bottom topography and surface elevation. It is also worth mentioning that a positive bottom topography makes a positive contribution, whereas a positive surface elevation makes a negative contribution to the quasi-geostrophic potential vorticity. The first and the third terms are related to the flow and the sum is therefore total fluid vorticity. The second and third terms are independent of the flow and are therefore called *ambient potential vorticity*.

Exercises

Exercise 1

Consider plane-wave solutions

$$\psi = Ae^{i(\omega t - kx - ly)}$$

to the quasi-geostrophic vorticity equation

$$\frac{\partial}{\partial t}\left[\nabla^2\psi - \delta\psi\right] + \beta\frac{\partial\psi}{\partial x} + J[\psi, (\nabla^2\psi - \delta\psi + \eta_b)] = 0.$$

Could this nonlinear equation have a plane-wave solution as proposed?

What would be the dispersion relationship $\omega(k, l)$ in this case? What is the difference from Rossby waves reviewed in Chap. 6?

Would these plane waves deserve to be called divergent, combined planetary-topographic Rossby waves? What would be the roles of the divergence and topographic terms? How would the phase and group velocities in the x and y directions and therefore the orientation of the phase and energy propagation directions differ from the regular Rossby waves of Chap. 6?

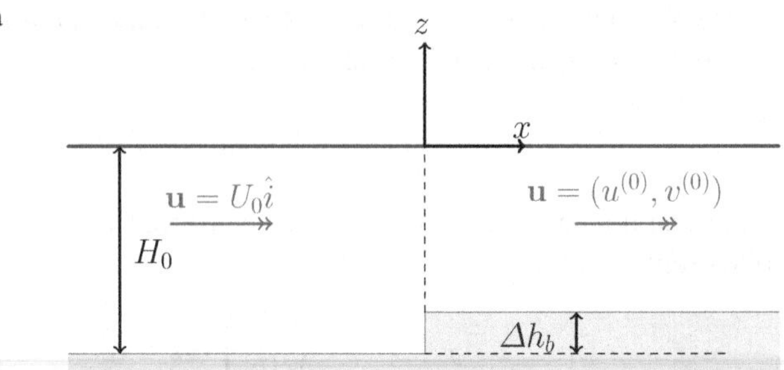

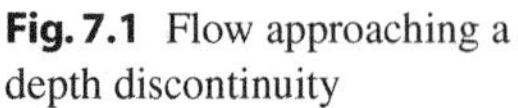
Fig. 7.1 Flow approaching a depth discontinuity

Exercise 2

An abrupt change of depth specified as

$$H(x) = \begin{cases} H_0 & \text{in region } x < 0 \\ H_0 - \Delta h_b & \text{in region } x > 0 \end{cases}$$

occurs along the path of a uniform current U_0 along the x-axis, as shown in Fig. 7.1.

Re-consider the β-plane problem of a uniform flow encountering a depth discontinuity studied in Sect. 6.5.3, and obtain an alternative solution using the quasi-geostrophic theory.

Sketch this quasi-geostrophic solution and compare with the nonlinear solution obtained in Sect. 6.5.3.

Elements of Ocean General Circulation 8

8.1 The Ocean Circulation at Global Scale

In this chapter we seek to develop the basic theory of ocean general circulation, i.e. persistent current patterns in world's oceans at global scales.

Considering a deep ocean basin that is governed by forcing at large-scale, we stipulate that the behavior of this immense body of water obeys a balance of the basic forces of fluid dynamics reviewed in the last few chapters.

Foresight based on the experience of human civilization navigating the oceans for ages tells us that there are certain patterns of ocean currents that seem to be permanent. For instance, the Gulf Stream and Kuroshio have been a part of collective experience for ages. As part of our basic knowledge, steady currents are often implied in most ocean atlases that have been constructed in the last centuries. Our collective memory based on ages of navigating and observing the oceans also would indicate certain patterns of currents established somehow in relation to the persistent wind patterns at large scales. Columbus, crossing the Atlantic ocean, as well as ancient seamen must have recognized these basic elements of nature as experienced observers.

By experience gained from typical maps of ocean currents, we then realize that various smaller parts of the ocean have quite different characteristics from the rest. These confined regions of special current or sea water characteristics often can be characterized as *boundary layers*. The coastal and continental shelf regions of the *coastal boundary layer* display complex patterns compared to the smoother and more persistent currents evident in the deep ocean regions. The Gulf Stream and the Kuroshio are typical examples of *western boundary currents* constituting the oceanic *western boundary layer*. *Upwelling boundary layers* such as near the coasts of West Africa or California are other examples of these confined regions of greater complexity.

© Springer Nature Switzerland AG 2020

E. Özsoy, *Geophysical Fluid Dynamics I*, Springer Textbooks in Earth Sciences, Geography and Environment, https://doi.org/10.1007/978-3-030-16973-2_8

Furthermore, experimental results also suggest *surface and bottom boundary layers*. Near the surface, water properties and currents are less uniform compared to the deep ocean, suggesting that the surface interaction with the atmosphere must be responsible for this confined region near the surface. In fact, we have seen in Chaps. 4 and 5 examples of how momentum flux is transferred between fluid elements by viscous effects, i.e. by frictional forces in the surface and bottom Ekman layers studied in Sect. 4.4. We have also found that the Ekman boundary layer thickness typically to be very small in comparison with the great depth of the ocean.

The basic theory used to construct the oceanic general circulation at large has to ignore the complex coastal boundary layers of incompatible horizontal scales, but has to consider the surface and bottom boundary layers responsible for transferring momentum fluxes at the surface and bottom. The vertical boundary layers at the top and bottom surfaces of the ocean are essential for the wind-driven, frictional ocean circulation. It will also be shown that the dynamics at global/regional scales necessitates the consideration of variation in the Coriolis parameter (the β effect), which essentially will bring into focus rather special types of boundary layers of somewhat larger scale, especially at the western boundaries.

The basic theory of ocean circulation has been developed mainly in the second half of twentieth century to expose the scientific basis of the observed patterns. Numerous contributions by leading ocean scientists to the solution of this true geophysical fluid dynamics problem have progressively revealed and explained the simple-looking but as yet unexplained features of the circulation.

Because of the fundamental nature of the problem, initial theoretical developments had to consider simple geometry and an idealized ocean at its simplest form. Only later developments had to take on increased complexity of the ocean and atmosphere, including the coupling between them, approaching the current age of numerical models using ever increasing levels of computational power.

We will not follow all the above developments here. Instead, will rest content by introducing a simple version of the most essential elements.

8.2 Elements of the Steady, Rigid Lid Ocean Circulation (f-Plane)

We denote the top and bottom surfaces respectively as S^t and S^b, assuming the ocean sandwiched between these two flat and rigid surfaces. Evidently, at this scale, a free surface is no longer relevant, allowing for the *rigid-lid* assumption.

8.2.1 Governing Equations in the f-Plane

We first return to the three dimensional equations of motion on the tangent plane (Sect. 5.1),

$$\frac{D\mathbf{q}}{Dt} + f\mathbf{k} \times \mathbf{q} = -\frac{1}{\rho}\nabla p + \nu\nabla^2\mathbf{q} \qquad (8.1a)[3.56]$$

and

$$\nabla \cdot \mathbf{q} = 0, \tag{8.1b)[3.1.b]}$$

where $\mathbf{q} = (u, v, w)$ denotes three-dimensional velocity, $\nabla = \frac{\partial}{\partial x_i}\mathbf{e}_i$ three-dimensional vector gradient operator and $f = f_0 + \beta_0 y'$ Coriolis parameter in the β-plane.

We will be using the equations in non-dimensional form, to make the analysis simpler and more tractable. Using the scales

$$\mathbf{q} \sim U_0, \quad f \sim \frac{f_0}{2} = \Omega \sin\phi_0, \quad p \sim \rho\frac{f_0 U_0 L_0}{2}, \quad (x, y) \sim L_0, \quad z \sim H_0, \quad t \sim \frac{f_0}{2} \tag{8.2a}$$

and the dimensionless numbers

$$R = \frac{2U_0}{\Omega_0 L_0}, \quad E = \frac{2\nu}{f_0 L_0}, \quad \bar{f} = \frac{f}{f_0}, \quad \bar{\beta} = \frac{\beta_0 L_0}{f_0}, \tag{8.2b}$$

where the first two respectively are the Rossby and Ekman numbers, the equations of motion can be made dimensionless. Here, we note that

$$\bar{f} = \frac{f}{f_0} = \frac{f_0 + \beta_0 y'}{f_0} = 1 + \frac{\beta_0 L_0 y}{f_0} = 1 + \bar{\beta} y, \tag{8.3a}$$

denotes the non-dimensional form of the Coriolis factor, where the coordinate y has been made non-dimensional. We leave out a coefficient of 2 for the Coriolis term to provide simpler arithmetic later.

Because we use two different scales L_0 and H_0 respectively for the horizontal and vertical coordinates the non-dimensional version of the three-dimensional gradient operator is

$$\nabla = \mathbf{i}\frac{\partial}{\partial x'/L_0} + \mathbf{j}\frac{\partial}{\partial y'/L_0} + \mathbf{k}\frac{\partial}{\partial z'/H_0} = \nabla_h + \mathbf{k}\frac{\partial}{\partial z} \tag{8.3b}$$

where the primed are dimensional coordinates and ∇_h and $\frac{\partial}{\partial z}$ respectively are the non-dimensional horizontal and vertical gradient operators.

With the above considerations, non-dimensional equations of motion become

$$\begin{gathered} \frac{\partial \mathbf{q}}{\partial t} + R(\mathbf{q}\cdot\nabla)\mathbf{q} + 2\bar{f}\mathbf{k}\times\mathbf{q} = -\nabla p + E\nabla^2\mathbf{q}, \\ \nabla\cdot\mathbf{q} = 0. \end{gathered} \tag{8.4a, b}$$

The third term with $\bar{f} = 1 + \bar{\beta} y$ is in fact important for the general circulation of the atmosphere and the ocean as we have seen in sections of the last two chapters. However, for simplicity in the initial part of the following sections, we will temporarily ignore this effect by setting $\bar{\beta} = 0$, or $\bar{f} = 1$ in the f-plane approximation. Then, in later sections we will recover the latitudinal variation of the Coriolis parameter (the β effect), and develop the corresponding solutions in the β-plane approximation.

We will set the unsteady term $\frac{\partial \mathbf{q}}{\partial t} = 0$ at the outset because we are interested in the steady circulation at large scale where only very slow variations are allowed in time.

Secondly, because these equations are non-linear, it is worthwhile to see if they could be linearized to make the analysis relatively simpler and more comprehensible. With the selected scales that apply to the general circulation, we expect that the Rossby number to be small, $R \ll 1$, so that it is also justified to drop the nonlinear term in the momentum equation.

With the above approximations, the flow satisfies

$$2\mathbf{k} \times \mathbf{q} = -\nabla p + E\nabla^2 \mathbf{q},$$
$$\nabla \cdot \mathbf{q} = 0. \tag{8.5a, b}$$

8.2.2 The Ocean Interior (f-Plane)

With approximations already made in the above section, we continue to investigate what happens if we also neglect the viscous friction terms by setting $E = 0$. It is immediately seen that this would actually result in a return to the geostrophic flow approximation reviewed in Chap. 5.

Since the steadily flowing large-scale ocean currents, away from frictional effects of coastal boundaries have evolved over great time span, geostrophic flow could be a good approximation for the ocean interior, which we will begin to denote with subscript i

$$2\mathbf{k} \times \mathbf{q}_i = -\nabla p_i,$$
$$\nabla \cdot \mathbf{q}_i = 0. \tag{8.6a, b}$$

On the other hand, in Sect. 5.2.3, we have shown that the geostrophic approximation fails to provide a solution by itself, as it is proven to be degenerate, the above two equations shown not being independent of each other: Cross multiplying $\mathbf{k}\times$ (8.6a), substituting into (8.6b) and making use of (1.27c) it may be shown that the latter equation is readily satisfied, making it impossible to obtain a solution for the two variables $\mathbf{q}_i$ and p_i.

We can then ask ourselves how we could justify the use of geostrophic equations for the interior, if they are degenerate? We will only be able to use this approximation by complementing the interior equations with those of the frictional effects at boundary layers, as shown in the following. For simplicity of demonstration we will continue with the simplest possible geometry of an open ocean domain with a flat bottom and rigid-lid at the surface as shown in Fig. 8.1.

A further characteristic of geostrophic flow that would be applicable to the ocean interior at this level of approximation is described by the *Taylor–Proudman theorem*, which imposes strong constraints. As shown in Sect. 5.2.3, both the velocity and pressure fields have to be independent of depth, in consistency with geostrophic

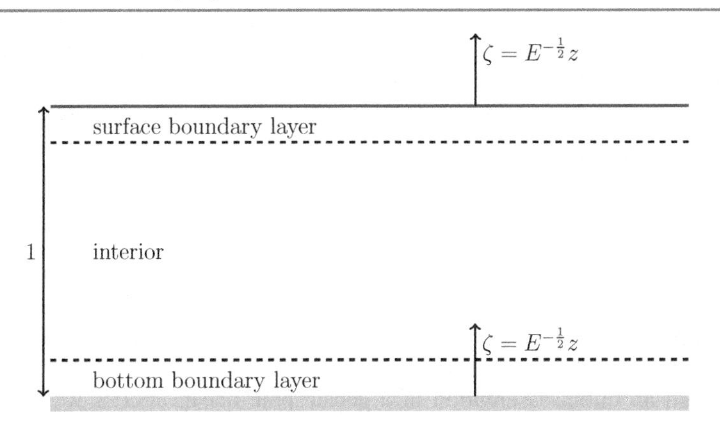

Fig. 8.1 Open ocean domain with a geostrophic interior and frictional surface and bottom boundary layers. The dotted lines suggest the separation between these solution regions

balance. This can be shown by taking $\nabla\times$ of (8.6a)

$$2\nabla \times \mathbf{k} \times \mathbf{q}_i \equiv -2(\mathbf{k}\cdot\nabla)\mathbf{q}_i = -\nabla \times \nabla p_i \equiv 0 \tag{8.7}$$

by making use of identities (1.27d) and (1.27i), which then yields

$$\frac{\partial \mathbf{q}_i}{\partial z} = 0. \tag{8.8}$$

But $\mathbf{q}_i = \mathbf{u}_i + w\mathbf{k}$, and therefore the horizontal and vertical components of the vector equation (8.8) separately imply

$$\frac{\partial \mathbf{u}_i}{\partial z} = 0, \quad \text{and} \quad \frac{\partial w_i}{\partial z} = 0, \tag{8.9a, b}$$

in other words, the horizontal and vertical velocity components $\mathbf{u}_i$ and w_i can only have two dimensional variations of (x, y); they are independent of depth. But then (8.6a) implies the same for pressure p_i, making it also independent of depth.

As a result, we can re-write (8.6a, b) as

$$\begin{aligned} \mathbf{u}_i &= \frac{1}{2}\mathbf{k} \times \nabla_h p_i, \\ \nabla_h\cdot\mathbf{u}_i &= 0. \end{aligned} \tag{8.10a, b}$$

where all variables are two-dimensional as well as the horizontal gradient operator ∇_h which applies to coordinates (x, y).

The reduction of original three-dimensional equations to strict two-dimensional form essentially contradicts with the momentum flux and no-slip boundary conditions to be satisfied respectively at the top and bottom boundaries. This is clear because a finite value is assigned to the horizontal velocity throughout the water column including the top to bottom boundaries S^t and S^b due to (8.6a), which contradicts these boundary conditions.

Further, equation (8.9b) says that the vertical velocity should be constant throughout depth and also at the top and bottom boundaries

$$w_i = w_i|_{S^t} = w_i|_{S^b} = \text{ constant}; \tag{8.11}$$

if some vertical velocity is introduced into the system for any reason at one of these boundaries (e.g. by bottom topography, coastal or surface flux effects), the flux have to be compensated by the other boundary. If both S^t and S^b respectively are rigid horizontal surfaces then the constant is zero.

Although an initial assumption of a geostrophic ocean interior was made and seemed acceptable, a solution does not seem possible without taking external factors and boundaries into consideration.

What this means is that the interior geostrophic flow cannot exist without having to adjust to momentum flux or no-slip boundary conditions at the vertical boundaries through viscous friction. These will be Ekman layers balancing Coriolis versus viscous forces as reviewed in Sect. 4.4, where the vertical extent of the Ekman boundary layers were estimated to be very thin compared to the ocean depth.

8.2.3 Ekman Boundary Layers (f-Plane)

We have remarked at the end of Sect. 5.2.3 that the removal of geostrophic degeneracy is possible by inclusion of some additional process such as due to frictional, nonlinear or unsteady terms neglected in the geostrophic equations. We are therefore forced to include frictional effects in the Ekman boundary layers to resolve the degeneracy of the interior equations.

We therefore resolve to separate the required solution into three parts

$$\begin{aligned} \mathbf{q} &= \mathbf{q}_i + \mathbf{q}_e^t + \mathbf{q}_e^b \\ p &= p_i + p_e^t + p_e^b \end{aligned} \tag{8.12a, b}$$

where the interior solution $\mathbf{q}_i$ is the geostrophic one described by (8.6a), (8.6b) and the top and bottom Ekman boundary layer solutions $\mathbf{q}_e^t$ and $\mathbf{q}_e^b$ respectively are the corrections that will be applied to it very near the top and bottom surfaces to match the boundary conditions. Because the top and bottom boundaries are far from each other and the Ekman boundary layer solutions decay fast as they join the interior solution, dropping the superscripts t and b we can write

$$\begin{aligned} \mathbf{q} &= \mathbf{q}_i + \mathbf{q}_e \\ p &= p_i + p_e \end{aligned} \tag{8.13a, b}$$

near each of the vertical boundaries, where the subscript e denotes the Ekman layer, irrespective of its being a top or bottom boundary.

The total solution variables (8.12) with the above approximations are required to satisfy (8.6a, b) with the interior terms replaced with their two dimensional forms as in (8.10a, b)

$$\begin{aligned} 2\mathbf{k} \times (\mathbf{u}_i + \mathbf{q}_e) &= -\nabla_h p_i - \nabla p_e + E\nabla_h^2 \mathbf{u}_i + E\nabla \mathbf{q}_e, \\ \nabla_h \cdot \mathbf{u}_i + \nabla \cdot \mathbf{q}_e &= 0. \end{aligned} \tag{8.14a, b}$$

and subtracting equations (8.10a, b) for the interior solution from the (8.14a, b) and making note of (8.12a, b), we have

$$\begin{aligned} 2\mathbf{k} \times \mathbf{q}_e &= -\nabla p_e + E\nabla_h^2 \mathbf{u}_i + E\nabla^2 \mathbf{q}_e, \\ \nabla \cdot \mathbf{q}_e &= 0. \end{aligned} \tag{8.15a, b}$$

We have seen in Chap. 4 that the Ekman boundary layer solutions (4.66a), (4.66b) and (4.74a), (4.74b) decay typically within a distance of $\sqrt{2\nu/f_0}$, which is the boundary layer thickness in dimensional coordinates. Considering the scalings of (8.2a) and definition in (8.2b), the corresponding thickness in dimensionless units would be $E^{\frac{1}{2}}$, a small number such that $E \ll E^{\frac{1}{2}} \ll 1$.

In review of the Ekman layer solutions of Sect. 4.4, we can therefore delegate an expanded boundary layer coordinate $\zeta = E^{-1/2}z$, which will enlarge the boundary layer domain. Transforming the vertical coordinate by substituting

$$z = E^{\frac{1}{2}}\zeta \tag{8.16}$$

will produce from (8.15a, b)

$$\begin{aligned} 2\mathbf{k} \times \mathbf{q}_e &= -\nabla_h p_e - \mathbf{k}E^{-1/2}\frac{\partial p_e}{\partial \zeta} + \frac{\partial^2 \mathbf{q}_e}{\partial \zeta^2} + E\nabla_h^2 \mathbf{q}_e + E\nabla_h^2 \mathbf{u}_i, \\ \frac{\partial w_e}{\partial \zeta} &= -E^{-1/2}\nabla_h \cdot \mathbf{u}_e = 0. \end{aligned} \tag{8.17a, b}$$

With $\mathbf{q}_e$ the three-dimensional and $\mathbf{u}_e = (u, v)$ the horizontal velocity component $\mathbf{q}_e = \mathbf{u}_e + w_e\mathbf{k}$, and ∇_h the two dimensional gradient, the vertical and horizontal terms are rearranged in (8.17a),

$$\begin{aligned} 2\mathbf{k} \times \mathbf{q}_e &= -\nabla_h p_e + \frac{\partial^2 \mathbf{q}_e}{\partial \zeta^2} + O(E) \\ 0 &= -E^{-1/2}\frac{\partial p_e}{\partial \zeta} + \frac{\partial^2 w_e}{\partial \zeta^2} + O(E) \end{aligned} \tag{8.18a, b}$$

where smaller terms of $O(E)$ are indicated on the right hand side.

Equation (8.17b) suggests that $w_e = O(E^{\frac{1}{2}})$ only, and comparing this with equation (8.18b), it is evident that the vertical gradient of Ekman layer pressure is small, of $O(E)$ at most.

Since there is no vertical gradient of the interior pressure p_i, and the gradient of the Ekman layer pressure being small, of $O(E)$, verifies that the pressure p_i in the interior is *impressed* on the boundary layer. Essentially $p_e \rightarrow 0$ outside the the Ekman layer and one can conclude that the correction is small, $p_e = 0(E)$ in the boundary layer. Consequently the horizontal pressure gradient can altogether be neglected in (8.18a). Dropping all $O(E)$ terms in (8.18a, b) and noting that $\mathbf{k} \times \mathbf{q}_e = \mathbf{k} \times \mathbf{u}_e$, the boundary later equations become:

$$2\mathbf{k} \times \mathbf{u}_e = \frac{\partial^2 \mathbf{u}_e}{\partial \zeta^2},$$
$$\frac{\partial w_e}{\partial \zeta} = -E^{-1/2} \nabla_h \cdot \mathbf{u}_e. \tag{8.19a, b}$$

8.2.4 Surface Ekman Layer (f-Plane)–Wind-Driven Flow

Since the upper ocean is mainly driven by the wind stress $\boldsymbol{\tau}^s$, shear at the surface must satisfy the dynamic boundary condition

$$\mu \frac{\partial \mathbf{u}_e}{\partial z} = \boldsymbol{\tau}^s \text{ at } z = 0 \tag{8.20}$$

in dimensional form, where where z is measured upwards from the surface and $\boldsymbol{\tau}^s = \boldsymbol{\tau}^s(x, y)$ in general. By selecting a wind stress scale

$$\boldsymbol{\tau}^s \sim \rho U_0 L_0 f_0 / 2, \tag{8.21}$$

the dimensionless form is

$$E \frac{\partial \mathbf{u}_e}{\partial z} = \boldsymbol{\tau}^s \text{ at } z = 0 \tag{8.22}$$

Transforming the vertical coordinate by (8.16) and defining

$$\mathbf{S} = E^{\frac{1}{2}} \boldsymbol{\tau}^s \tag{8.23}$$

we write (8.22) as

$$\frac{\partial \mathbf{u}_e}{\partial \zeta} = \mathbf{S} \text{ at } \zeta = 0 \tag{8.24a}$$

providing the surface boundary condition for (8.19a, b). The other boundary condition is that the Ekman layer velocity correction should vanish in the interior

$$\mathbf{u}_e \rightarrow 0 \text{ as } \zeta \rightarrow -\infty. \tag{8.24b}$$

To solve the system (8.19a) with (8.24a), (8.24b) we first form

$$2\mathbf{k} \times \mathbf{k} \times \mathbf{u}_e = \frac{\partial^2}{\partial \zeta^2}(\mathbf{k} \times \mathbf{u}_e) \tag{8.25}$$

by taking cross product of (8.19a) with $\mathbf{k}$. Then by multiplying (8.19a) with i and adding to (8.25) we obtain

$$\frac{\partial^2 \mathbf{Q}}{\partial \zeta^2} - 2i\mathbf{Q} = 0 \tag{8.26}$$

where we define

$$\mathbf{Q} \equiv \mathbf{k} \times \mathbf{u}_e + i\mathbf{u}_e \tag{8.27}$$

as a new variable.

The boundary conditions are similarly transformed into

$$\frac{\partial \mathbf{Q}}{\partial \zeta} = \mathbf{k} \times \mathbf{S} + i\mathbf{S} \text{ at } \zeta = 0, \tag{8.28a}$$

and

$$\mathbf{Q} \to 0 \text{ as } \zeta \to -\infty. \tag{8.28b}$$

The general solution to (8.27)

$$\mathbf{Q} = \mathbf{A}e^{(1+i)\zeta} + \mathbf{B}e^{-(1+i)\zeta},$$

with constants $\mathbf{A}$ and $\mathbf{B}$ evaluated with boundary conditions (8.28a), (8.28b) yields

$$\mathbf{Q} = \left\{ \frac{\mathbf{k} \times \mathbf{S} + i\mathbf{S}}{1+i} \right\} e^{(1+i)\zeta}, \tag{8.29}$$

so that we can recover $\mathbf{u}_e$ by taking the imaginary part

$$\begin{aligned} \mathbf{u}_e &= \Im\{\mathbf{Q}\} = \Im \left\{ \frac{\mathbf{k} \times \mathbf{S} + i\mathbf{S}}{1+i} e^{(1+i)\zeta} \right\} \\ &= \frac{1}{2}\{(\mathbf{k} \times \mathbf{S} + \mathbf{S}) \sin \zeta - (\mathbf{k} \times \mathbf{S} - \mathbf{S}) \cos \zeta\} e^{\zeta}. \end{aligned} \tag{8.30}$$

The surface value of the Ekman layer velocity is

$$\mathbf{u}_e(0) = \Im \left\{ \frac{\mathbf{k} \times \mathbf{S} + i\mathbf{S}}{1+i} \right\} = \frac{1}{2}\{\mathbf{S} - \mathbf{k} \times \mathbf{S}\}, \tag{8.31}$$

which makes an angle $\pi/4$ clockwise from the direction of the wind-stress vector.

The dimensional solution can be recovered by back substituting the various scales used up to the present, to show that the above solution is identical to that obtained in Sect. 4.4.1.

The net mass flux $\mathbf{M}'$ defined in dimensional form is

$$\mathbf{M}_e^{t'} \equiv \int_{-\infty}^{0} \mathbf{u}_e' dz$$

and its dimensionless form $\mathbf{M}$ can be calculated by integrating (8.31) as

$$\mathbf{M}_e^t = E^{\frac{1}{2}} \int_{-\infty}^{0} \mathbf{u}_e d\zeta = -\frac{E^{\frac{1}{2}}}{2} \mathbf{k} \times \mathbf{S} \tag{8.32}$$

at right angles and to the right of the wind-stress vector.

The vertical velocity can be integrated across the boundary layer by making use of (8.19b)

$$\begin{aligned} w_e(0) - w_e(-\infty) &= -E^{\frac{1}{2}} \nabla_h \cdot \int_{-\infty}^{0} \mathbf{u}_e d\zeta \\ &= -E^{\frac{1}{2}} \nabla_h \cdot \Im \left\{ \frac{\mathbf{k} \times \mathbf{S} + i\mathbf{S}}{(1+i)^2} \right\} \\ &= -\frac{E^{\frac{1}{2}}}{2} \nabla_h \cdot \Im \{ \mathbf{S} - i\mathbf{k} \times \mathbf{S} \} \end{aligned} \tag{8.33}$$

and since $w_e(-\infty) \to 0$, using (8.10a) and vector identity (1.27c) we find

$$w_e(0) = -\frac{E^{\frac{1}{2}}}{2} \mathbf{k} \cdot \nabla_h \times \mathbf{S}. \tag{8.34}$$

Now, to match the zero normal flux boundary condition at the ocean surface (top), in view of (8.13a) we write

$$w(0) = w_e(0) + w_i^t = 0 \text{ on } z = 0 \tag{8.35}$$

which gives

$$w_i^t = -w_e(0) = \frac{E^{\frac{1}{2}}}{2} \mathbf{k} \cdot \nabla_h \times \mathbf{S}. \tag{8.36}$$

The generation of vertical velocity in the interior by combined action of the wind-stress and vertical transfer of momentum by the surface Ekman layer is appropriately called *Ekman pumping*. It is this vertical flux of volume that actually creates vorticity in the interior and sets it into motion.

We also notice with the help of (8.32) and (8.36) that

$$\nabla_h \cdot \mathbf{M}_e^t + w_i^t = 0, \tag{8.37}$$

stands as a continuity equation for the surface Ekman layer, showing the divergence of the Ekman layer horizontal volume flux creating the vertical flux. This is why *Ekman pumping* is sometimes referred to as *Ekman flux divergence*.

8.2.5 Bottom Ekman Layer (f-Plane)–Interior-Driven Flow

For the bottom Ekman layer, the governing equations are the same, (8.19a, b), and the only difference is in the boundary conditions. At the bottom, the total flow velocity should satisfy the no-slip condition:

$$\mathbf{u}(0) = \mathbf{u}_e(0) + \mathbf{u}_i^t = 0 \text{ on } z = 0 \tag{8.38}$$

where z is measured upwards from the bottom; equivalently with the transformation $z = E^{\frac{1}{2}}\zeta$

$$\mathbf{u}_e(0) = -\mathbf{u}_i^t \text{ on } \zeta = 0 \tag{8.39a}$$

and that the Ekman velocity should decay away from the boundary

$$\mathbf{u}_e \to 0 \text{ as } \zeta \to \infty. \tag{8.39b}$$

Equations (8.39a), (8.39b) show that the bottom Ekman layer is driven by the interior flow, balanced by the bottom friction. The solution for $\mathbf{u}_e$ from equations (8.38) and (8.39a), (8.39b) is very similar to that for the surface layer demonstrated in the last section,

$$\mathbf{Q} \equiv \mathbf{k} \times \mathbf{u}_e + i\mathbf{u}_e = -\{\mathbf{k} \times \mathbf{u}_i + i\mathbf{u}_i\}e^{-(1+i)\zeta}, \tag{8.40}$$

so that

$$\begin{aligned} \mathbf{u}_e &= -\Im\{[\mathbf{k} \times \mathbf{u}_i + i\mathbf{u}_i]e^{-(1+i)\zeta}\} \\ &= \{\mathbf{k} \times \mathbf{u}_i \sin\zeta - \mathbf{u}_i \cos\zeta\}e^{-\zeta}. \end{aligned} \tag{8.41}$$

The mass flux is

$$\mathbf{M}_e^b = E^{\frac{1}{2}} \int_0^\infty \mathbf{u}_e d\zeta = \frac{E^{\frac{1}{2}}}{2}[\mathbf{k} \times \mathbf{u}_i - \mathbf{u}_i] \tag{8.42}$$

and the vertical velocity

$$\begin{aligned} w_e(\infty) - w_e(0) &= -E^{\frac{1}{2}}\nabla_h \cdot \int_0^\infty \mathbf{u}_e d\zeta \\ &= E^{\frac{1}{2}}\nabla_h \cdot \Im\left\{\frac{\mathbf{k} \times \mathbf{u}_i + i\mathbf{u}_i}{1+i}\right\} \\ &= -\frac{E^{\frac{1}{2}}}{2}\nabla_h \cdot \{-\mathbf{k} \times \mathbf{u}_i + i\mathbf{u}_i\} \end{aligned} \tag{8.43}$$

and since $w_e(\infty) \to 0$, using (8.10b) and vector identity (1.27c) we find

$$w_e(0) = -\frac{E^{\frac{1}{2}}}{2}\mathbf{k} \cdot \nabla_h \times \mathbf{u}_i. \tag{8.44}$$

Now, to match the zero normal flux boundary condition at the ocean surface (top), in view of (8.13a) we write

$$w(0) = w_e(0) + w_i^b = 0 \text{ on } z = 0 \tag{8.45}$$

which gives

$$w_i^b = -w_e(0) = \frac{E^{\frac{1}{2}}}{2}\mathbf{k} \cdot \nabla_h \times \mathbf{u}_i. \tag{8.46}$$

8.2.6 Ocean Circulation on the f-Plane

We developed in the last two sections the vertical volume fluxes due to the surface and bottom Ekman boundary layers as follows:

$$w_i^t = \frac{E^{\frac{1}{2}}}{2}\mathbf{k} \cdot \nabla_h \times \mathbf{S}. \tag{8.36}$$

$$w_i^b = \frac{E^{\frac{1}{2}}}{2}\mathbf{k} \cdot \nabla_h \times \mathbf{u}_i. \tag{8.46}$$

The surface and bottom fluxes of (8.36) and (8.46) appear different from each other. On the other hand, as we have shown in Sect. 8.2.2, any difference between the top and bottom values of the vertical velocity of the interior flow can not be allowed, by (8.9b) and (8.11),

$$w_i^t = w_i^b, \tag{8.47}$$

and therefore substituting (8.36) and (8.46) in (8.47) gives

$$\zeta \equiv \mathbf{k} \cdot \nabla_h \times \mathbf{u}_i = \mathbf{k} \cdot \nabla_h \times \mathbf{S}. \tag{8.48}$$

The above equation states the direct conversion of the torque applied by the wind stress curl into vorticity of the interior flow. The wind-driven circulation owes its existence to this source of vorticity transferred into the deep ocean by the surface Ekman layer through what we call as *Ekman pumping*.

Finally, by making use of (8.10a) and (1.27d) in (8.48), the vorticity equation for f-plane ocean circulation takes the following form:

$$\nabla_h^2 p_i = 2\mathbf{k} \cdot \nabla_h \times \mathbf{S}. \tag{8.49}$$

Alternatively, defining a stream-function

$$\psi = -p_i/2, \tag{8.50}$$

and inserting the wind-stress $\mathbf{S} = S^x\mathbf{i} + S^y\mathbf{j}$ and its curl

$$\nabla_h \times \mathbf{S} = \begin{vmatrix} \mathbf{i} & \mathbf{j} & \mathbf{k} \\ \frac{\partial}{\partial x} & \frac{\partial}{\partial y} & \frac{\partial}{\partial z} \\ S^x & S^y & 0 \end{vmatrix},$$

the vorticity equation (8.49) becomes

$$\nabla_h^2 \psi = -\mathbf{k} \cdot \nabla_h \times \mathbf{S} = \frac{\partial S^x}{\partial y} - \frac{\partial S^y}{\partial x}. \tag{8.51}$$

The circulation generated in a rectangular ocean basin with a non-dimensional coordinate system centered with respect to boundaries at $x = -\ell/2, \ell/2$ and $y = -1/2, 1/2$ can be calculated with the present model.

For instance, assuming the typical ocean basin to be analogous of the North Atlantic Ocean, the wind regime can be simplified to consist of mid-latitude westerlies and sub-tropical easterlies (trade winds) described by

$$S^x = \sin \pi y, \tag{8.52}$$

while we let $S^y = 0$, yielding a sinusoidal forcing function

$$\nabla_h^2 \psi = \frac{\partial S^x}{\partial y} = \pi \cos \pi y, \tag{8.53}$$

with the boundary conditions stating a constant value, selected to be zero, along all four coasts

$$\begin{aligned} \psi = 0 \text{ on } & -\frac{\ell}{2} < x < \frac{\ell}{2}, \quad y = -\frac{1}{2}, \frac{1}{2}, \\ & \text{and} \\ \psi = 0 \text{ on } & -\frac{1}{2} < y < \frac{1}{2}, \quad x = -\frac{\ell}{2}, \frac{\ell}{2}. \end{aligned} \tag{8.54a, b}$$

A separable solution

$$\psi = [G(x) + H] \cos \pi y \tag{8.55}$$

is suggested for the above equation, which upon substituting into (8.53),

$$\frac{\partial^2 G(x)}{\partial x^2} - \pi^2 [G(x) + H] = \pi, \tag{8.56}$$

yields

$$\frac{\partial^2 G}{\partial x^2} - \pi^2 G = 0 \text{ and } H = -\frac{1}{\pi}. \tag{8.57a, b}$$

A solution for $G(x)$ in the form

$$G(x) = Ae^{\pi x} + Be^{-\pi x} \tag{8.58}$$

must satisfy the boundary conditions (8.54). This requires

$$\begin{aligned} Ae^{-\pi \ell} + Be^{\pi \ell} &= \frac{1}{\pi} \\ Ae^{\pi \ell} + Be^{-\pi \ell} &= \frac{1}{\pi} \end{aligned} \tag{8.59a, b}$$

yielding

$$A = B = \frac{1}{2\pi \cosh(\pi \ell/2)} \tag{8.60}$$

so that the solution is given as

$$\psi = \left\{ \frac{\cosh(\pi x)}{\pi \cosh(\pi \ell/2)} - \frac{1}{\pi} \right\} \cos \pi y. \tag{8.61}$$

The solution displayed in Fig. 8.2 indicates an anticyclonic major circulation of the entire basin. However, it can be observed that this solution has east-west and north-south symmetry with respect to the origin, which is at the center of the basin.

This is in obvious disagreement with the observed circulation in ocean basins. For instance in the North Atlantic Ocean, the anticyclonic general circulation is observed to have an asymmetric structure, especially in the east-west direction, with currents intensified along the western side, where the Gulf Stream transport confined near the American coast balances the return flow of the ocean-wide anticyclonic cell. It seems there is certainly something ging wrong at this level of approximation we have so far used. There has been a number of simplifications along the way, that we could blame, but really the most important element that we have ignored so far turns out to be the β effect, which could account for asymmetry, for instance as we have observed in the cases of Rossby waves, based on vorticity balance arguments of the earlier chapters. We therefore give greater respect to this effect in remedy to the situation.

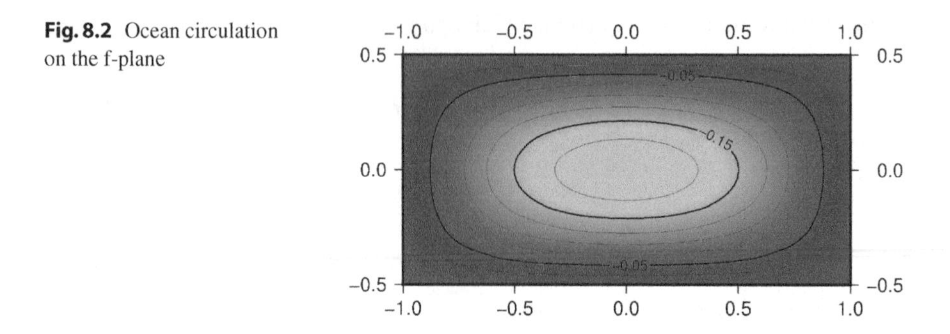

Fig. 8.2 Ocean circulation on the f-plane

8.3 Ocean Circulation on the β-Plane

8.3.1 Development of the Model

In the following we start by rewriting the interior geostrophic equations,

$$\begin{aligned} 2\bar{f}\mathbf{k} \times \mathbf{q}_i &= -\nabla p_i, \\ \nabla \cdot \mathbf{q}_i &= 0. \end{aligned} \tag{8.62a, b}$$

where ∇ is once again the three dimensional gradient operator defined in (8.3b) and now we approximate the Coriolis parameter on the β-plane as a function of the meridional distance, as described by $\bar{f} = 1 + \bar{\beta} y$ on the β-plane, according to (8.2a), (8.2b) and (8.2.3).

Taking $\nabla\times$ of (8.62) in this case yields

$$\nabla \times 2\bar{f}\mathbf{k} \times \mathbf{q}_i \equiv (\mathbf{q}_i \cdot \nabla)2\bar{f}\mathbf{k} - (2\bar{f}\mathbf{k} \cdot \nabla)\mathbf{q}_i = \nabla \times \nabla p_i \equiv 0 \tag{8.63}$$

which yields

$$2\bar{f}\frac{\partial \mathbf{q}_i}{\partial z} = 2v_i \frac{\partial \bar{f}}{\partial y}\mathbf{k} = 2\bar{\beta} v_i \mathbf{k} \tag{8.64}$$

This is a vector equation that can be decomposed into the following horizontal and vertical components:

$$\frac{\partial \mathbf{u}_i}{\partial z} = 0 \quad \text{and} \quad \frac{\partial w_i}{\partial z} = \frac{\bar{\beta}}{\bar{f}} v_i, \tag{8.65a, b}$$

so that the horizontal velocity $\mathbf{u}_i$ is depth independent, but the vertical velocity w_i is not. The latter has a linear variation on depth, proportional to the meridional velocity, and when (8.65b) is integrated from the bottom to top of the ocean, i.e. from $z = 0$ to $z = 1$ in non-dimensional vertical coordinates, yields

$$w_i^t - w_i^b = \frac{\bar{\beta}}{\bar{f}} v_i. \tag{8.66}$$

We have seen that the interior flow velocity $\mathbf{q}_i = \mathbf{u}_i + w_i \mathbf{k}$ is three-dimensional, but by (8.65a), the horizontal component $\mathbf{u}_i$ is two-dimensional, $\mathbf{u}_i = \mathbf{u}_i(x, y)$, which by (8.62a) necessitates that the pressure is also two-dimensional, $p_i = p_i(x, y)$. Then we can re-write (8.62a, b) as

$$\begin{aligned} 2\bar{f}\mathbf{k} \times \mathbf{u}_i &= -\nabla_h p_i, \\ \nabla_h \cdot \mathbf{u}_i + \frac{\partial w_i}{\partial z} &= 0, \end{aligned} \tag{8.67a, b}$$

where ∇_h is the horizontal gradient operator.

The first equation is cross multiplied by $\mathbf{k}/(2\bar{f})$, and assigning a suitable stream-function

$$\psi = -\frac{p_i}{2} \tag{8.68}$$

gives

$$\mathbf{u}_i = \frac{1}{2\bar{f}}\mathbf{k} \times \nabla_h p_i = -\frac{1}{\bar{f}}\mathbf{k} \times \nabla_h \psi. \tag{8.69}$$

It now remains to see if the continuity equation is satisfied. Noting that $\bar{f} = 1 + \bar{\beta} y$ and

$$v_i = -\frac{1}{\bar{f}}\frac{\partial \psi}{\partial x}, \tag{8.70}$$

by substituting (8.69), (8.70) and (8.65b), and using vector identities (1.27c), (1.27i), we evaluate the left hand side of (8.67b)

$$\begin{aligned}
\nabla_h \cdot \mathbf{u}_i + \frac{\partial w_i}{\partial z} &= -\nabla_h \cdot \mathbf{k} \times \frac{\nabla_h \psi}{\bar{f}} + \frac{\bar{\beta}}{\bar{f}} v_i = -\mathbf{k} \cdot \nabla_h \times \frac{\nabla_h \psi}{\bar{f}} - \frac{\bar{\beta}}{\bar{f}^2}\frac{\partial \psi}{\partial x} \\
&= -\mathbf{k} \cdot \nabla_h \frac{1}{\bar{f}} \times \nabla_h \psi - \frac{\bar{\beta}}{\bar{f}^2}\frac{\partial \psi}{\partial x} \\
&= \frac{\bar{\beta}}{\bar{f}^2}\frac{\partial \psi}{\partial x} - \frac{\bar{\beta}}{\bar{f}^2}\frac{\partial \psi}{\partial x} \\
&\equiv 0
\end{aligned} \tag{8.71}$$

which shows that the continuity equation is identically satisfied by the stream-function. This in turn is a repeated verification of the geostrophic indeterminacy in the β-plane, because it shows once again that the members of the equation set (8.67a, b) are not independent. In other words, (8.67b) does not provide any new information other then confirming (8.67a) to be true.

We then review the Ekman boundary layers, which are described by

$$\begin{aligned}
2\bar{f}\mathbf{k} \times \mathbf{u}_e &= \frac{\partial^2 \mathbf{u}_e}{\partial \zeta^2}, \\
\frac{\partial w_e}{\partial \zeta} &= -E^{-1/2}\nabla_h \cdot \mathbf{u}_e.
\end{aligned} \tag{8.72a, b}$$

The solution for the β-plane can be obtained similar to the f-plane case, for the surface and bottom Ekman layers.

By manipulating (8.72a) we obtain

$$\frac{\partial^2 \mathbf{Q}}{\partial \zeta^2} - 2i\bar{f}\mathbf{Q} = 0 \tag{8.73}$$

where we define $\mathbf{Q} \equiv \mathbf{k} \times \mathbf{u}_e + i\mathbf{u}_e$ as in (8.27) and obtain a solution

$$\mathbf{Q} \equiv \mathbf{k} \times \mathbf{u}_e + i\mathbf{u}_e = \mathbf{A}e^{(1+i)\bar{f}^{\frac{1}{2}}\zeta}, \tag{8.74}$$

where $\mathbf{A}$ is an appropriate constant that would satisfy the boundary conditions. The solution for the Ekman layer horizontal velocity is therefore obtained from $\mathbf{u}_e = \Im\{\mathbf{Q}\}$ as done in the earlier sections.

The solutions satisfying boundary conditions (8.28a), (8.28b) for the surface Ekman layer and the same satisfying boundary conditions (8.39a), (8.39b) for the bottom Ekman layer respectively are

$$\begin{aligned} \mathbf{u}_e^t &= \Im\left\{\frac{\mathbf{k} \times \mathbf{S} + i\mathbf{S}}{(1+i)\bar{f}^{\frac{1}{2}}} e^{(1+i)\bar{f}^{\frac{1}{2}}\zeta}\right\} \\ &= \frac{1}{2\bar{f}^{\frac{1}{2}}}\{(\mathbf{k} \times \mathbf{S} + \mathbf{S})\sin(\bar{f}^{\frac{1}{2}}\zeta) - (\mathbf{k} \times \mathbf{S} - \mathbf{S})\cos(\bar{f}^{\frac{1}{2}}\zeta)\}e^{\bar{f}^{\frac{1}{2}}\zeta}. \end{aligned} \tag{8.75}$$

and

$$\begin{aligned} \mathbf{u}_e^b &= -\Im\left\{[\mathbf{k} \times \mathbf{u}_i + i\mathbf{u}_i]e^{-(1+i)\bar{f}^{\frac{1}{2}}\zeta}\right\} \\ &= \{\mathbf{k} \times \mathbf{u}_i \sin(\bar{f}^{\frac{1}{2}}\zeta) - \mathbf{u}_i \cos(\bar{f}^{\frac{1}{2}}\zeta)\}e^{-\bar{f}^{\frac{1}{2}}\zeta}. \end{aligned} \tag{8.76}$$

Carrying out steps (8.33)–(8.36) for the surface and (8.43)–(8.46) for the bottom Ekman layers respectively, we obtain the vertical velocity at the top and bottom of the interior ocean as

$$w_i^t = -w_e(0) = \frac{E^{\frac{1}{2}}}{2}\mathbf{k} \cdot \nabla_h \times \left(\frac{\mathbf{S}}{\bar{f}}\right). \tag{8.77}$$

$$w_i^b = -w_e(0) = \frac{E^{\frac{1}{2}}}{2}\left\{\mathbf{k} \cdot \nabla_h \times \left(\frac{\mathbf{u}_i}{\bar{f}^{\frac{1}{2}}}\right) + \nabla_h \cdot \left(\frac{\mathbf{u}_i}{\bar{f}^{\frac{1}{2}}}\right)\right\}. \tag{8.78}$$

and since

$$w_i^t - w_i^b = \frac{\bar{\beta}}{\bar{f}}v_i \tag{8.79}$$

the equation is reduced to

$$\frac{E^{\frac{1}{2}}}{2}\left\{\mathbf{k} \cdot \nabla_h \times \left(\frac{\mathbf{u}_i}{\bar{f}^{\frac{1}{2}}}\right) + \nabla_h \cdot \left(\frac{\mathbf{u}_i}{\bar{f}^{\frac{1}{2}}}\right)\right\} + \left(\frac{\bar{\beta}}{\bar{f}}\right)v_i = \frac{E^{\frac{1}{2}}}{2}\mathbf{k} \cdot \nabla_h \times \left(\frac{\mathbf{S}}{\bar{f}}\right). \tag{8.80}$$

One should note that this is a new vorticity equation for the interior, balancing the fluid and planetary vorticity components with the wind-stress torque applied at the surface. We then note by substituting (8.69) and expanding each of the terms

$$\begin{aligned}\mathbf{k}\cdot\nabla\times\left(\frac{\mathbf{u}_i}{\bar{f}^{\frac{1}{2}}}\right) &= -\mathbf{k}\cdot\nabla_h\times\mathbf{k}\times\frac{\nabla_h\psi}{\bar{f}^{\frac{3}{2}}} = -\nabla_h\cdot\frac{\nabla_h\psi}{\bar{f}^{\frac{3}{2}}} \\ &= -\frac{1}{\bar{f}^{\frac{3}{2}}}\nabla_h^2\psi - \nabla_h\frac{1}{\bar{f}^{\frac{3}{2}}}\cdot\nabla_h\psi = -\frac{1}{\bar{f}^{\frac{3}{2}}}\nabla_h^2\psi + \frac{3\bar{\beta}}{2\bar{f}^{\frac{5}{2}}}\frac{\partial\psi}{\partial y}\end{aligned} \tag{8.81a}$$

and

$$\begin{aligned}\nabla_h\cdot\left(\frac{\mathbf{u}_i}{\bar{f}^{\frac{1}{2}}}\right) &= -\nabla_h\cdot\mathbf{k}\times\frac{\nabla_h\psi}{\bar{f}^{\frac{3}{2}}} = -\mathbf{k}\cdot\nabla_h\times\frac{\nabla_h\psi}{\bar{f}^{\frac{3}{2}}} \\ &= -\nabla_h\frac{1}{\bar{f}^{\frac{3}{2}}}\times\nabla_h\psi = \frac{3\bar{\beta}}{2\bar{f}^{\frac{5}{2}}}\frac{\partial\psi}{\partial x},\end{aligned} \tag{8.81b}$$

leads to a new form of the vorticity equation (8.80):

$$\frac{E^{\frac{1}{2}}}{2}\left\{-\frac{1}{\bar{f}^{\frac{3}{2}}}\nabla_h^2\psi + \frac{3\bar{\beta}}{2\bar{f}^{\frac{5}{2}}}\left(\frac{\partial\psi}{\partial x}+\frac{\partial\psi}{\partial y}\right)\right\} - \frac{\bar{\beta}}{\bar{f}^2}\frac{\partial\psi}{\partial x} = \frac{E^{\frac{1}{2}}}{2}\mathbf{k}\cdot\nabla_h\times\frac{\mathbf{S}}{\bar{f}}. \tag{8.82}$$

Re-arranging terms, we have the following β-plane vorticity equation for the ocean interior:

$$\nabla_h^2\psi - \frac{3\bar{\beta}}{2\bar{f}}\left(\frac{\partial\psi}{\partial x}+\frac{\partial\psi}{\partial y}\right) + \frac{2\bar{\beta}}{E^{\frac{1}{2}}\bar{f}^{\frac{1}{2}}}\frac{\partial\psi}{\partial x} = -\bar{f}^{\frac{3}{2}}\mathbf{k}\cdot\nabla_h\times\frac{\mathbf{S}}{\bar{f}}. \tag{8.83}$$

This equation involves more terms than the f-plane case in (8.51), but can similarly be solved subject to lateral boundary conditions to determine the ocean circulation on the β-plane. Although numerical solutions would be in order, we note that this equation is not so much amenable to an analytical solution as has been carried out in the f-plane case, because of the additional terms with variable coefficients $\bar{f}$.

We can return to (8.82) or (8.83) later, but at this stage, the developed theory for the β-plane wind-driven ocean circulation already has important differences from the f-plane version we have considered in the last section. We notice that equation (8.79) and the consequent (8.80) represent a significant change in the interior flow. Since the horizontal component of the velocity is independent of z by (8.65a), the vertical velocity given by (8.65b) has a linear dependence on z. By taking the z-coordinate pointing upwards at the surface ($z=0$), the integration of (8.65b) from the bottom ($z=-1$) upwards yields

$$w_i(z) - w_i^b = \frac{\bar{\beta}}{\bar{f}}v_i(x,y)(1+z) \tag{8.84}$$

and its complete integration to the surface ($z = 0$) would then result in (8.79). One also notes that (8.84) implies $w_i^b \ll w_i^t$ and possibly $w_i^b/w_i^t = O(E^{\frac{1}{2}})$ or smaller.

We also note that the problem is not closed for the total ocean domain. The wind-driven surface layer dynamics is determined by the dynamic boundary condition at the top, producing an Ekman pumping volume flux which drives the interior flow, which in turn drives the bottom Ekman layer which on the other hand has to satisfy the no-slip bottom boundary condition.

Yet, it is not clear how the various flux components are finally balanced, for instance, in a closed basin, the net horizontal flux integrated around the lateral boundaries of the basin has to vanish. This would then require that the three-dimensional scheme of circulation partitioned between horizontal and vertical flows of the interior and Ekman layers have to be balanced. The net balance actually depends on the complex processes at coastal and shelf areas, which we have so far avoided to include in the development.

8.3.2 The Sverdrup Interior and an Approximate Solution for the Circulation

For the time being, from the developed equations we can observe that the Ekman pumping vertical velocity near the surface is small, at $O(E^{\frac{1}{2}})$, as dictated by (8.77). As a consequence of (8.82), we also expect the interior flow horizontal velocity to be of the same order, $O(E^{\frac{1}{2}})$. The decreasing of the vertical flux towards the bottom by (8.84) would then possibly lead to the the bottom boundary layer velocities to be much smaller, of $O(E)$. Then, in accordance with the earlier approximations, we can assume $w_i^b = O(E)$ in (8.79) to leave out the bottom boundary layer due to its small contribution to the circulation, in which case we only keep the largest terms in the balance to approximate (8.80) as

$$\left(\frac{\bar{\beta}}{\bar{f}}\right) v_i = \frac{E^{\frac{1}{2}}}{2}\mathbf{k} \cdot \nabla_h \times \frac{\mathbf{S}}{\bar{f}}. \qquad (8.85)$$

The above equation is known as the *Sverdrup relation*, implying a direct effect of the wind-stress curl on creating a meridional velocity in the interior. This balance has been used by Sverdrup to construct a first approximation to the ocean circulation.

The above approximate dynamics of Sverdrup amounts to the following: (*i*) planetary-vorticity (β effect) signified by the last term on the left hand side of (8.82) has a greater role than the fluid vorticity represented by the first two terms, (*ii*) the bottom Ekman pumping (again the first two terms) is much smaller than the surface Ekman pumping terms on the right hand side of (8.82).

We can better understand the same balance from the alternative form in (8.83), where only the third term appears sufficiently large so as to balance the wind-stress curl on the right hand side, yielding the Sverdrup relation

$$\frac{\partial \psi}{\partial x} = -\frac{E^{\frac{1}{2}} \bar{f}^2}{2\bar{\beta}} \mathbf{k} \cdot \nabla \times \frac{\mathbf{S}}{\bar{f}}. \tag{8.86}$$

In this equation the right hand side is expanded as

$$\frac{\partial \psi}{\partial x} = -\frac{E^{\frac{1}{2}} \bar{f}}{2\bar{\beta}} \left\{ \frac{\partial S^x}{\partial y} - \frac{\partial S^y}{\partial x} - \frac{\bar{\beta}}{\bar{f}} S^x \right\}. \tag{8.87}$$

Setting up a case of demonstration at this stage, we introduce the familiar wind-stress distribution

$$S^x = \sin \pi y, \qquad S^y = 0, \tag{8.88a, b}$$

to obtain a direct solution to (8.86). Remembering also (8.70), the meridional velocity is obtained as

$$v_i = -\frac{1}{\bar{f}} \frac{\partial \psi}{\partial x} = \frac{E^{\frac{1}{2}}}{2\bar{\beta}} \left\{ \pi \cos \pi y - \frac{\bar{\beta}}{\bar{f}} \sin \pi y \right\}. \tag{8.89}$$

As a result of the approximations made, it should have been noticed that the to order of the original equation (8.83) has been reduced from second-order to first-order, resulting in (8.87). Yet this reduced-order system involving only one of the coordinates has been expected to yield a miraculous solution for the two-dimensional field $\psi(x, y)$, subject to the same boundary conditions as (8.54a, b)

$$\begin{aligned} &\psi = 0 \text{ on } \quad -\frac{\ell}{2} < x < \frac{\ell}{2}, \quad y = -\frac{1}{2}, \frac{1}{2}, \\ &\qquad \text{and} \\ &\psi = 0 \text{ on } \quad -\frac{1}{2} < y < \frac{1}{2}, \quad x = -\frac{\ell}{2}, \frac{\ell}{2}. \end{aligned} \tag{8.90a, b}$$

We still need to integrate (8.89) once to yield

$$\psi = -\frac{E^{\frac{1}{2}} \bar{f}}{2\bar{\beta}} \left\{ \pi \cos \pi y - \frac{\bar{\beta}}{\bar{f}} \sin \pi y \right\} x + h(y), \tag{8.91}$$

where $h(y)$ is a function, arising from integration in x of the partial derivative.

It is obvious that the solution to the first order equation (8.86) or (8.89), displayed in Fig. 8.3 can not satisfy the boundary conditions (8.90a, b) in general, except choosing an appropriate $h(y)$ to satisfy either of the eastern or western boundary conditions. If we choose the eastern boundary condition to be satisfied, we obtain the approximate solution

$$\psi = -\frac{E^{\frac{1}{2}}}{2} \left\{ \frac{\pi \bar{f}}{\bar{\beta}} \cos \pi y - \sin \pi y \right\} \left(x - \frac{\ell}{2} \right). \tag{8.92}$$

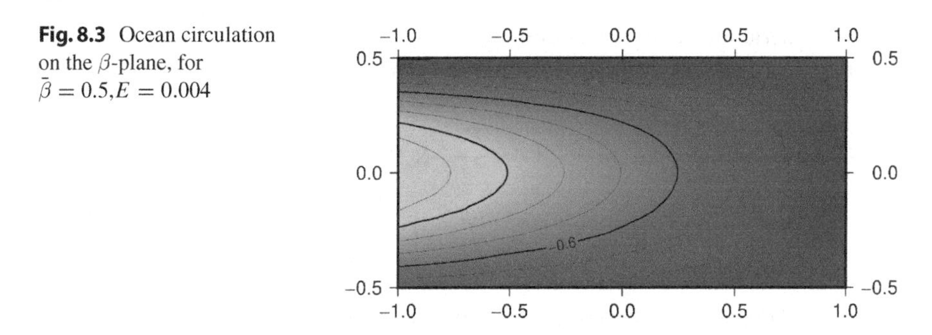

Fig. 8.3 Ocean circulation on the β-plane, for $\bar{\beta} = 0.5, E = 0.004$

We note that there are no more free constants to be determined that would satisfy all of the boundary conditions (8.90a, b), namely on the western, northern or southern boundaries.

Clearly this low order solution is insufficient, establishing within errors of $O(E^{\frac{1}{2}})$, an approximate relationship of the ocean interior meridional velocity as a function of the surface wind-stress by (8.89), failing to give correct information especially near the boundaries.

The solution is asymmetric about the origin, in both the x and y directions. On the centerline along the x-axis, where the second term of (8.89) disappears, by virtue of (8.85) we see that the meridional velocity is always negative, $v_i < 0$. It is also evident from (8.84) that the Ekman pumping produces negative vertical velocity $w_i < 0$ in the central part of the interior, i.e. downwards from the surface. Because the solution (8.92) performs well near the eastern boundary and it produces negative meridional velocity at the east-west centerline, we can try to sketch the circulation that would be expected despite the present imperfect solution. The fact that the the northern and southern boundaries are streamlines, connected by the eastern boundary segment, with linearly spaced streamlines at the center, we arrive at an incomplete sketch of an anticyclonic gyre as shown in Fig. 8.3, with eastward flow on the northern boundary joined to the westward flow on the southern boundary by the southward meridional flow of the interior. However, the streamlines are disconnected on the western boundary, giving incorrect fluxes in and out of the boundary.

Because the Sverdrup interior solution appears to be better near the east-west centerline, we can take a closer look at the solution near this axis, by comparing the transport components in the meridional direction.

The surface Ekman layer transport, in analogy to (8.32) can be obtained as

$$\mathbf{M}_e = E^{\frac{1}{2}} \int_{-\infty}^{0} \mathbf{u}_e d\zeta = -\frac{E^{\frac{1}{2}}}{2\bar{f}} \mathbf{k} \times \mathbf{S}, \tag{8.93a}$$

and its meridional component is

$$M_e^y = \mathbf{M}_e \cdot \mathbf{j} = -\frac{E^{\frac{1}{2}}}{2\bar{f}} S^x, \tag{8.93b}$$

while the meridional component of the interior transport is simply the interior horizontal velocity multiplied with the non-dimensional depth 1,

$$\begin{aligned} M_i^y =& \mathbf{M}_i \cdot \mathbf{j} = \int_{-1}^{0} v_i dz = v_i\,(1) = \frac{E^{\frac{1}{2}}\bar{f}}{2\bar{\beta}}\mathbf{k}\cdot\nabla_h \times \frac{\mathbf{S}}{\bar{f}} \\ =& \frac{E^{\frac{1}{2}}\bar{f}}{2\bar{\beta}}\mathbf{k}\cdot\left\{\frac{1}{\bar{f}}\nabla_h \times \mathbf{S} + \nabla_h\frac{1}{\bar{f}}\times\mathbf{S}\right\} = \frac{E^{\frac{1}{2}}}{2\bar{\beta}}\mathbf{k}\cdot\nabla_h \times \mathbf{S} + \frac{\bar{\beta}}{\bar{f}}S^x. \end{aligned} \tag{8.94}$$

Adding (8.93a), (8.93b) together gives the other form of the Sverdrup relation

$$M_t^y = M_e^y + M_i^y = \frac{E^{\frac{1}{2}}}{2\bar{\beta}}\mathbf{k}\cdot\nabla_h \times \mathbf{S} = \frac{E^{\frac{1}{2}}}{2\bar{\beta}}\left\{\frac{\partial S^y}{\partial x} - \frac{\partial S^x}{\partial y}\right\}. \tag{8.95}$$

Combining (8.93b) and (8.95) gives

$$M_i^y = v_i = M_t^y + \frac{E^{\frac{1}{2}}}{2\bar{f}}S^x. \tag{8.96}$$

8.3.3 Approximate Solution for the Circulation with a Western Boundary Layer

We have seen that the Sverdrup model of ocean circulation is incomplete, because of it fails to satisfy the boundary conditions. In developing this solution we have neglected O(E) terms representing fluid vorticity and the bottom boundary layer contributions in (8.82).

Now we can try to make the solution better by including some of the terms we neglected. In fact, the incomplete solution seemed to perform better near the basin center as compared to near the boundaries. In fact, the circulation sketched in Fig. 8.3 suggests that there should be a rather intense northward flow along the western boundary to balance the southerly flow in the Sverdrup interior, that is not handled by the approximate equations so far. Therefore we attempt to improve the solution near the western boundary.

Because an intense return flow seems to be needed close to the western boundary, coordinate stretching near there seems to be in order. Stretching out the zonal coordinate to scale as $x = \delta\xi$ where δ is a small parameter that is on the order of the other small parameter, $O(E^{\frac{1}{2}})$, we can rewrite some terms in (8.83) in terms of the stretched coordinate

$$\nabla_h^2 = \frac{1}{\delta^2}\frac{\partial^2}{\partial^2\xi} + \frac{\partial^2}{\partial y^2}, \qquad \frac{\partial}{\partial x} = \frac{1}{\delta}\frac{\partial}{\partial\xi} \quad \text{etc.,}$$

observing that the terms that survive in (8.83) to $O(1)$ are

$$\frac{1}{\delta^2}\frac{\partial^2\psi}{\partial\xi^2} + \frac{2\bar{\beta}}{\delta E^{\frac{1}{2}}\bar{f}^{-\frac{1}{2}}}\frac{\partial\psi}{\partial\xi} = -\bar{f}^{\frac{3}{2}}\mathbf{k}\cdot\nabla_h \times \frac{\mathbf{S}}{\bar{f}} \tag{8.97}$$

and consequently, rewriting in the original coordinates, we have

$$\frac{\partial^2\psi}{\partial x^2} + \frac{2\bar{\beta}}{E^{\frac{1}{2}}\bar{f}^{-\frac{1}{2}}}\frac{\partial\psi}{\partial x} = -\bar{f}^{\frac{3}{2}}\mathbf{k}\cdot\nabla_h \times \frac{\mathbf{S}}{\bar{f}}. \tag{8.98}$$

We can re-define the variable coefficient of the second term

$$a(y) = \frac{2\bar{\beta}}{E^{\frac{1}{2}}\bar{f}^{-\frac{1}{2}}} = \frac{2\bar{\beta}}{E^{\frac{1}{2}}(1+\bar{\beta}y)^{\frac{1}{2}}} \tag{8.99a}$$

and the right hand side term

$$g(x,y) = -\bar{f}^{\frac{1}{2}}\left\{\frac{\partial S^x}{\partial y} - \frac{\partial S^y}{\partial x} - \frac{\bar{\beta}}{\bar{f}}S^x\right\}, \tag{8.99b}$$

to rewrite (8.98) as

$$\frac{\partial^2\psi}{\partial x^2} + a(y)\frac{\partial\psi}{\partial x} = g(x,y). \tag{8.100}$$

If we accept also the simple wind-stress distribution of (8.88a, b), then we can write the right hand side of (8.100) as

$$g(y) = -\bar{f}^{\frac{1}{2}}\left\{\pi\cos\pi y - \frac{\bar{\beta}}{\bar{f}}\sin\pi y\right\}. \tag{8.101}$$

It remains to obtain a solution, first by putting (8.100) in the form

$$e^{-a(y)x}\frac{\partial}{\partial x}\left(e^{a(y)x}\frac{\partial\psi}{\partial x}\right) = g(y). \tag{8.102}$$

This can then be readily integrated to give

$$e^{a(y)x}\frac{\partial\psi}{\partial x} = \frac{g(y)}{a(y)}e^{a(y)x} + c(y), \tag{8.103}$$

where $c(y)$ is a constant of integration in x. Integrating once again and choosing a form compatible with the coordinate geometry, we have

$$\psi = \frac{g(y)}{a(y)}\left(x+\frac{\ell}{2}\right) - \frac{c_1}{a(y)}e^{-a(y)(x+\frac{\ell}{2})} + c_2, \tag{8.104}$$

where, now, c_1 and c_2 are constants.

The western and eastern boundary conditions, requiring

$$\psi = 0 \text{ at } x = -\frac{\ell}{2}, \quad \text{and} \quad \psi = 0 \text{ at } x = +\frac{\ell}{2} \tag{8.105}$$

are then used to set the two constants, whereby the solution becomes

$$\begin{aligned} \psi &= \frac{g(y)}{a(y)} \left[\left(x + \frac{\ell}{2} \right) - \ell \left\{ \frac{1 - e^{a(y)(x+\frac{\ell}{2})}}{1 - e^{a(y)\ell}} \right\} \right] \\ &= -\frac{E^{\frac{1}{2}}}{2} \left\{ \frac{\pi(1+\bar{\beta}y)}{\bar{\beta}} cos\pi y - sin\pi y \right\} \left[\left(x + \frac{\ell}{2} \right) - \ell \left\{ \frac{1 - e^{a(y)\left(x+\frac{\ell}{2}\right)}}{1 - e^{a(y)\ell}} \right\} \right] \end{aligned} \tag{8.106}$$

The circulation corresponding to the solution (8.106) is shown in Fig. 8.4. Both of the approximate solutions in Figs. 8.3 and 8.4 have been generated using (8.92) and (8.106) respectively, selecting dimensional parameters, viscosity $\nu = 0.2\,\text{m}^2\text{s}^{-1}$, horizontal size $L_0 = 1.5 \times 10^6$, earth's radius $r_0 = 6.4 \times 10^6\,\text{m}$, center latitude $\phi_0 = 30°$ N, $\beta = 2^{-11}\,\text{m}^{-1}\text{s}^{-1}$ yielding $\bar{\beta} = 0.5$, $E = 0.004$.

The solution (8.106) allowing adjustment of currents at the western boundary layer correctly connects the streamlines originally disconnected in the Sverdrup solution (8.92). The correction is made with an intense northward current at the boundary, which may roughly correspond to currents typically observed along the western sides of the world's oceans. We note however, the great difference in the intensity of circulation between the two solutions respectively in Figs. 8.3 and 8.4, although the same parameters are used to make the two solutions comparable. Yet, both of solutions are approximate, and while the Sverdrup solution fails to satisfy the boundary conditions on three of the boundaries, the new solution in the present section fails to satisfy only on two of them, the north and south boundaries, where the solution seems to be shifted slightly to the south, possibly as result of the small y-derivative terms in (8.83) being neglected.

In adding the functional western boundary layer in the present section to have closed cell circulation, only the boundary conditions on the western and eastern

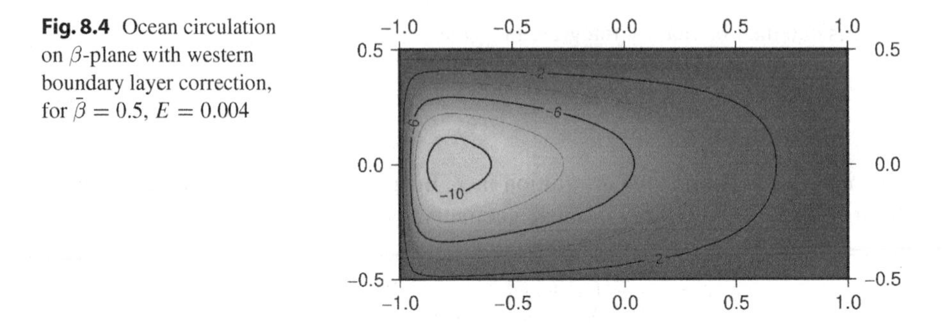

Fig. 8.4 Ocean circulation on β-plane with western boundary layer correction, for $\bar{\beta} = 0.5$, $E = 0.004$

boundaries have been implemented. One should note however, that the streamfunction that we choose to have closed cells is for the interior horizontal current $\mathbf{u}_i$, excluding the total horizontal components of the respective surface and bottom Ekman boundary layer currents $\mathbf{u}_e^t$ and $\mathbf{u}_e^b$ which would add up to the horizontal current averaged over depth. The other way, that could ensure western boundary current returning the meridional flow of the Sverdrup interior, is to check if the total mass meridional flux is balanced.

Noting (8.70) we can also write (8.96) as

$$\frac{E^{\frac{1}{2}}}{2\bar{f}^{\frac{1}{2}}}\frac{\partial v_i}{\partial x}+\frac{\bar{\beta}}{\bar{f}}v_i=\frac{E^{\frac{1}{2}}}{2}\mathbf{k}\cdot\nabla_h\times\frac{\mathbf{S}}{\bar{f}}. \tag{8.107}$$

Substituting into (8.107) from (8.95) gives

$$\frac{E^{\frac{1}{2}}}{2\bar{f}^{\frac{1}{2}}}\frac{\partial M_t^y}{\partial x}+\frac{\bar{\beta}}{\bar{f}}M_t^y=\frac{E^{\frac{1}{2}}}{2\bar{f}}\mathbf{k}\cdot\nabla_h\times\mathbf{S}. \tag{8.108}$$

The solution to (8.108) is composed of homogeneous and particular solutions

$$M_t^y=\hat{M}_t^y+\frac{E^{\frac{1}{2}}}{2\bar{f}}\mathbf{k}\cdot\nabla_h\times\mathbf{S} \tag{8.109}$$

where $\hat{M}_t^y$ is a solution of

$$\frac{E^{\frac{1}{2}}}{2\bar{f}^{\frac{1}{2}}}\frac{\partial \hat{M}_t^y}{\partial x}+\frac{\bar{\beta}}{\bar{f}}\hat{M}_t^y=0. \tag{8.110}$$

and solving (8.110),

$$\hat{M}_t^y=A\exp\left\{-\frac{2\bar{\beta}}{E^{\frac{1}{2}}\bar{f}^{\frac{1}{2}}}x\right\} \tag{8.111}$$

and therefore

$$M_t^y=A\exp\left\{-\frac{2\bar{\beta}}{E^{\frac{1}{2}}\bar{f}^{\frac{1}{2}}}x\right\}+\frac{E^{\frac{1}{2}}}{2\bar{\beta}}\mathbf{k}\cdot\nabla_h\times\mathbf{S}. \tag{8.112}$$

The constant A is obtained by requiring that

$$\int_0^{\ell} M_t^y dx=0 \tag{8.113}$$

and therefore the meridional volume flux becomes

$$M_t^y=\frac{E^{\frac{1}{2}}}{2\bar{\beta}}\mathbf{k}\cdot\nabla_h\times\mathbf{S}-\frac{1}{\bar{f}^{\frac{1}{2}}}\left(\int_0^{\ell}\mathbf{k}\cdot\nabla_h\times\mathbf{S}dx\right)\exp\left\{-\frac{2\bar{\beta}}{E^{\frac{1}{2}}\bar{f}^{\frac{1}{2}}}x\right\}. \tag{8.114}$$

If the wind stress is given by (8.88a, b), then

$$M_t^y = \left(\frac{\pi E^{\frac{1}{2}}}{2\bar{\beta}} - \frac{\pi \ell}{\bar{f}^{\frac{1}{2}}} \exp\left\{ -\frac{2\bar{\beta}}{E^{\frac{1}{2}} \bar{f}^{\frac{1}{2}}} x \right\} \right) \cos \pi y. \tag{8.115}$$

We have developed the simplest possible model of circulation in an ocean basin, using the essential elements to demonstrate basic observed features. One should note however, the approximations and simplifications often have been too generous. The analysis is limited to an ocean with uniform density within the context of the present volume. We have aimed for the simplest demonstration model, leaving further discussion and improved models to expert texts such as by Pedlosky (1998).

8.3.3.1 Application: Global Ocean Circulation

Various important physical factors such as non-linearity, secular time-dependence, turbulence, horizontal friction effects, 3D coastal and shelf dynamics, topographic variations inherent in the equations that could be of critical importance for the circulation, and these have not been taken into account. The significant changes that occur in the circulation as we switch between different levels of approximation in the last few sections stand for the possible changes that could be expected if further improvements were to be seeked. In particular, the real global ocean basin has complicated geometry and topography, in addition to the all too important interaction with the equally complex atmospheric dynamics and thermodynamics that increasingly has become an immense computational as well as observational task.

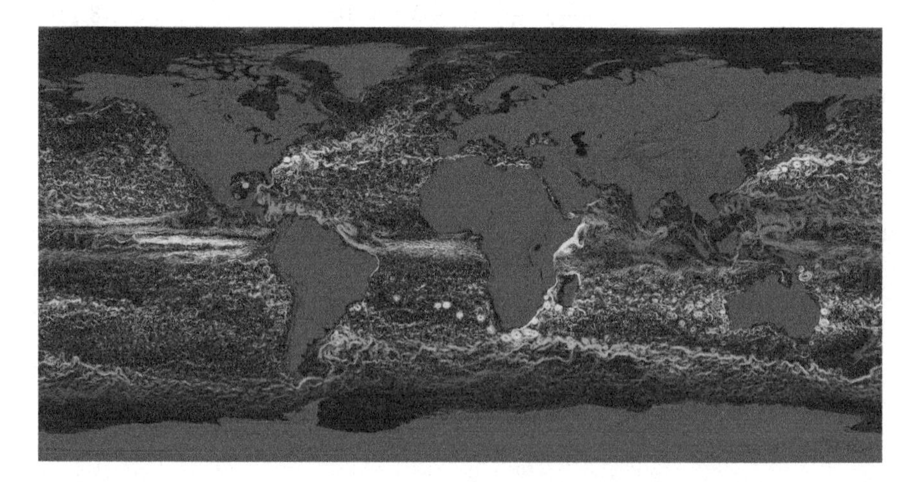

Fig. 8.5 High resolution ocean currents simulated by MITgcm model based on satellite altimetry and in situ data. Credits: MIT/JPL ECCO2 project, made available under NASA Goddard Scientific Visualization Studio policy guidelines (https://svs.gsfc.nasa.gov/)

An example of realistic estimates using most recent syntheses of global ocean circulation in Fig. 8.5 is based on extensive networks of observation and simulation, ultimately providing an extremely detailed picture of global ocean circulation.

Exercises

`Exercise 1`

Consider a hypothetical flow between two plates with horizontal surfaces S^t and S^b in a rotating frame of reference such as on earth, where the f-plane approximation is applied. This could be a model of hypothetical fluid of infinite horizontal extent, with constant Coriolis parameter, rotating about the vertical axis sandwiched between two moving horizontal plates generating Ekman layers at the top and bottom due to viscous forces and a geostrophic interior between them, as we have reviewed in Sect. 8.1.

In fact, this problem has some relationship to the “Einstein's tea leaves” problem explored in Exercise 2 of Chap. 2, with only conditional similarity because Einstein's experiment in a tea cup does not have to take Coriolis effect into account. Consequently, the sense of rotation of the tea spoon does not change the result in that experiment, while in the ocean and atmosphere where the present problem could arise, there is rotation in the background imposed by the inertial frame of reference.

`(i)` Describe the geostrophic interior and the Ekman layers near the top and bottom boundaries S^t and S^b moving with horizontal velocities $\mathbf{U}^t$ and $\mathbf{U}^b$, to develop respective equations and corresponding boundary conditions. Obtain separate solutions for the Ekman boundary layers and the interior and match their fluxes as we have done for the ocean circulation problem.

Obtain a vorticity equation for the interior and interpret this new form of the vorticity equation for the given problem, discussing effects of the individual terms.

`(ii)` Consider a case in which the top plate is stationary and the bottom plate has pure rotation specified as

$$\mathbf{U}^b = \omega_0 \mathbf{k} \times \mathbf{r} = \omega_0 r \mathbf{e}_\theta$$

where ω_0 is the dimensionless constant angular velocity of the plate ($\omega_0'/(f/2)$ with ω_0' in dimensional units) and $\mathbf{e}_\theta$ is the unit vector in the angular direction in cylindrical coordinates. Perform the Ekman boundary layer analysis in the relevant coordinate system to find the 3D flow velocities in the Ekman layer.

What would be the functional expressions for the total fluid horizontal and vertical velocities, respectively $\mathbf{u} = \mathbf{u}_i + \mathbf{u}_e^b$ and $w = w_i + w_e^b$ near the bottom boundary layer, and for pressure $p = p_i$ of the interior? Sketch the 3D flow near the bottom.

Finally, obtain the vertically integrated horizontal mass transport of the Ekman layer to show the magnitude and direction of this transport. In which direction is the transport, relative to the bottom surface? In which direction is it relative to the center of rotation?

(iii) What would happen if the angular velocity of the rotating bottom surface is reversed to take on a negative value $-\omega_0$? How would the 3D structure of the flow near the plate boundary and the mass flux $\mathbf{M}_e$ change?

Two-Layer Models

9

9.1 Introduction

The subjects covered so far in this volume have been limited to fluids with homogeneous properties. Specifically, the density has always been taken as constant, although we know this is not quite true for either the ocean or atmosphere, our main areas of study in GFD. The reader will remember that we have derived the Navier–Stokes equations for earthly fluids assumed to have well-behaved solutions with a minimum of simplifying assumptions, while possible further complications have carefully been excluded.

Following Chap. 3, we have purposefully limited our attention to homogeneous fluids, and in Chap. 4 we have made even further simplifications in adopting the shallow water theory, all in order to focus our attention to the still too great series of possible earthly motions that we try to understand using simple tools. One would heuristically expect much greater variety when inhomogeneous effects are taken into account, although homogeneous fluids have already appeared sufficiently complex.

In transition to studying inhomogeneous fluids, we provide in this chapter, a simple extension of the shallow water theory to two-layer stratified case, which will still be homogeneous within each layer and provide some first impression of what can be expected in the more general cases of multi-layered, continuously stratified and inhomogeneous fluids.

9.2 Two-Layer Shallow Water Equations

We assume two immiscible layers of fluids with different density, with the lighter fluid of density ρ_1 overlying fluid of higher density $\rho_2 > \rho_1$. The upper layer has a free surface with displacement η_1 from equilibrium, while the interface has displacement

© Springer Nature Switzerland AG 2020

E. Özsoy, *Geophysical Fluid Dynamics I*, Springer Textbooks in Earth Sciences, Geography and Environment, https://doi.org/10.1007/978-3-030-16973-2_9

η_2 between the two fluids. Equations of momentum and continuity for the layers can be written as

$$\frac{\partial \mathbf{u}_1}{\partial t} + f\hat{k} \times \mathbf{u}_1 = -\frac{1}{\rho_1}\nabla p_1 + \frac{\boldsymbol{\tau}^s}{\rho_1 H_1} \tag{9.1a}$$

$$\frac{\partial \mathbf{u}_2}{\partial t} + f\hat{k} \times \mathbf{u}_2 = -\frac{1}{\rho_2}\nabla p_2 \tag{9.1b}$$

$$\frac{\partial(\eta_1 - \eta_2)}{\partial t} + \nabla \cdot H_1\mathbf{u}_1 = 0 \tag{9.1c}$$

$$\frac{\partial \eta_2}{\partial t} + \nabla \cdot H_2\mathbf{u}_2 = 0, \tag{9.1d}$$

where pressure is assumed to be hydrostatic in each layer, consistent with the shallow water theory,

$$p_1 = \rho_1 g(\eta_1 - z) + p_a \tag{9.1e}$$

$$p_2 = \rho_1 g(\eta_1 - \eta_2 + h_1) + \rho_2 g[\eta_2 - (z - h_1)] + p_a \tag{9.1f}$$

such that $H_1 = \eta_1 + h_1 - \eta_2$, $H_2 = \eta_2 + h_2 - h_b$, and h_b is the bottom topography. Wind stress at the surface is $\boldsymbol{\tau}^s$ applied to the upper layer, while the atmospheric pressure at the surface is p_a.

We define volume transports in each layer as

$$\mathbf{U}_1 = \int_{-h_1+\eta_2}^{\eta_1} \mathbf{u}_1 dz = H_1\mathbf{u}_1 \tag{9.2a}$$

$$\mathbf{U}_2 = \int_{-h_1-h_2+h_b}^{-h_1+\eta_2} \mathbf{u}_2 dz = H_2\mathbf{u}_2, \tag{9.2b}$$

and set

$$\epsilon = \frac{\Delta\rho}{\rho_2} = \frac{\rho_2 - \rho_1}{\rho_2} \tag{9.2c}$$

or

$$\frac{\rho_1}{\rho_2} = 1 - \epsilon. \tag{9.2d}$$

By substituting hydrostatic pressure p_1 and p_2 (9.1e,f), approximating $H_1 \approx h_1$ and $H_2 \approx h_2 - h_b \equiv \tilde{h}_2$, i.e. assuming the surface and interface displacements to be small, to linearize respective terms, the vertically integrated momentum and continuity equations become

$$\frac{\partial \mathbf{U}_1}{\partial t} + f\hat{k} \times \mathbf{U}_1 = -gh_1\nabla\eta_1 + \frac{h_1}{\rho_1}\nabla p_a + \frac{\boldsymbol{\tau}^s}{\rho_1} \tag{9.3a}$$

$$\frac{\partial \mathbf{U}_2}{\partial t} + f\hat{k} \times \mathbf{U}_2 = -g\tilde{h}_2\nabla[r\eta_1 + \epsilon\eta_2] + \frac{\tilde{h}_2}{\rho_2}\nabla p_a \tag{9.3b}$$

$$\frac{\partial(\eta_1 - \eta_2)}{\partial t} + \nabla \cdot \mathbf{U}_1 = 0 \tag{9.3c}$$

$$\frac{\partial \eta_2}{\partial t} + \nabla \cdot \mathbf{U}_2 = 0 \tag{9.3d}$$

where we re-define $r \equiv 1 - \epsilon$ and $\tilde{h}_2(x, y) \equiv h_2 - h_b(x, y)$.

We have, in this treatment, included surface forcing by wind-stress and barometric pressure imparted by the overlying atmosphere, but have excluded internal friction between the layers and at the bottom.

We can symbolically write the above equations as

$$\frac{\partial \mathbf{U}_1}{\partial t} + f\hat{k} \times \mathbf{U}_1 = \mathbf{X}_1 + \frac{\boldsymbol{\tau}^s}{\rho_1} \tag{9.4a}$$

$$\frac{\partial \mathbf{U}_2}{\partial t} + f\hat{k} \times \mathbf{U}_2 = \mathbf{X}_2 \tag{9.4b}$$

$$\frac{\partial N_1}{\partial t} + \nabla \cdot \mathbf{U}_1 = 0 \tag{9.4c}$$

$$\frac{\partial N_2}{\partial t} + \nabla \cdot \mathbf{U}_2 = 0 \tag{9.4d}$$

where $\mathbf{X}_1$ and $\mathbf{X}_2$ include those terms which are written as gradient of a scalar

$$\mathbf{X}_1 = -gh_1 \nabla \eta_1 + \frac{h_1}{\rho_1} \nabla p_a \tag{9.5a}$$

$$\mathbf{X}_2 = -g\tilde{h}_2 \nabla[r\eta_1 + \epsilon\eta_2] + \frac{\tilde{h}_2}{\rho_2} \nabla p_a \tag{9.5b}$$

and

$$N_1 = \eta_1 - \eta_2 \tag{9.5c}$$

$$N_2 = \eta_2. \tag{9.5d}$$

We further manipulate the system of equations by first taking time derivative $\frac{\partial \eta_2}{\partial t}$, secondly multiplying with $-f\hat{k}\times$ each of the two momentum equations above and adding together, to obtain:

$$\left(\frac{\partial^2}{\partial t^2} + f^2\right) \mathbf{U}_1 = \left(\frac{\partial \mathbf{X}_1}{\partial t} + f\hat{k} \times \mathbf{X}_1\right) - \frac{1}{\rho_1}\left(\frac{\partial \boldsymbol{\tau}}{\partial t} + f\hat{k} \times \boldsymbol{\tau}\right) \tag{9.6a}$$

$$\left(\frac{\partial^2}{\partial t^2} + f^2\right) \mathbf{U}_2 = \left(\frac{\partial \mathbf{X}_2}{\partial t} + f\hat{k} \times \mathbf{X}_2\right). \tag{9.6b}$$

Taking divergence of both sides of (9.6a,b), we have:

$$\left(\frac{\partial^2}{\partial t^2}+f^2\right)\nabla\cdot\mathbf{U}_1=\nabla\cdot\left(\frac{\partial\mathbf{X}_1}{\partial t}+f\hat{k}\times\mathbf{X}_1\right)-\frac{1}{\rho_1}\left(\frac{\partial}{\partial t}\nabla\cdot\boldsymbol{\tau}^s+f\hat{k}\cdot\nabla\times\boldsymbol{\tau}^s\right) \tag{9.7a}$$

$$\left(\frac{\partial^2}{\partial t^2}+f^2\right)\nabla\cdot\mathbf{U}_2=\nabla\cdot\left(\frac{\partial\mathbf{X}_2}{\partial t}+f\hat{k}\times\mathbf{X}_2\right). \tag{9.7b}$$

The second parts of the terms including $\mathbf{X}_1$ and $\mathbf{X}_2$ vanish as a consequence of the vector identity

$$\nabla\cdot(\hat{k}\times\nabla s)=-\hat{k}\cdot\nabla\times\nabla s=0, \tag{9.8}$$

yielding

$$\left(\frac{\partial^2}{\partial t^2}+f^2\right)\nabla\cdot\mathbf{U}_1=\frac{\partial}{\partial t}\nabla\cdot\mathbf{X}_1-\frac{1}{\rho_1}\left(\frac{\partial}{\partial t}\nabla\cdot\boldsymbol{\tau}^s+f\hat{k}\cdot\nabla\times\boldsymbol{\tau}^s\right) \tag{9.9a}$$

$$\left(\frac{\partial^2}{\partial t^2}+f^2\right)\nabla\cdot\mathbf{U}_2=\frac{\partial}{\partial t}\nabla\cdot\mathbf{X}_2. \tag{9.9b}$$

Finally, replacing the divergence terms from continuity Eqs. (9.4c,d), we have

$$\frac{\partial}{\partial t}\left(\frac{\partial^2}{\partial t^2}+f^2\right)[\eta_1-\eta_2]=\frac{\partial}{\partial t}gh_1\nabla^2\eta_1+\frac{h_1}{\rho_1}\frac{\partial}{\partial t}\nabla^2p_a-\frac{1}{\rho_1}\left(\frac{\partial}{\partial t}\nabla\cdot\boldsymbol{\tau}^s+f\hat{k}\cdot\nabla\times\boldsymbol{\tau}^s\right) \tag{9.10a}$$

$$\frac{\partial}{\partial t}\left(\frac{\partial^2}{\partial t^2}+f^2\right)\eta_2=\frac{\partial}{\partial t}[gh_2r\nabla^2\eta_1+gh_2\epsilon\nabla^2\eta_2+gr\nabla\cdot h_b\nabla\eta_1+gfr\hat{k}\cdot\nabla\times h_b\nabla\eta_1]+\frac{\tilde{h}_2}{\rho_2}\frac{\partial}{\partial t}\nabla^2p_a. \tag{9.10b}$$

The above system of coupled Eqs. (9.10a,b) can be solved for η_1 and η_2, later to determine $\mathbf{U}_1$ and $\mathbf{U}_2$ from the elementary forms of the momentum Eqs. (9.3a,b).

We also need boundary conditions to supplement the governing equations. The no flux boundary conditions at the coast are specified as

$$\mathbf{n}\cdot\left(\frac{\partial\mathbf{X}_1}{\partial t}+f\hat{k}\times\mathbf{X}_1\right)-\frac{1}{\rho_1}\mathbf{n}\cdot\left(\frac{\partial\boldsymbol{\tau}^s}{\partial t}+f\hat{k}\times\boldsymbol{\tau}^s\right)=\left(\frac{\partial^2}{\partial t^2}+f^2\right)\mathbf{U}_1\cdot\mathbf{n}=0 \tag{9.11a}$$

$$\mathbf{n}\cdot\left(\frac{\partial\mathbf{X}_2}{\partial t}+f\hat{k}\times\mathbf{X}_2\right)=\left(\frac{\partial^2}{\partial t^2}+f^2\right)\mathbf{U}_2\cdot\mathbf{n}=0 \tag{9.11b}$$

where $\mathbf{X}_1$ and $\mathbf{X}_2$ are appropriately defined by (9.5a,b), in terms η_1 and η_2, added with the barometric pressure term, yielding mixed boundary conditions in terms of these variables.

9.3 Free Periodic Motions

Free motions are those without any forcing, i.e. when the elements of atmospheric forcing vanish, $\tau^s = 0$ and $p_a = 0$. Furthermore, if the motions are assumed to be periodic, solutions could be obtained directly from the coupled vorticity Eqs. (9.10a,b) with applied boundary conditions (9.11a,b).

For periodic solutions we can assume

$$\eta_i = N_i(x, y)e^{i\omega t} i = 1, 2 \tag{9.12}$$

upon which the governing equations are reduced to

$$\nabla^2 N_1 + \left(\frac{\omega^2 - f^2}{gh_1}\right)[N_1 - N_2] = 0 \tag{9.13a}$$

$$\nabla^2 N_2 + \left(\frac{\omega^2 - f^2}{g\epsilon h_2}\right) N_2 + \left(\frac{1-\epsilon}{\epsilon}\right)\nabla \cdot \left(\frac{h_2 - h_b}{h_2}\right)\nabla N_1 = 0 \tag{9.13b}$$

with the no flux boundary conditions

$$\frac{\partial N_1}{\partial s} + i\left(\frac{\omega}{f}\right)\frac{\partial N_1}{\partial n} = 0 \tag{9.14a}$$

$$\frac{\partial}{\partial s}[rN_1 + \epsilon N_2] + i\left(\frac{\omega}{f}\right)\frac{\partial}{\partial n}[(rN_1 + \epsilon N_2] = 0. \tag{9.14b}$$

with n being the coordinate normal to the boundary and s the coordinate along the boundary.

These are coupled Helmholz equations that would allow free wave motions along coasts or within closed basins or around islands. The main difficulty however lies in the mixed boundary conditions which imply asymmetrical reflection properties at lateral boundaries. When external forcing terms are included in general, the solution would be a suitable sum of free modes.

9.4 Normal Modes of Oscillation

The coupled system (9.10a,b) and 9.11a,b) implies that motion generated in one of the layers would essentially involve the other layer as well. However as in other coupled mechanical systems, there may be a way to decouple the system into separate "normal modes", which would be the basis set to make up the solution.

We therefore seek decomposition of the above two-layer system into normal modes, which would enable better understanding of the coupling between motions of the layers.

The ultimate aim is to recover original forms of the shallow water equations for each of the modes, 'as if' for a single layer:

$$\frac{\partial \mathbf{U}}{\partial t} + f\hat{k} \times \mathbf{U} = -g\nabla h_e \zeta + \mathbf{F} \tag{9.15a}$$

$$\frac{\partial \zeta}{\partial t} + \nabla \cdot \mathbf{U} = 0. \tag{9.15b}$$

Towards this end, we first make linear combinations of the momentum and continuity equations of each layer subject to a constant α

$$\mathbf{U} \equiv \alpha \mathbf{U}_1 + \mathbf{U}_2 \tag{9.16a}$$

$$\zeta \equiv \alpha N_1 + N_2 = \alpha(\eta_1 - \eta_2) + \eta_2 \tag{9.16b}$$

and similarly for the forcing

$$\mathbf{F} \equiv \left(\alpha \frac{h_1}{\rho_1} + \frac{\tilde{h}_2}{\rho_2} \right) \nabla p_a + \alpha \frac{\boldsymbol{\tau}^s}{\rho_1}. \tag{9.16c}$$

Accordingly, forming the linear combination of Eqs. (9.10a,b), we have

$$\frac{\partial}{\partial t}[\alpha \mathbf{U}_1 + \mathbf{U}_2] + f\hat{k} \times [\alpha \mathbf{U}_1 + \mathbf{U}_2] = -g\nabla[\alpha h_1 \eta_1 + \tilde{h}_2 r \eta_1 + \tilde{h}_2 \epsilon \eta_2] + \left(\alpha \frac{h_1}{\rho_1} + \frac{\tilde{h}_2}{\rho_2} \right) \nabla p_a + \alpha \frac{\boldsymbol{\tau}^s}{\rho_1} \tag{9.17a}$$

$$\frac{\partial}{\partial t}[\alpha(\eta_1 - \eta_2) + \eta_2] + \nabla \cdot [\alpha \mathbf{U}_1 + \mathbf{U}_2] = 0. \tag{9.17b}$$

In order that we can reduce the system to (9.15a,b) we need also to set

$$g h_e \zeta = g[\alpha h_1 \eta_1 + \tilde{h}_2 r \eta_1 + \tilde{h}_2 \epsilon \eta_2], \tag{9.17c}$$

where $h_e(x, y)$ is an "equivalent depth". We also note that the bottom topography $h_b = h_b(x, y)$, $\tilde{h}_2 = h_2 - h_b(x, y)$, while h_1 and h_2 are constants. In the present case with variable bottom topography, it is therefore implied that $h_e = h_e(x, y)$ to be determined from the necessary condition

$$h_e(x, y)[\alpha(\eta_1 - \eta_2) + \eta_2] = [\alpha h_1 \eta_1 + r\tilde{h}_2(x, y)\eta_1 + \epsilon \tilde{h}_2(x, y)\eta_2]. \tag{9.18}$$

By equating the coefficients of η_1 and η_2, i.e. we find:

$$\alpha(x, y) h_e(x, y) = \alpha(x, y) h_1 + r\tilde{h}_2(x, y) \tag{9.19a}$$

$$[1 - \alpha(x, y)] h_e(x, y) = \epsilon \tilde{h}_2(x, y), \tag{9.19b}$$

which forces that $\alpha = \alpha(x, y)$ is also a horizontally varying coefficient.

We have to solve these two equations for $h_e(x, y)$ and $\alpha(x, y)$, which appear as variable coefficients, in order to be compatible with the horizontal variations of $\tilde{h}_2(x, y) = h_2 - h_b(x, y)$ on the right hand sides of the equations.

9.4.1 Case I: Constant Depth

The problem will be more easily resolvable if we assume constant depth, $h_b = 0$. Then solving for both α and h_e as constants is possible:

$$\alpha h_e = \alpha h_1 + r h_2 \tag{9.20a}$$

$$(1 - \alpha) h_e = \epsilon h_2 \tag{9.20b}$$

by adding the two together, and noting $r + \epsilon = 1$

$$h_e = \alpha h_1 + h_2 \tag{9.21}$$

and eliminating α between the equations gives

$$h_e^2 - (h_1 + h_2) h_e + \epsilon h_1 h_2 = 0 \tag{9.22}$$

with two roots corresponding respectively to barotropic (superscript t) and the baroclinic (superscript c) modes:

$$h_e^{t,c} = \frac{h_1 + h_2}{2} \left(1 \pm \sqrt{1 - \frac{4 \epsilon h_1 h_2}{(h_1 + h_2)^2}} \right). \tag{9.23}$$

The roots of this quadratic equation must satisfy

$$h_e^t h_e^c = \epsilon h_1 h_2 \tag{9.24}$$

and making use of (9.21b)

$$\alpha = 1 - \frac{\epsilon h_2}{h_e}, \tag{9.25}$$

so that we obtain

$$\alpha = 1 - \frac{h_e^t h_e^c}{h_1 h_e}. \tag{9.26}$$

Making the appropriate selections for either the barotropic or baroclinic modes, we obtain the corresponding roots for α as follows

$$\alpha^t = 1 - \frac{h_e^c}{h_1} \tag{9.27a}$$

$$\alpha^c = 1 - \frac{h_e^t}{h_1}. \tag{9.27b}$$

9.4.2 Case II: Variable Bottom Topography

For the variable depth case $h_b(x, y)$ we can study the case where the upper layer depth does not intersect the bottom so that h_1 = constant, while the lower layer depth $\tilde{h}_2(x, y) = h_2 - h_b(x, y)$ is variable. Following the same steps as in the constant depth case,

$$\alpha(x, y)h_e(x, y) = \alpha h_1 + r\tilde{h}_2(x, y) \tag{9.28a}$$

$$(1 - \alpha(x, y))h_e(x, y) = \epsilon\tilde{h}_2(x, y). \tag{9.28b}$$

Adding the two together, and noting $r + \epsilon = 1$

$$h_e = \alpha h_1 + \tilde{h}_2. \tag{9.29}$$

Eliminating α between these equations gives

$$h_e^2 - (h_1 + \tilde{h}_2)h_e + \epsilon h_1\tilde{h}_2 = 0 \tag{9.30}$$

with the two roots

$$h_e^{t,c}(x, y) = \frac{h_1 + \tilde{h}_2(x, y)}{2}\left(1 \pm \sqrt{1 - \frac{4\epsilon h_1\tilde{h}_2(x, y)}{(h_1 + \tilde{h}_2(x, y))^2}}\right). \tag{9.31}$$

Similarly with the constant depth case, the coefficients α are given by

$$\alpha^t(x, y) = 1 - \frac{h_e^c(x, y)}{h_1}, \tag{9.32a}$$

$$\alpha^c(x, y) = 1 - \frac{h_e^t(x, y)}{h_1}. \tag{9.32b}$$

Following the same methodology as earlier, we propose the reduced system:

$$\frac{\partial \mathbf{U}}{\partial t} + f\hat{k} \times \mathbf{U} = -g\nabla h_e\zeta + \mathbf{F} \tag{9.33a}$$

$$\frac{\partial \zeta}{\partial t} + \nabla \cdot \mathbf{U} = 0. \tag{9.33b}$$

where $h_e(x, y)$ is a coefficient. By eliminating $\mathbf{U}$ between these equations, the following closed form is obtained

$$\frac{\partial}{\partial t}(\frac{\partial^2}{\partial t^2} + f^2)\zeta = \frac{\partial}{\partial t}g\nabla^2 h_e\zeta - \frac{\partial}{\partial t}\nabla \cdot \mathbf{F} - \nabla \times \mathbf{F}. \tag{9.34}$$

The above equation represents both of the two modes and has to be solved separately inserting the appropriate function $h_e^{t,c}(x, y)$ for the barotropic or baroclinic modes, remembering that the corresponding variables were defined as

$$\mathbf{U} = \alpha \mathbf{U}_1 + \mathbf{U}_2 \tag{9.35a}$$

$$\zeta = \alpha \eta_1 + (1 - \alpha)\eta_2. \tag{9.35b}$$

where the functions $\alpha^{t,c}(x, y)$ corresponding to the appropriate $h_e^{t,c}(x, y)$ for the barotropic (superscript t) and the baroclinic (superscript c) modes. In all, the above equations hold for each mode contributing to the layer motions.

The modal contributions for the barotropic mode are then

$$\mathbf{U}^t = \alpha^t \mathbf{U}_1 + \mathbf{U}_2 \tag{9.36a}$$

$$\zeta^t = \alpha^t \eta_1 + (1 - \alpha^t)\eta_2 \tag{9.36b}$$

satisfying

$$\frac{\partial \mathbf{U}^t}{\partial t} + f\hat{k} \times \mathbf{U}^t = -g\nabla h_e^t \zeta^t + \mathbf{F}^t \tag{9.37a}$$

$$\frac{\partial \zeta^t}{\partial t} + \nabla \cdot \mathbf{U}^t = 0. \tag{9.37b}$$

or

$$\frac{\partial}{\partial t}\left(\frac{\partial^2}{\partial t^2} + f^2\right)\zeta^t = \frac{\partial}{\partial t} g\nabla^2 h_e^t \zeta^t - \frac{\partial}{\partial t}\nabla \cdot \mathbf{F}^t - \nabla \times \mathbf{F}^t. \tag{9.38}$$

The modal contributions of the baroclinic mode are

$$\mathbf{U}^c = \alpha^c \mathbf{U}_1 + \mathbf{U}_2 \tag{9.39a}$$

$$\zeta^c = \alpha^c \eta_1 + (1 - \alpha^c)\eta_2 \tag{9.39b}$$

satisfying

$$\frac{\partial \mathbf{U}^c}{\partial t} + f\hat{k} \times \mathbf{U}^c = -g\nabla h_e^c \zeta^c \tag{9.40a}$$

$$\frac{\partial \zeta^c}{\partial t} + \nabla \cdot \mathbf{U}^c = 0. \tag{9.40b}$$

or

$$\frac{\partial}{\partial t}\left(\frac{\partial^2}{\partial t^2} + f^2\right)\zeta^c = \frac{\partial}{\partial t} g\nabla^2 h_e^c \zeta^c - \frac{\partial}{\partial t}\nabla \cdot \mathbf{F}^c - \nabla \times \mathbf{F}^c. \tag{9.41}$$

In the modal equations developed above, note that $\mathbf{F}^t$ and $\mathbf{F}^c$ respectively are the forcing terms partitioned with barotropic and baroclinic components of atmospheric pressure and wind stress by applying the respective coefficients α^t and α^c according to (9.16c).

Once the barotropic solutions ζ^t, $\mathbf{U}^t$ and the baroclinic solutions ζ^c, $\mathbf{U}^c$ are determined from the governing equations for each mode, their layer contributions are found by solving from Eqs. (9.16a,b), (9.35a,b) and (9.39a,b) which were used to sum up modal contributions as follows:

$$\eta_1 = \frac{(1-\alpha^c)\zeta^t - (1-\alpha^t)\zeta^c}{\alpha^t - \alpha^c} \tag{9.42a}$$

$$\eta_2 = \frac{-\alpha^c\zeta^t + \alpha^t\zeta^c}{\alpha^t - \alpha^c} \tag{9.42b}$$

and

$$\mathbf{U}_1 = \frac{\mathbf{U}^t - \mathbf{U}^c}{\alpha^t - \alpha^c} \tag{9.42c}$$

$$\mathbf{U}_2 = \frac{-\alpha^c\mathbf{U}^t + \alpha^t\mathbf{U}^c}{\alpha^t - \alpha^c} \tag{9.42d}$$

For any given forcing $\mathbf{F}$, both the barotropic and baroclinic modes will be present at the same time, their relative contributions determined by the effectiveness of the forcing. The solution in the forced case will be a combination of the free modes, which are studied in the next section.

9.5 Equations for Normal Modes

We can determine the properties of the normal modes in the absence of forcing, by looking at the 'pure' barotropic and baroclinic motions in the absence of atmospheric forcing (i.e. free motion).

9.5.1 Approximate Modes

Since ϵ is a small parameter $\epsilon \ll 1$, we can approximate the above results for the normal modes. Making use of the Taylor Series expansion

$$\sqrt{1-u^2} \approx 1 - \frac{u^2}{2} + O(x^4) \quad \text{near} \quad u = 0, \tag{9.43}$$

we can approximate the borotropic and baroclinic coefficients h_e and α respectively as

$$h_e^t(x,y) = h_1 + \tilde{h}_2(x,y) - \frac{\epsilon h_1 \tilde{h}_2(x,y)}{(h_1 + \tilde{h}_2(x,y))} + O(\epsilon^2) \tag{9.44a}$$

$$h_e^c(x,y) = \frac{\epsilon h_1 \tilde{h}_2(x,y)}{(h_1 + \tilde{h}_2(x,y))} + O(\epsilon^2) \tag{9.44b}$$

and

$$\alpha^t(x, y) = 1 - \frac{\epsilon \tilde{h}_2(x, y)}{(h_1 + \tilde{h}_2(x, y))} + O(\epsilon^2) \tag{9.44c}$$

$$\alpha^c(x, y) = -\frac{\tilde{h}_2(x, y)}{h_1} + \frac{\epsilon \tilde{h}_2(x, y)}{(h_1 + \tilde{h}_2(x, y))} + O(\epsilon^2). \tag{9.44d}$$

To leading order we can approximate, for the barotropic mode:

$$h_e^t(x, y) = h_1 + \tilde{h}_2(x, y) \equiv \tilde{h}(x, y) \tag{9.45a}$$

$$\alpha^t(x, y) = 1 \tag{9.45b}$$

(where $h(x, y)$ defines the total depth) and for the baroclinic mode:

$$h_e^c(x, y) = \frac{\epsilon h_1 \tilde{h}_2(x, y)}{(h_1 + \tilde{h}_2(x, y))} = \frac{\epsilon h_1 \tilde{h}_2(x, y)}{\tilde{h}(x, y)} \tag{9.46a}$$

$$\alpha^c(x, y) = -\frac{\tilde{h}_2(x, y)}{h_1}. \tag{9.46b}$$

Substituting for the barotropic mode yields

$$\mathbf{U}^t = \mathbf{U}_1 + \mathbf{U}_2 \tag{9.47a}$$

$$\zeta^t = \eta_1, \tag{9.47b}$$

which indicate that the transports of layers are additive and the surface displacement only includes the upper layer contribution, i.e. there is no influence of baroclinic contribution to this order.

For the baroclinic mode, substitution yields

$$\mathbf{U}^c = \mathbf{U}_2 - \frac{\tilde{h}_2(x, y)}{h_1}\mathbf{U}_1 \tag{9.48a}$$

$$\zeta^c = \frac{\tilde{h}(x, y)}{h_1}\eta_2 - \frac{\tilde{h}_2(x, y)\eta_1}{h_1}. \tag{9.48b}$$

9.5.2 Barotropic Normal Mode—Leading Order

By setting the baroclinic part of the solution, ζ^c and $\mathbf{U}^c$ to zero in Eqs. (9.42a–9.42d) and utilizing the leading order approximations for α^t and α^c, the pure barotropic mode is described by

$$\eta_1 = \zeta^t \tag{9.49a}$$

$$\eta_2 = \frac{\tilde{h}_2(x, y)}{\tilde{h}(x, y)}\zeta^t \tag{9.49b}$$

and

$$\mathbf{U}_1 = \frac{h_1}{\tilde{h}(x, y)}\mathbf{U}^t \tag{9.49a}$$

$$\mathbf{U}_2 = \frac{\tilde{h}_2(x, y)}{\tilde{h}(x, y)}\mathbf{U}^t \tag{9.49b}$$

and therefore,

$$\mathbf{U}_1 + \mathbf{U}_2 = \mathbf{U}^t \tag{9.50}$$

i.e. the barotropic velocity is the total transport velocity of both layers.

The structure of the barotropic mode to leading order is clear: in pure barotropic motion, the surface elevation is equal to the barotropic mode elevation, which is the only mode present. The interface elevation proportionally follows the surface elevation without any phase shift. The integrated velocities of each layer add up to the total transport velocity of the entire water column following the surface elevation without any phase shift.

We have for the effective depth:

$$h_e^t(x, y) = \tilde{h}(x, y), \tag{9.51}$$

and the equations we solve to determine the barotropic normal mode are therefore

$$\frac{\partial \mathbf{U}^t}{\partial t} + f\hat{k} \times \mathbf{U}^t = -g\nabla\tilde{h}(x, y)\zeta^t \tag{9.52a}$$

$$\frac{\partial \zeta^t}{\partial t} + \nabla \cdot \mathbf{U}^t = 0. \tag{9.52b}$$

or alternatively,

$$\nabla^2\left(\tilde{h}(x, y)\zeta^t\right) - \frac{1}{g}\left(\frac{\partial^2}{\partial t^2} + f^2\right)\zeta^t = 0. \tag{9.53}$$

9.5.3 Baroclinic Normal Mode—Leading Order

By setting the barotropic part of the solution, ζ^t and $\mathbf{U}^t$, to zero in the above equations and utilizing the leading order approximations for α^t and α^c in (9.42a–9.42d) the barotropic normal mode is described by

$$\eta_1 = 0 \tag{9.54a}$$

$$\eta_2 = \frac{h_1}{\tilde{h}(x, y)} \zeta^c \tag{9.54b}$$

and

$$\mathbf{U}_1 = -\frac{h_1}{\tilde{h}(x, y)} \mathbf{U}^c \tag{9.54c}$$

$$\mathbf{U}_2 = \frac{h_1}{\tilde{h}(x, y)} \mathbf{U}^c \tag{9.54d}$$

and therefore,

$$\mathbf{U}_1 + \mathbf{U}_2 = 0 \tag{9.55}$$

i.e. the net transport for both layers is zero, the transport in each layer being opposite to the other.

We have for the effective depth:

$$h_e^c(x, y) = \frac{\epsilon h_1 \tilde{h}_2(x, y)}{\tilde{h}(x, y)} \tag{9.56}$$

and the equations we solve to determine the barotropic normal mode are therefore

$$\frac{\partial \mathbf{U}^c}{\partial t} + f\hat{k} \times \mathbf{U}^c = -g\nabla \left(\frac{\epsilon h_1 \tilde{h}_2(x, y)}{\tilde{h}(x, y)} \right) \zeta^c \tag{9.57a}$$

$$\frac{\partial \zeta^c}{\partial t} + \nabla \cdot \mathbf{U}^c = 0. \tag{9.57b}$$

or alternatively,

$$\nabla^2 \left(\frac{\epsilon h_1 \tilde{h}_2(x, y)}{\tilde{h}(x, y)} \right) \zeta^c - \frac{1}{g} \left(\frac{\partial^2}{\partial t^2} + f^2 \right) \zeta^c = 0. \tag{9.58}$$

9.5.4 Basin Lateral Boundary Conditions

For both the barotropic or the baroclinic modes, the normal transport velocity at basin lateral boundaries must vanish:

$$\mathbf{n} \cdot \left(\frac{\partial \nabla h_e^{t,c} \zeta^{t,c}}{\partial t} + f\hat{k} \times \nabla h_e^{t,c} \zeta^{t,c} \right) = (\frac{\partial^2}{\partial t^2} + f^2) \mathbf{U}^{t,c} \cdot \mathbf{n} = 0 \tag{9.59}$$

which reduces to the mixed boundary condition

$$\frac{\partial}{\partial t}\frac{\partial}{\partial n}h_e^{t,c}\zeta^{t,c} + f\frac{\partial}{\partial s}h_e^{t,c}\zeta^{t,c} = 0 \tag{9.60}$$

where n is the normal coordinate and s is the tangent coordinate along the lateral boundary.

9.5.5 Periodic Motions in a Basin

Assuming periodic motions

$$\zeta^{t,c} = \psi^{t,c}(x, y)e^{i\omega t} \tag{9.61}$$

the governing equations are reduced to

$$\nabla^2\left(\tilde{h}(x, y)\psi^t\right) + \frac{1}{g}\left(\omega^2 - f^2\right)\psi^t = 0 \tag{9.62}$$

for the barotropic mode and

$$\nabla^2\left(\frac{\epsilon h_1\tilde{h}_2(x, y)}{\tilde{h}(x, y)}\psi^c\right) + \frac{1}{g}\left(\omega^2 - f^2\right)\psi^c = 0 \tag{9.63}$$

for the baroclinic mode, with corresponding basin boundary conditions

$$i\omega\frac{\partial}{\partial n}h_e^{t,c}\psi^{t,c} + f\frac{\partial}{\partial s}h_e^{t,c}\psi^{t,c} = 0. \tag{9.64}$$

The solution of the above eigenvalue problem for closed, variable topography basins can hopefully yield normal mode structures and eigenfrequencies for Kelvin and topographic basin modes.

Exercises

10

Chapter 1, **Exercise 1**

$$\mathbf{a} \cdot \mathbf{b} = a_i \hat{e}_i b_j \hat{e}_j = a_j \hat{e}_j b_i \hat{e}_i = b_i \hat{e}_i a_j \hat{e}_j = \mathbf{b} \cdot \mathbf{a}$$

$$\mathbf{a} \times \mathbf{b} = \epsilon_{ijk} a_j b_k \hat{e}_i = \epsilon_{ikj} a_k b_j \hat{e}_i = -\epsilon_{ijk} b_j a_k \hat{e}_i = -\mathbf{b} \times \mathbf{a}$$

$$\begin{aligned}(\mathbf{a} \times \mathbf{b}) \cdot \mathbf{c} &= \epsilon_{ijk} a_j b_k \hat{e}_i c_i \hat{e}_i = \epsilon_{kij} a_i b_j \hat{e}_k c_k \hat{e}_k \\ &= \epsilon_{kij} b_j c_k \hat{e}_k a_i \hat{e}_k = \epsilon_{ijk} b_j c_k \hat{e}_i a_i \hat{e}_i = (\mathbf{b} \times \mathbf{c}) \cdot \mathbf{a}\end{aligned}$$

Chapter 1, **Exercise 2**

Preparing a table of a_j, b_k ϵ_{ijk} and c_i for all possible combinations of i, j and k over the range 1, 2, 3 will show that $\mathbf{c} = \mathbf{a} \times \mathbf{b}$.

Chapter 1, **Exercise 3**

$$dW = \mathbf{f} \cdot d\mathbf{r} = m \frac{d^2\mathbf{r}}{dt^2} \cdot d\mathbf{r} = d\left(\frac{1}{2} m \frac{d\mathbf{r}}{dt} \cdot \frac{d\mathbf{r}}{dt}\right) = d\left(\frac{1}{2} m \mathbf{u} \cdot \mathbf{u}\right) \equiv dT$$

The work done by moving the particle from position $\mathbf{r}_1$ to $\mathbf{r}_2$ against the force field $\mathbf{f}$ is equal to the change in its kinetic energy.

$$W_{12} = \int_1^2 dW = \int_1^2 dT = \Delta T = T_2 - T_1 = T(\mathbf{r}_2) - T(\mathbf{r}_1)$$

.

(ii) Work done around the closed orbit C is equivalent to

$$W_{00} = \oint_C \mathbf{f} \cdot d\mathbf{r}.$$

© Springer Nature Switzerland AG 2020

E. Özsoy, *Geophysical Fluid Dynamics I*, Springer Textbooks in Earth Sciences, Geography and Environment, https://doi.org/10.1007/978-3-030-16973-2_10

Since $T = \frac{1}{2}m\mathbf{u}\cdot\mathbf{u}$, the choices are: (a) if $W_{00} = 0$, $|\mathbf{u}(t_0 + \Delta t)| = |\mathbf{u}(t_0)|$ and no work is done, which means that the force field $\mathbf{f}(\mathbf{r}(t), t)$ is *conservative*, (b) $W_{00} > 0$, work is done against the force field with a gain in kinetic energy, i.e. $|\mathbf{u}(t_0 + \Delta t)| > |\mathbf{u}(t_0)|$, (c) $W_{00} < 0$, work is done by the force field at the expense of kinetic energy, $|\mathbf{u}(t_0 + \Delta t)| < |\mathbf{u}(t_0)|$.

(iii) A conservative force field implies

$$W_{00} = \oint_C \mathbf{f}\cdot d\mathbf{r} = 0,$$

requiring the integration of a *total derivative*. This is only true if the force $\mathbf{f}$ can be expressed as a gradient of a scalar function V

$$\mathbf{f} = -\nabla V$$

where V is only a function of position,

$$V = V(\mathbf{r}),$$

without any dependence on time t. On the other hand, by Stokes' theorem (1.35), we can write

$$W_{00} = \oint_C \mathbf{f}\cdot d\mathbf{r} \equiv \int_S \nabla\times\mathbf{f}\cdot\mathbf{n}\, dS = 0,$$

where S is any surface encircled by the closed curve C and $\mathbf{n}$ is the unit normal vector to this surface. Since the closed curve C and the surface S are totally arbitrary, a conservative force field, $W_{00} = 0$, implies that the curl of vector force field has to vanish everywhere,

$$\nabla\times\mathbf{f} = 0,$$

for a conservative force, which can be used to check if the force field $\mathbf{f}$ is conservative or not. Note also that this would readily be satisfied by a conservative force field $\mathbf{f} = -\nabla V$, due to vector identity (1.27i):

$$\nabla\times\mathbf{f} = -\nabla\times\nabla V \equiv 0.$$

(iv) Based on the above developments, the integral around closed orbit C can be broken into two parts for a conservative force field:

$$\oint_C \mathbf{f}\cdot d\mathbf{r} = \int_{\mathbf{r}_0}^{\mathbf{r}} \mathbf{f}\cdot d\mathbf{r} + \int_{\mathbf{r}}^{\mathbf{r}_0} \mathbf{f}\cdot d\mathbf{r} = 0$$

where $\mathbf{r}_0 = \mathbf{r}(t_0)$ is the initial position at time $t = t_0$ and $\mathbf{r}(t)$ is any position at a later time t along the closed path.

$V(\mathbf{r})$ is the *potential function* that we recognize from the last part, and we can change one part of the above closed path integral to the other side of the equality to obtain

$$V(\mathbf{r}) = -\int_{\mathbf{r}_0}^{\mathbf{r}} \mathbf{f} \cdot d\mathbf{r} = \int_{\mathbf{r}}^{\mathbf{r}_0} \mathbf{f} \cdot d\mathbf{r}$$

proving that

$$dW = dT = \mathbf{f} \cdot d\mathbf{r} \equiv -dV$$

This result shows that $V(\mathbf{r})$ can be interpreted as *potential energy*, converted from *kinetic energy* that is spent (*or vica versa*) as the mass is moved from an initial position $\mathbf{r}_0 = \mathbf{r}(t_0)$ to $\mathbf{r}(t)$. The change in energy with position does not depend on the path, because the conservative process is *reversible* as the breakup of integral along the closed path C demonstrates.

(v) For a conservative force field $\mathbf{f}$ that is constant in time we can form

$$\frac{dT}{dt} = \mathbf{f} \cdot \frac{d\mathbf{r}}{dt} = \mathbf{f} \cdot \mathbf{u}$$

and similarly

$$\frac{dV}{dt} = -\mathbf{f} \cdot \frac{d\mathbf{r}}{dt} = -\mathbf{f} \cdot \mathbf{u},$$

resulting in

$$\frac{dE}{dt} \equiv \frac{d}{dt}(T + V) = 0$$

where we define $E = T + V$ as *total energy* (sum of kinetic and potential), and $\mathbf{u} = \frac{d\mathbf{r}}{dt}$ as the particle velocity. This result shows that in a conservative system kinetic energy can be converted into potential (and *or vica versa*), with the total energy remaining constant.

(vi) A force field that varies with time, $\mathbf{f} = \mathbf{f}(r(t), t)$, can not be a conservative field because it could not be a gradient of a potential that would yield a total derivative in the energy contour integral.

(vii) Setting $\mathbf{f} = \mathbf{f}_c + \mathbf{f}_n$, the rate of change of kinetic energy is

$$\frac{dT}{dt} = \mathbf{f}_c \cdot \mathbf{u} + \mathbf{f}_n \cdot \mathbf{u}.$$

whereas, the form of potential energy equation is the same as in part (vi) since it represents the conservative potential,

$$\frac{dV}{dt} = -\mathbf{f}_c \cdot \mathbf{u}$$

so that the first terms on the right hand side are cancelled upon adding these equations, to yield

$$\frac{dE}{dt} \equiv \frac{d}{dt}(T + V) = \mathbf{f}_n \cdot \mathbf{u}.$$

It is obvious that total energy can not be conserved in this general case, since the right hand side represents energy loss.

(a) For instance if the non-conservative force component $\mathbf{f}_n$ is a *frictional force*, assumed to be linearly parameterized as

$$\mathbf{f}_n = -\mu \mathbf{u}$$

acting in a direction opposite to that of the particle motion, the energy equation would be

$$\frac{dE}{dt} = \mathbf{f}_n \cdot \mathbf{u} = -\mu \mathbf{u} \cdot \mathbf{u} = -\mu |\mathbf{u}|^2$$

which would indicate a loss of total energy.

(b) In the case that non-conservative force component $\mathbf{f}_n$ is the fictitious *Coriolis force* (due to earth's rotation, see next chapter),

$$\mathbf{f}_n = -f \mathbf{k} \times \mathbf{u}$$

where f is defined as the *Coriolis parameter* and $\mathbf{k}$ is the unit vector normal to the plane of horizontal motion on the earth's surface. Coriolis force acts in the direction normal to velocity, which therefore gives

$$\frac{dE}{dt} = \mathbf{f}_n \cdot \mathbf{u} = -f(\mathbf{k} \times \mathbf{u}) \cdot \mathbf{u} = 0.$$

This is a surprising result. In this case, total energy is still conserved, in spite of the fact that Coriolis force is non-conservative, i.e. it can not be derived from gradient of a scalar potential.

Chapter 1, Exercise 4

`(i)` Since the motion is only in one direction, we replace the line integral around a closed orbit by what happens when the mass is displaced from rest to a maximum position x_m and brought back

$$\begin{aligned} W_{00} &= \int_0^{x_m} f\,dx + \int_{x_m}^0 f\,dx = -\int_0^{x_m} kx\,dx - \int_{x_m}^0 kx\,dx \\ &= -\frac{1}{2}kx_m^2 + \frac{1}{2}kx_m^2 = 0, \end{aligned}$$

which shows that the restoring force $f = -kx$ of the spring is conservative.

Making use of Newton's Second Law of motion $f = ma$, where the acceleration a is

$$a = \frac{du}{dt} = \frac{d}{dt}\frac{dx}{dt} = \frac{d^2x}{dt^2},$$

the equation of motion is written as

$$m\frac{d^2x}{dt^2} + kx = 0.$$

With the motion started from rest, with initial velocity u_0, given as initial conditions

$$x(0) = 0; \quad u(0) = u_0,$$

the solution is

$$x = c_1 \sin\left(\sqrt{\frac{k}{m}}t\right) + c_2 \cos\left(\sqrt{\frac{k}{m}}t\right)$$

where c_1 and c_2 are constants that determined from the initial conditions. Solution is then

$$x = \frac{u_0}{\omega_0}\sin(\omega_0 t)$$

where

$$\omega_0 = \sqrt{\frac{k}{m}}$$

is the natural frequency of the free oscillation. Motion is oscillatory and will be sustained for eternity since the system is conservative.

The maximum displacements are

$$x_m = \pm\frac{u_0}{\omega_0} = \pm u_0\sqrt{\frac{m}{k}}$$

and the values of velocity at these maximum displacements are $u_m = 0$.

`(iii)` For thr conservative spring force, the 1D analogue of the potential is

$$f = -kx = -\frac{dV}{dx}; \qquad V = \frac{1}{2}kx^2 = \frac{1}{2}mu_0^2\sin^2(\omega_0 t)$$

which is the potential energy of the system at any time. The kinetic energy is

$$T = \frac{1}{2}mu^2 = \frac{1}{2}mu_0^2\cos^2(\omega_0 t)$$

Total energy is conserved:

$$E = T + V = mu_0^2 \qquad \frac{dE}{dt} = 0$$

`(iv)` Adding friction force

$$f = -kx - \mu u = -kx - \mu\frac{dx}{dt},$$

we can confirm that the frictional term is not conservative, since the integral from rest to maximum displacement and back, in comparison to part `(ii)` will give

$$W_{00} = \int_0^{x_m} f\,dx + \int_{x_m}^0 f\,dx = -\int_0^{x_m}\left(kx + \mu\frac{dx}{dt}\right)dx - \int_{x_m}^0\left(kx + \mu\frac{dx}{dt}\right)dx$$
$$= -\int_0^{x_m}\mu\left(\frac{dx}{dt}\right)^2 dt - \int_{x_m}^0\mu\left(\frac{dx}{dt}\right)^2 dt < 0,$$

which due to positive integrand at all times, indicates negative work done against the force, or a positive loss of energy.

The equation of motion in this case is

$$m\frac{d^2x}{dt^2} + \mu\frac{dx}{dt} + kx = 0,$$

which has the following damped solution with the same initial conditions as in part `(ii)`:

$$x = \frac{u_0}{\omega_0}e^{-\frac{\mu}{2m}t}\sin(\omega_0 t)$$

where

$$\omega_0 = \sqrt{\frac{k}{m} - \frac{\mu^2}{4m^2}}$$

provided that $\mu < 2\sqrt{mk}$. For higher values of friction $\mu > 2\sqrt{mk}$ an over-damped solution is obtained as

$$x = -\frac{u_0}{\alpha}e^{-\alpha t}$$

where

$$\alpha = \frac{\mu}{2m} + \sqrt{\frac{\mu^2}{4m^2} - \frac{k}{m}}.$$

The kinetic energy at any time for the damped case is

$$T = \frac{1}{2}mu^2 = \frac{1}{2}mu_0^2 e^{-\frac{\mu}{m}t}\cos^2(\omega_0 t)$$

or for the over-damped case,

$$T = \frac{1}{2}mu_0^2 e^{-2\alpha t}.$$

Potential energy is due to the conservative force of the spring and the same as in part `(ii)`

$$V = \frac{1}{2}kx^2 = \frac{1}{2}k\frac{u_0^2}{\omega_0^2}e^{-\frac{\mu}{m}t}\sin^2(\omega_0 t) = \frac{1}{2}mu_0^2\left(\frac{1}{1 - \frac{\mu^2}{4mk}}\right)e^{-\frac{\mu}{m}t}\sin^2(\omega_0 t)$$

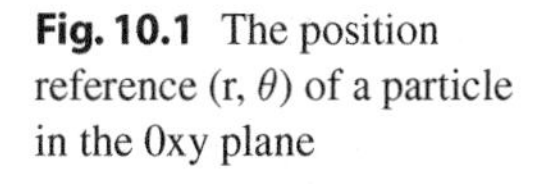

Fig. 10.1 The position reference (r, θ) of a particle in the 0xy plane

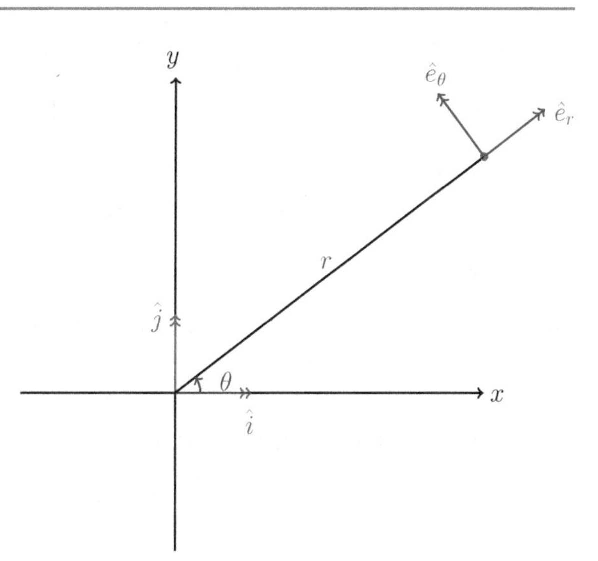

yielding

$$E = T + V = \frac{1}{2} m u_0^2 e^{-\frac{\mu}{m} t} + m u_0^2 \left(\frac{1 + \frac{\mu^2}{4mk}}{1 - \frac{\mu^2}{4mk}} \right) e^{-\frac{\mu}{m} t} \sin^2(\omega_0 t).$$

For the over-damped case,

$$E = T + V = \frac{1}{2} m u_0^2 \left(\frac{1}{\alpha^2 + 1} \right) e^{-\alpha t}$$

Chapter 2, Exercise 1

To set an example of *simple shearing motion* as a superposition of *pure-straining* and *rigid-body rotation* modes, we start with rigid body rotation with angular velocity $-\frac{1}{2}\omega\hat{e}\theta$ (in the clockwise sense) perpendicular to the $0xy$ plane (or Or_1r_2) shown in Fig. 10.1, which would have velocity components

$$\delta \mathbf{u}_\theta = -\frac{1}{2}\omega r \; \hat{e}_\theta \quad \text{and} \quad \delta \mathbf{u}_r = 0$$

in polar coordinates (r, θ), with unit vectors $\hat{e}_r$, $\hat{e}_\theta$, transformed into cartesian coordinates Oxy with unit vectors $\hat{i}$, $\hat{j}$ along respective axes.

The unit vectors transformed to cartesian coordinates are

$$\hat{e}_r = \hat{i} \cos\theta + \hat{j} \sin\theta \quad \text{and} \quad \hat{e}_\theta = -\hat{i} \sin\theta + \hat{j} \cos\theta$$

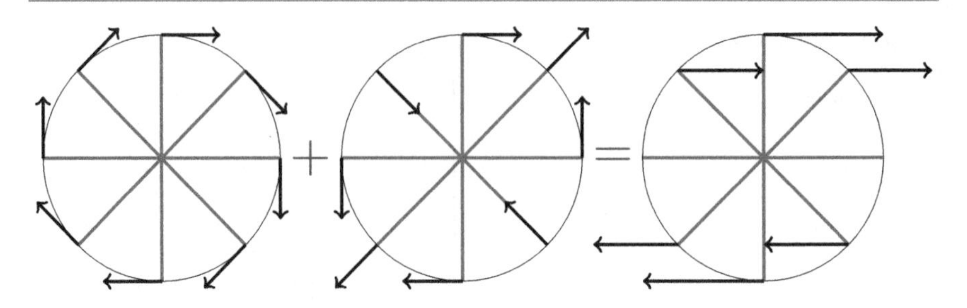

Fig. 10.2 The sum of *rigid-body rotation* and *pure-straining* modes resulting in *simple shearing* motion

so that the the rigid-body rotation mode becomes

$$\delta \mathbf{u}^r = -\frac{1}{2}\omega r\ \hat{e}_\theta = \frac{1}{2}\omega r[\hat{\imath}\sin\theta - \hat{\jmath}\cos\theta].$$

The pure straining motion (without volume change), along principal axes oriented 45° to the x, y axes is proposed to be

$$\delta \mathbf{u}^s = \frac{1}{2}\omega(x\hat{\jmath} + y\hat{\imath}) = \frac{1}{2}\omega r[\hat{\imath}\sin\theta + \hat{\jmath}\cos\theta].$$

As shown in Fig. 10.2 the sum gives rise to a simple shearing motion

$$\delta \mathbf{u} = \delta \mathbf{u}^r + \delta \mathbf{u}^s = \omega r \sin\theta\ \hat{\imath} = \omega y\ \hat{\imath}.$$

Chapter 2, **Exercise 2**

The continuity equation $\nabla \cdot \mathbf{u} = 0.$ is satisfied by the velocity components $u_r = -a/r,\ u_\theta = -a/r$. The stream-function ψ is obtained by integrating

$$-\frac{\partial \psi}{\partial r} = -\frac{a}{r}, \quad \text{and} \quad \frac{1}{r}\frac{\partial \psi}{\partial \theta} = -\frac{a}{r}$$

to yield

$$\psi = a \ln r + c_1(\theta)$$
$$\psi = -a\theta + c_2(r)$$

or

$$\psi = a(\ln r - \theta),$$

which gives a vorticity field of $\boldsymbol{\omega} = 0$, indicating an irrotational velocity field. For constant stream-function values of $\psi = \psi_0$, the streamlines are spirals

$$r = e^{\frac{\psi_0}{a}} e^\theta = C e^\theta$$

where C is a constant.

Chapter 3, **Exercise 1**

Angular velocity of earth's rotation is $\Omega = 2\pi/86400 = 0.727\ 10^{-4}$ and the Coriolis parameter is $f = 2\Omega \sin\phi$, where ϕ is the latitude of the location, where the cannonball is assumed to be fired towards the north, with horizontal velocity $v = 120\,\text{m/s}$ to last for $\Delta t = 25\,\text{s}$. The lateral deflection of the cannonball relative to the planned course is calculated in the following table.

location	latitude	f	$\frac{du}{dt} = fv$	$u = fv\Delta t$	$x = fv(\Delta t)^2/2$
-	ϕ	$(10^{-4}s^{-1})$	(ms^{-2})	(ms^{-1})	(m)
Paris	$48°52'N$	0.642	1.31	32.86	411
Quito	$0°13'S$	−0.00548	−0.0065	−0.16	−20
Johannesburg	$26°12'S$	−1.095	−0.77	−19.26	−240

Chapter 3, **Exercise 2**

The Great Red Spot of Jupiter is centered at latitude of $\phi = 22°$ and spans $12° - 25°$ of arc on the planet with a equatorial radius of $71,400\,\text{km}$, from which the horizontal scale of the motion can roughly be estimated as $L_0 = 71,000 * tan(20°) = 24,420\,\text{km}$. With angular speed of rotation $1.763 \times 10^{-4}\,\text{s}^{-1}$, the Coriolis parameter for the given latitude is $f = 2\Omega sin\phi = 1.321 \times 10^{-4}\,\text{s}^{-1}$, and for typical wind speeds of about $100\,\text{m/s}$, the Rossby number is estimated as

$$Ro = \frac{U_0}{f_0 L_0} = 0.03,$$

which shows that the rotation effects are significant in shaping the Red Spot, since $1/Ro$ measures the relative importance of Coriolis to inertia terms in the equations of motion.

Chapter 3, **Exercise 3**

Continuity equation between the two sections is

$$Q = \frac{\pi}{4} D_1^2 U_1 = \frac{\pi}{4} D_2^2 U_2.$$

The other equation we need is the Bernouilli equation (see preface photo)

$$\Delta p = p_1 - p_2 = \frac{1}{2}\rho(U_2^2 - U_1^2).$$

Solving for the discharge Q between these equations yield

$$Q = \frac{\pi}{4} D_1^2 \left(\frac{2\Delta p}{\rho\left[(\frac{D_1^4}{D_2^4} - 1) \right]} \right)^{\frac{1}{2}}$$

This configuration of Venturi can be used to measure the total discharge through the pipe, using geometry and measured values of Δp.

Chapter 3, **Exercise 4**

Using the Bernouilli equation (3.83b) and (3.84), and setting $\mathbf{u} = 0$ and $p = 0$ for this steady and inviscid equilibrium state, we have

$$h(r) - \frac{1}{2g}\Omega^2 r^2 = h_c$$

where $h(r)$ is the free surface height that varies with radial distance, and $h_c = h(0)$ the height at the center ($r = 0$).

The constant height at the center h_c can be found by equating fluid volume before and after rotation,

$$2\pi \int_0^R h(r)\, r\, dr = h_0 \pi R^2$$

where R is the radius of the cylindrical container. This yields

$$2\pi \left(\frac{1}{2g}\Omega^2 \frac{R^4}{4} + h_c \frac{R^2}{2} \right) = h_0 \pi R^2$$

or

$$h_c = h_0 - \frac{1}{4g}\Omega^2 R^2$$

so that the surface shape is given by

$$h(r) - h_0 = \frac{1}{2g}\Omega^2 \left(r^2 - \frac{1}{2}R^2 \right).$$

Chapter 4, **Exercise 1**

At initial time, hydrostatic pressure along the tube is given by

$$p(z) = \rho g(\ell - z) + p_a, \text{ vertical leg, } z > 0$$

$$p(x) = \rho g \ell + p_a, \text{ horizontal leg. } x > 0$$

Pressure distribution immediately after opening the lower end at time $t = t_{0+} > t_0$ can be obtained from elementary forms of the linear, inviscid equations of motion for each leg

$$-\frac{dU}{dt} = -\frac{1}{\rho}\frac{dp}{dz} - g, \text{ vertical leg}, z > 0, \text{ at time } t = t_{0+},$$
$$\frac{dU}{dt} = -\frac{1}{\rho}\frac{dp}{dx} \text{ horizontal leg}, x > 0, \text{ at time } t = t_{0+},$$

which are integrated to give

$$p(z) = -\rho g z + \rho\left(\frac{dU}{dt}\right) z + c_1, \text{ vertical leg}, \ z > 0, \text{ at time } t = t_{0+},$$
$$p(x) = -\rho\left(\frac{dU}{dt}\right) x + c_2, \text{ horizontal leg}, \ x > 0, \text{ at time } t = t_{0+}.$$

since the pressure in both legs have to be equal at the corner (B),

$$p(z = 0) = c_1 = p(x = 0) = c_2.$$

At the two open ends (A and C) pressure is equalized to atmospheric value p_a,

$$p(\ell) = -\rho g \ell + \rho\left(\frac{dU}{dt}\right)\ell + c_1 = p_a, \text{ at point C}, t = t_{0+},$$
$$p(\ell) = -\rho\left(\frac{dU}{dt}\right)\ell + c_2 = p_a, \text{ at point } A, t = t_{0+}.$$

Adding the two equations together

$$-\rho g \ell + c_1 + c_2 = 2p_a$$

and noting $c_1 = c_2$ from the above,

$$c_1 = c_2 = p_a + \frac{1}{2}\rho g \ell.$$

Subtracting yields the fluid acceleration

$$\left(\frac{dU}{dt}\right) = \frac{1}{2}g,$$

so that the pressure at the corner (B) is calculated as

$$p(x = 0, z = 0) = \frac{1}{2}g\ell$$

and therefore the pressure distribution in both legs are obtained as follows

$$p(z) = \frac{1}{2}\rho g(\ell - z) + p_a, \text{ vertical leg, } z > 0, \text{ at time } t = t_{0+},$$

$$p(x) = \frac{1}{2}\rho g(\ell - x) + p_a, \text{ horizontal leg, } x > 0, \text{ at time } t = t_{0+}.$$

Chapter 4, **Exercise 2**

For physical effects leading to the "Einstein's tea leaves" phenomenon, see the article.

Tandon, A. and J. C. Marshall, 2010. Einstein's Tea Leaves and Pressure Systems in the Atmosphere, The Physics Teacher 48(5):292-295, DOI: 10.1119/1.3393055.

Chapter 4, **Exercise 3**

The proposed solution

$$u = A + By + Cz + Dy^2 + Ez^2$$

has to satisfy the uniform viscous flow equation

$$\frac{\partial^2 u}{\partial y^2} + \frac{\partial^2 u}{\partial z^2} = -\frac{G}{\mu},$$

subject to a pressure gradient $G = -\frac{dp}{dx}$, as well as the no slip boundary condition

$$u(\text{on boundary } C : (y_0, z_0)) = 0,$$

the boundary C of the pipe being described by an ellipse in the (y, z)-plane

$$\frac{y_0^2}{b^2} + \frac{z_0^2}{c^2} = 1.$$

First we consider the symmetrical geometry of the problem and observe that for the trial solution not to violate symmetry some terms should vanish, requiring

$$B = 0, \qquad C = 0.$$

On the other hand, evaluating the trial solution first at $y = 0, z = b$ and later at $y = y_0, z = 0$, we obtain

$$u(y = b, z = 0) = A + Db^2 = 0, \qquad u(y = 0, z = c) = A + Ec^2 = 0.$$

On the other hand, substituting the trial solution in the governing equation yields

$$2D + 2E = -\frac{G}{\mu}.$$

Solving for the coefficients A, D, E from the above algebraic equations gives the solution

$$u = \frac{G}{2\mu}\left[\frac{1-\left(\frac{y^2}{b^2}+\frac{z^2}{c^2}\right)}{\left(\frac{1}{b^2}+\frac{1}{c^2}\right)}\right]$$

Chapter 4, **Exercise 4**

The two-layer flow is driven by a constant pressure gradient $\frac{\partial p}{\partial x}$. For each layer i, the governing uniform flow equations are

$$\frac{\partial^2 u_i}{\partial z^2} = \frac{1}{\mu_i}\frac{\partial p}{\partial x}, \text{ for } i = 1, 2,$$

with boundary and jump conditions

$$u_1 = 0 \text{ at } z = a, \qquad u_1 = 0 \text{ at } z = -a,$$

$$u_1 = u_2 \text{ at } z = 0,$$

$$\mu_1\frac{\partial u_1}{\partial z} = \mu_2\frac{\partial u_2}{\partial z} \text{ at } z = 0.$$

Solutions that can be tried are

$$u_1 = \frac{1}{2\mu_1}\frac{\partial p}{\partial x}z^2 + A_1 z + A_2,$$

$$u = 2 = \frac{1}{\mu_2}\frac{\partial p}{\partial x}z^2 + B_1 z + B_2.$$

Boundary and jump conditions require

$$\frac{1}{2\mu_1}\frac{\partial p}{\partial x}a^2 + A_1 a + A_2 = 0,$$

$$\frac{1}{\mu_2}\frac{\partial p}{\partial x}a^2 - B_1 a + B_2 = 0.$$

$$A_2 = B_2$$

$$\mu_1 A_1 = \mu_2 A_2$$

Solving for the constants A_1, A_2, B_1, B_2 yields

$$u_1 = \frac{a^2}{2\mu_1}\left[\left(\frac{z}{a}\right)^2 - \frac{\mu_2-\mu_1}{\mu_1+\mu_2}\left(\frac{z}{a}\right) - \frac{2\mu_1}{\mu_1+\mu_2}\right]\frac{\partial p}{\partial x}$$

$$u_2 = \frac{a^2}{2\mu_2}\left[\left(\frac{z}{a}\right)^2 - \frac{\mu_2-\mu_1}{\mu_1+\mu_2}\left(\frac{z}{a}\right) - \frac{2\mu_2}{\mu_1+\mu_2}\right]\frac{\partial p}{\partial x}$$

Chapter 4, Exercise 5

The flow is described by

$$-V_0 \frac{\partial U}{\partial y} = \nu \frac{\partial^2 U}{\partial y^2}$$

with boundary conditions

$$U = 0 \text{ at } y = 0$$

$$U \to U_0 \text{ as } y \to \infty.$$

The solution has the form

$$U(y) = A + Be^{-\frac{V_0}{\nu} y}$$

and determining constants A, B from boundary conditions gives

$$U(y) = U_0(1 - e^{-\frac{V_0}{\nu} y}).$$

The momentum flux across any horizontal surface S at level y, with a normal vector $\mathbf{n}$ pointing down in the $-y$ direction is

$$\mathbf{M} = \int_S \rho \mathbf{u}(\mathbf{u} \cdot \mathbf{n}) ds$$

and we can define $\mathbf{m}$ as the momentum flux calculated per unit area ds

$$\mathbf{m} = \rho \mathbf{u}(\mathbf{u} \cdot \mathbf{n}) = \rho(-uv\mathbf{i} - vv\mathbf{j})$$

evaluated for the above solution as

$$\mathbf{m} = \rho U_0 V_0 (1 - e^{-\frac{V_0}{\nu} y})\mathbf{i} - \rho V_0^2 \mathbf{j}$$

At the surface,

$$\mathbf{m}(0) = -\rho V_0^2 \mathbf{j}$$

and at large distance from the surface,

$$\mathbf{m} \to \rho U_0 V_0 \mathbf{i} - \rho V_0^2 \mathbf{j} \quad \text{as y} \to \infty.$$

The shear stress at any horizontal surface in the fluid is

$$\tau = \mu \frac{\partial u}{\partial y} = \mu \frac{\partial}{\partial y} U(y) = \rho U_0 V_0 e^{-\frac{V_0}{\nu} y}$$

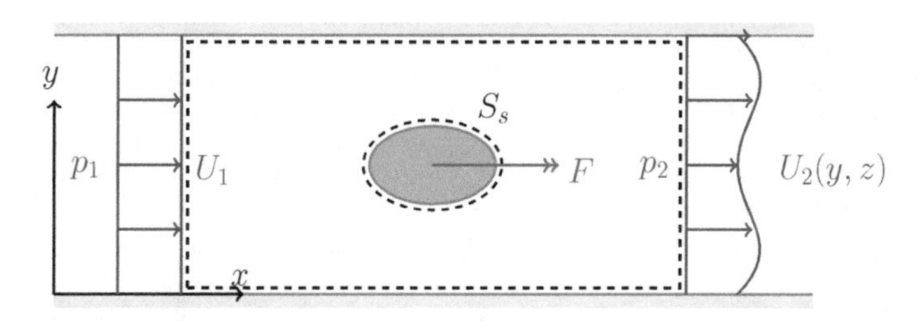

Fig. 10.3 Flow past a rigid body in a channel

and the force applied per unit area at the surface is the shear stress at the boundary

$$\tau = \rho U_0 V_0$$

Chapter 4, Exercise 6

The situation could arise in a wind-tunnel test where measurements in the fore and aft of a tested object could be used to obtain remote information on the object's behavior. We consider the volume shown with dashed lines, enclosed by the fluid exterior domain and containing the tested object as shown in Fig 10.3. For simplicity we will assume that there is no friction on the outer surface along the wind tunnel boundaries, and pressure distributions on the wind tunnel are either negligible or symmetrical to cancel out.

The integral momentum equation (3.12) can be used for the fluid volume of which the outer boundary is shown by dashed lines, to write

$$\int_S \rho \mathbf{u}\mathbf{u} \cdot \mathbf{n}\, dS = \int_S (\rho\psi\mathbf{n} + \hat{\sigma} \cdot \mathbf{n})\, dS.$$

Using the notation of Chap. 2 and remembering the stress tensor

$$\hat{\sigma} = \sigma_{ij} = -p\delta_{ij} + d_{ij}$$

evaluated in the above equation is

$$\hat{\sigma} \cdot \mathbf{n} = \sigma_{ij} n_j = -p n_i + d_{ij} n_j,$$

the integral momentum equation takes the form

$$\int_S \rho \mathbf{u}\mathbf{u} \cdot \mathbf{n}\, dS = \int_S \rho\psi\mathbf{n}\, dS - \int_S p\mathbf{n}\, dS + \int_S \hat{d} \cdot \mathbf{n}\, dS.$$

We note that the first term on the right hand side is negligible as the scale of the experiment is too small to consider any changes in the gravity potential, and applying the integral momentum equation with the various other assumptions we have made,

Fig. 10.4 Upwelling on the eastern shores of the Aegean Sea on 9 July 2017. Credits: NASA Worldview (https://worldview.earthdata.nasa.gov/)

the force $\mathbf{F}_s$ i.e. sum of frictional and pressure forces applied on the body surface S_s is obtained as

$$\mathbf{F}_s = \int_{S_s} \hat{d} \cdot \mathbf{n}\, dS - \int_{S_s} p\mathbf{n}\, dS \simeq \rho U_1^2 A - \rho \int_A U_2^2(y,z)\, dy\, dz + (p_1 - p_2)A.$$

Chapter 4, **Exercise 7**

This exercise is for the reader to develop a general understanding on coastal upwelling processes based on literature research. We only provide an example of upwelling in the Aegean Sea in Fig. 10.4, forced by the persistent northerly *Etesian* wind regime that typically intensifies in summer.

Chapter 5, **Exercise 1**

The motion is governed by

$$\frac{du}{dt} - fv = g\, \sin\alpha - ku$$

$$\frac{dv}{dt} + fu = -kv$$

The Coriolis term is a cross product in the vector form of the equations

$$\frac{d\mathbf{u}}{dt} + f\mathbf{k} \times \mathbf{u} = g_s\, \imath - k\mathbf{u}$$

where $g_s = g\ \sin\alpha$ is the short for the slope component of gravity. The initial condition is given as $\mathbf{u}(t = 0) = 0$.

Assigning vector components of the horizontal velocity to real and imaginary parts of a complex variable w

$$\mathbf{u} = (u, v) = u\imath + v\jmath \rightarrow w = u + iv$$

the vector equation is converted to a single equation for w

$$\frac{dw}{dt} + (k + if)w = g_s$$

with initial conditions, chosen to be

$$w(0) = 0.$$

The solution follows as

$$w = \frac{g_s}{k + if}\left[1 - e^{(k+if)t}\right].$$

The position of the ball can be estimated by integrating velocity using the Euler approximation, although used to replace actual Lagrangian dynamics as

$$\frac{dz}{dt} = w$$

where

$$z = x + iy = re^{i\theta}$$

is the complex representation of the position in the coordinate system Oxy, to yield

$$\begin{aligned} z =& \frac{g_s}{k + if}\left[t + \frac{1}{k + if}\left(e^{(k+if)t} - 1\right)\right] \\ =& \frac{g_s}{(k + if)^2}\left[(k + if)t + \left(e^{(k+if)t} - 1\right)\right] \end{aligned}$$

We define the non-dimensional complex position

$$Z = X + iY \equiv \frac{(k + if)^2}{g_s} z = \rho e^{i\delta} z = \rho e^{i\delta} r e^{i\theta}$$

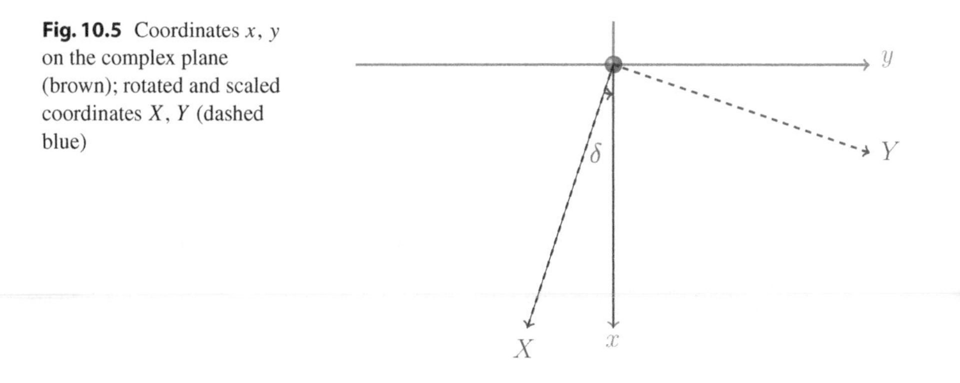

Fig. 10.5 Coordinates x, y on the complex plane (brown); rotated and scaled coordinates X, Y (dashed blue)

where

$$\rho = \frac{[(f^2 - k^2)^2 + (2kf)^2]^{\frac{1}{2}}}{g_s}; \qquad \delta = -tan^{-1}\frac{2kf}{f^2 - k^2}.$$

The complex coordinate Z is a non-dimensional form of the original coordinates z scaled by a ratio ρ and rotated by the angle δ. The original and rotated coordinates are shown in Fig. 10.5 Also defining a non-dimensional time variable

$$T = ft, \qquad \nu = \frac{k}{f},$$

the solution in rotated coordinates are

$$Z = (1 - e^{-(\nu T + iT)}) - (\nu T + iT)$$

so that

$$X = 1 - e^{-\nu T}\cos T - \nu T \quad \text{and} \quad Y = e^{-\nu T}\sin T - T.$$

Combining these parametric equations gives the trajectory

$$(X - 1 + \nu T)^2 + (Y + T)^2 = 1$$

which is a cnoidal functional form.

With realistic values of parameters, assuming that the experiment is done at latitude $\phi = 30° N$, with a slope angle of $\alpha = 10°$ and $k = 0.01$, we find $f = 7.27 \times 10^{-5}\,\text{s}^{-1}$ (corresponding to an inertial period of $T_i = 24\,\text{h}$), the slope gravity component of $g_s = 1.7\,\text{ms}^{-2}$.

To see if in reality we could perform an experiment as imagined we can calculate the scaling factor $\rho = 3.1 \times 10^{-9}$ so that an oscillation cycle which completes in $Y = 2\pi$ in the dimensionless scale would amount to a horizontal distance of 2×10^6 km !!!—hence not very realistic. For this experiment to be realistically performed, rotation rate should be increased by more than a million times, most likely to be realized on on a rotating table. In the ocean and atmosphere inertial oscillations

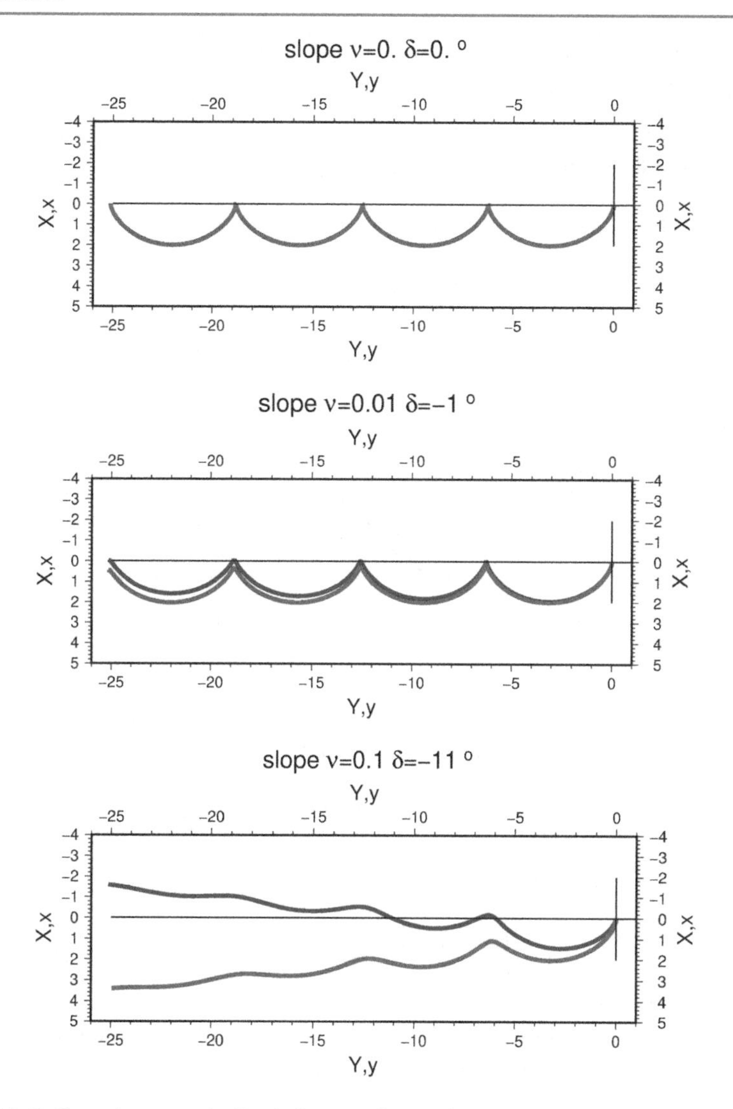

Fig. 10.6 Ball moving on an inclined plane on the rotating earth, (top) without friction $\nu = 0$, (middle) $\nu = 0.01$, (bottom) $\nu = 0.1$. Blue line: with respect to rotated coordinates and scaled X, Y; red line: with respect to original but scaled coordinates x, y

are commonly observed but with much smaller horizontal scales of motion, as a result of quite different forcing and dissipation mechanisms in comparison to the present hypothetical slope experiment.

(i) The general case $f \neq 0,\ k \neq 0$ has been analyzed above with examples displayed in Figs. 10.6 and 10.7,

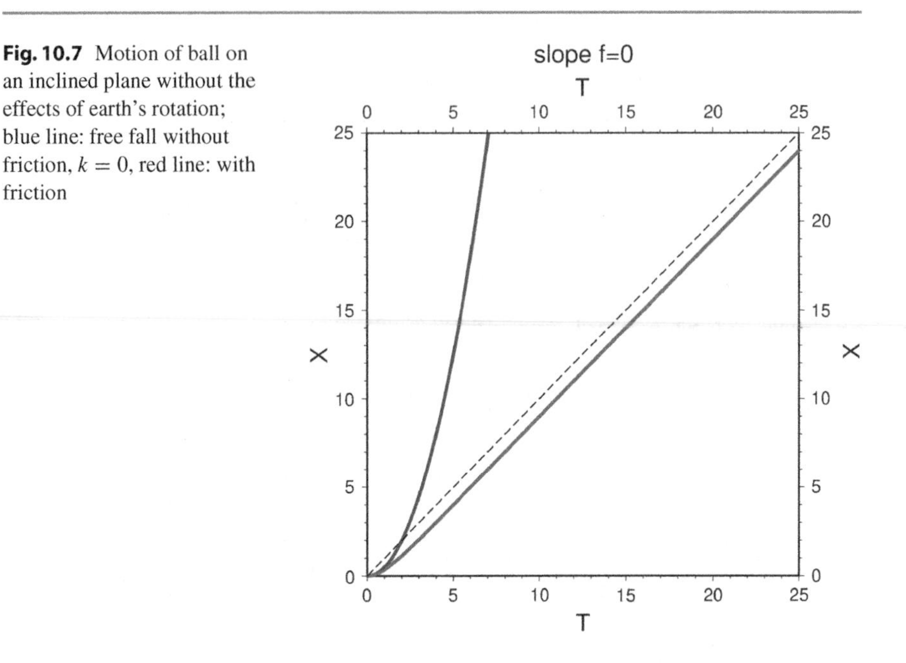

Fig. 10.7 Motion of ball on an inclined plane without the effects of earth's rotation; blue line: free fall without friction, $k = 0$, red line: with friction

(ii) the special case without friction is the first panel of the plot in Fig. 10.6 calculated by setting $\nu = 0$, yielding the solution

$$X = 1 - \cos T \quad \text{and} \quad Y = \sin T - T$$

with the cnoidal trajectory

$$(X - 1)^2 + (Y + T)^2 = 1.$$

shown in the top panel.

(iii) The case without rotation $f = 0$, $k \neq 0$ is obtained by setting $f = 0$ which necessitates that $T = ft = 0$ and letting $\tilde{T} = kt$

$$X = \tilde{T} - (e^{-\tilde{T}} - 1) \quad \text{and} \quad Y = 0,$$

which in dimensional terms gives

$$x = \frac{g_s}{(k)^2}[kt + (e^{-kt} - 1)].$$

Note also that the sign of X is reversed in the definition $X = \frac{(k)^2}{g_s}x$. This is a damped motion along the x direction, with the time history given in Fig. 10.7.

(iv) The case for $f = 0,\ k = 0$; results in a straight line motion along x-axis with velocity $u = g_s t$, making the distance traveled $x = \frac{1}{2} g_s t^2$. This can be seen from the solution limit for $f = 0, k \to 0$ $(\nu \to 0)$

$$
\begin{aligned}
x &= lim\{k \to 0\} \frac{g_s}{k^2}[kt + e^{-kt} - 1] \\
&= lim\{k \to 0\} \frac{g_s}{k^2}[kt + 1 - kt + \frac{(kt)^2}{2!} - \ldots - 1] \\
&= \frac{1}{2} g_s t^2
\end{aligned}
$$

which is a familiar result for free-fall.

(v) If an initial acceleration a_0 is applied,

$$\frac{dw}{dz} = a_0$$

and therefore the solution for velocity would become

$$w = \frac{1}{k + if}\left[g_s - a_0 e^{(k+if)t}\right]$$

and the position of the ball is

$$z = \frac{1}{k + if}\left[g_s t + \frac{a_0}{k + if}\left(e^{(k+if)t} - 1\right)\right].$$

Letting

$$\gamma = \gamma^R + i\gamma^I = \frac{a_0}{g_s}$$

where both γ and a_0 can be complex constants, and persisting with the earlier non-dimensional variables

$$Z = \gamma(1 - e^{-(\nu+i)T}) - \nu T - iT$$

and

$$
\begin{aligned}
X &= \gamma^R - \nu T - \gamma^R e^{-\nu T} \cos T - \gamma^I e^{-\nu T} \sin T \\
Y &= \gamma^I - T - \gamma^I e^{-\nu T} \cos T + \gamma^R e^{-\nu T} \sin T.
\end{aligned}
$$

These solutions are plotted in Fig. 10.8 for positive initial acceleration $\gamma = 2 + 2i$ in the X and Y directions, with larger cnoidal loop trajectories as displayed.

Chapter 5, Exercise 2

Continuity equation for the fluid is

$$U_0 H_0 = U(x) H(x).$$

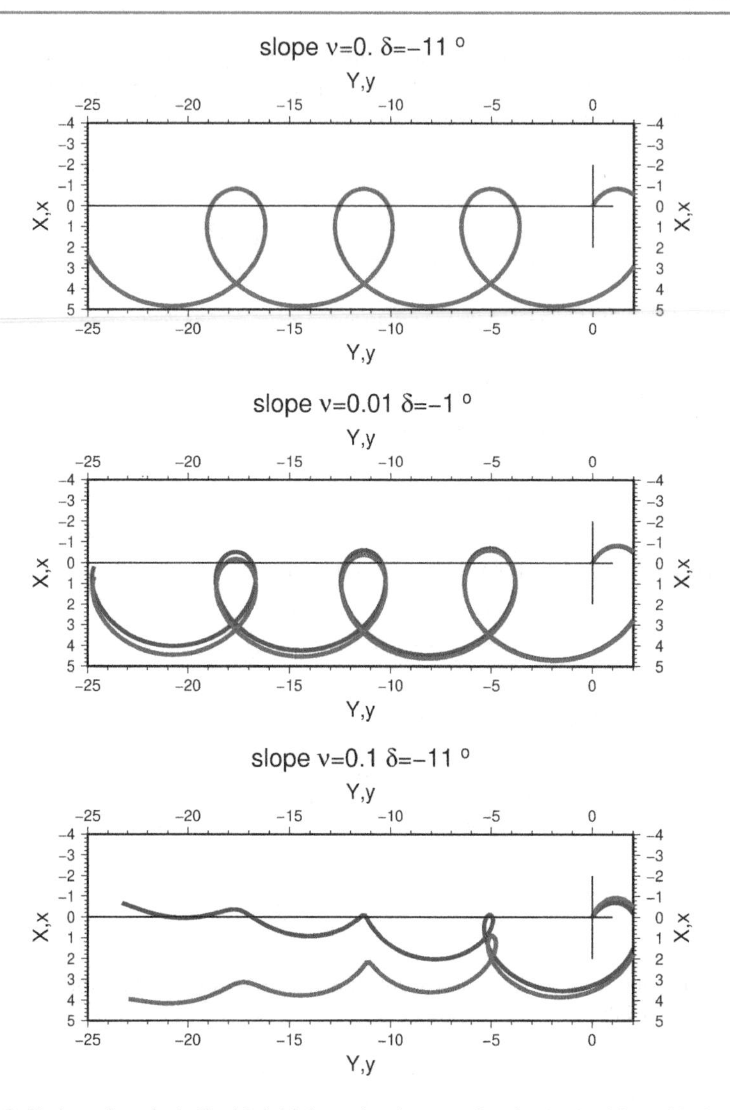

Fig. 10.8 Trajectories of a ball with initial acceleration $\gamma = 2 + 2i$. (top) without friction $\nu = 0$, (middle) $\nu = 0.01$, (bottom) $\nu = 0.1$. Blue line: with respect to rotated coordinates and scaled X, Y; red line: with respect to original but scaled coordinates x, y

On the other hand, over the mound, the kinematic boundary condition at the bottom (5.32) requires that the vertical velocity w_b at the bottom should be

$$w_b = -U(x)\frac{dH(x)}{dx} \quad \text{on} \quad y = H(x) \quad \text{on} \quad \ell < x < \ell,$$

which is non-zero only in the region of the mound, and $w_b = 0$ elsewhere on the bottom.

Combining these two equations, we have

$$w_b = \mathbf{u}_h \cdot \nabla h = -\frac{U_0 H_0}{H(x)} \frac{dH(x)}{dx} = -U_0 H_0 \frac{d}{dx} \ln H(x) \quad \text{on} \quad y = H(x) \quad \text{on} \quad -\ell < x < \ell.$$

For the inviscid fluid flow regime at small Rossby number, the equations approach the geostrophic approximation of Eqs. (5.27) and (5.28) given in the text

$$2\Omega \mathbf{k} \times \mathbf{u}_h = -\frac{1}{\rho} \nabla_h p$$

$$\nabla_h \cdot \mathbf{u}_h = 0,$$

where the subscript $()_h$ refers to purely horizontal variables and operators, with Ω being the speed of rotation of the earth. We have shown that the geostrophic equations require that the horizontal velocity $\mathbf{u}_h = 0$ and pressure p can not have dependence on the vertical z. This embodies the fact that we can not have any variation of vertical velocity w with z, since the original three dimensional continuity equation

$$\frac{\partial w}{\partial z} + \nabla_h \cdot \mathbf{u}_h = 0$$

subtracted from the two-dimensional geostrophic version gives

$$\frac{\partial w}{\partial z} = 0$$

requiring the top and bottom vertical velocity components to be equal

$$w_t = w_b.$$

Therefore it is not possible for the flow to pass over the mound unless adjustment is made to suck or inject fluid at the top boundary to give

$$w_t = -U_0 H_0 \frac{d}{dx} \ln H(x) \quad \text{on} \quad y = H_0 \quad \text{on} \quad -\ell < x < \ell.$$

It is therefore required that the fluid is sucked out for $-\ell < x < 0$ and injected in for $0 > x > \ell$.

Chapter 5, Exercise 3

(i) The governing equation for damped inertial motions

$$\frac{\partial \mathbf{u}_h}{\partial t} + 2\Omega \mathbf{k} \times \mathbf{u}_h = -\mu \mathbf{u}_h$$

can be differentiated with respect to time t and added to cross product $2\Omega\times$ of itself to yield

$$\frac{\partial^2 \mathbf{u}_h}{\partial t^2} + 2\mu \frac{\partial \mathbf{u}_h}{\partial t} + (4\Omega^2 + \mu^2)\mathbf{u}_h = 0.$$

Assuming a periodic solution

$$\mathbf{u}_h \sim e^{i\omega t}$$

gives the characteristic equation

$$\omega^2 - 2i\mu\omega + (4\Omega^2 + \mu^2) = 0$$

with roots

$$\omega_{1,2} = i\mu \pm \sqrt{(i\mu)^2 + (4\Omega^2 + \mu^2)} = i\mu \pm 2\Omega.$$

A solution to the governing equation is therefore

$$\mathbf{u}_h = \mathbf{a}_0 e^{-\mu t} e^{-i2\Omega t} + \mathbf{b}_0 e^{-\mu t} e^{+i2\Omega t}$$

where $\mathbf{a}_0$ and $\mathbf{b}_0$ are constants.

We can make the (now customary) changes from a vector to a complex variable as $u_h \rightarrow w = u + iv$ so that

$$w = a_0 e^{-\mu t} e^{-i2\Omega t} + b_0 e^{-\mu t} e^{+i2\Omega t}$$

is the solution represented in the complex domain, where it is understood that a_0 and b_0 are complex constants.

We then consciously make the choice $b_0 = 0$ and reject the second term in the solution so that the velocity vector rotates in the clockwise sense in the northern hemisphere where $\Omega > 0$. The reason for this choice becomes clear if we notice that the original first order equation has been raised into second order producing two roots one of which is not realistic. In fact, it can be shown that the second term of the proposed solution does not satisfy the original first order equation.

With the initial condition at time $t = 0$

$$w(0) = U_0 = a_0$$

and therefore the solution becomes

$$w = U_0 e^{-\mu t} e^{-i2\Omega t}$$

(ii) The components u, v are

$$u = U_0 e^{-\mu t} \cos 2\omega t \quad \text{and} \quad v = -U_0 e^{-\mu t} \sin 2\Omega t$$

where U_0 is chosen as a real constant, without loss of generality, i.e. assuming initial velocity to be in the positive x-direction.

Assuming the Eulerian approximation to be valid for small displacements, the particle path $Z = x + iy$ can be integrated from the velocity field

$$\frac{dZ}{dt} = w$$

to give

$$Z = \frac{U_0}{-(\mu + i2\Omega)} e^{-\mu t} e^{-i2\Omega t}$$

or

$$x = \frac{U_0}{4\Omega^2 + \mu^2} e^{-\mu t} [2\Omega \sin 2\Omega t - \mu \cos 2\Omega t] + x_0$$
$$y = \frac{U_0}{4\Omega^2 + \mu^2} e^{-\mu t} [2\Omega \cos 2\Omega t + \mu \sin 2\Omega t] + y_0$$

The particle trajectory in the case with $\mu \neq 0$ is a spiral rotating in the clockwise sense that would converge to a single point as the motion decays, compared to the circular rotation reviewed in text for $\mu = 0$.

Chapter 6, **Exercise 1**

The function is a sum of two wave components with slightly differing frequency and wave-number. Making use of trigonometric identities and making use of

$$\begin{aligned} \omega_1 &= \omega - \Delta\omega/2 \\ \omega_2 &= \omega + \Delta\omega/2 \\ k_1 &= k - \Delta k/2 \\ k_2 &= k + \Delta k/2 \end{aligned}$$

we can write this waveform as

$$\begin{aligned} \eta &= \cos(k_1 x - \omega_1 t) + \cos(k_2 x - \omega_2 t) \\ &= \cos(\phi_1) + \cos(\phi_2) \\ &= 2\cos\left(\frac{\phi_1 + \phi_2}{2}\right) \cos\left(\frac{\phi_1 - \phi_2}{2}\right) \\ &= 2\cos\left(\frac{k_1 x - \omega_1 t + k_2 x - \omega_2 t}{2}\right) \cos\left(\frac{k_1 x - \omega_1 t - k_2 x + \omega_2 t}{2}\right) \\ &= 2\cos(kx - \omega t)\cos(\Delta k x - \Delta\omega t). \end{aligned}$$

Alternatively,

$$\eta = A(x, t)\cos(kx - \omega t)$$

where

$$A(x, t) = 2\cos(\Delta k x - \Delta\omega t)$$

is a slowly varying amplitude of the carrier wave, remembering that $\Delta\omega << \omega$ and $\Delta k << k$.

In this new form, the modulated waveform is an oscillating function with frequency ω and wave-number k, enveloped by a function with much smaller frequency $\Delta\omega$ and wave-number Δk.

Plotting and visualizing the function in this form can verify that the modulated carrier wave travels at the *phase speed*

$$c_p = \omega/k,$$

while the envelope with amplitude $A(x,t)$ travels with the *group speed*

$$c_g = \Delta\omega/\Delta k.$$

Chapter 6, Exercise 2

For small-amplitude motion, the following linear shallow water wave equation for sea surface height η applies:

$$\frac{\partial}{\partial t}\left\{\left(\frac{\partial^2}{\partial t^2}+f^2\right)\eta - gh_0\nabla^2\eta\right\} = 0$$

where h_0 is the constant depth, with the horizontal fluid velocity $\mathbf{u}$ given by

$$\left(\frac{\partial^2}{\partial t^2}+f^2\right)\mathbf{u} = -g\left\{\frac{\partial}{\partial t}\nabla\eta - f\mathbf{k}\times\nabla\eta\right\}$$

Let the waves generated by the wave-maker be of the following periodic form

$$\eta = \Re\{\hat{\eta}(x)e^{i\omega t}\}, \qquad u = \Re\{\hat{u}(x)e^{i\omega t}\} \text{ and } v = \Re\{\hat{v}(x)e^{i\omega t}\}.$$

Substituting in the wave equation, we have for the amplitude

$$\frac{\partial^2\hat{\eta}}{\partial x^2} - \left(\frac{f^2-\omega^2}{gh_0}\right)\hat{\eta} = 0$$

and the amplitude of velocity components are given as

$$(f^2-\omega^2)\hat{u} = -g(i\omega)\frac{\partial\hat{\eta}}{\partial x}$$

$$(f^2-\omega^2)\hat{v} = gf\frac{\partial\hat{\eta}}{\partial x}.$$

(i) The solution for $\omega > f$ is obtained by letting

$$k^2 = \frac{\omega^2 - f^2}{gh_0} > 0,$$

so that the solution is obtained as

$$\eta = \Re\{Ae^{i(kx+\omega t)} + Be^{-i(kx-\omega t)}\}.$$

In order that the wave would propagate out in the $+x$ direction, we must have $A = 0$. It remains to obtain constant B from the boundary condition at the wave-maker,

$$u = \mathbf{u} \cdot n = \frac{dx_0}{dt} = \Re\{i\omega a_0 e^{i\omega t}\} = \Re\{\hat{u}(0)e^{i\omega t}\} \text{at} x = 0,$$

or

$$\hat{u}(0) = i\omega a_0.$$

Substituting in the equation for $\hat{u}$ velocity component

$$\frac{\partial \hat{\eta}(0)}{\partial x} = \frac{a_0}{g}(\omega^2 - f^2),$$

and checking against the solution for η, we have

$$B = i(kh_0)a_0$$

and the solution becomes

$$\eta = \Re\{i(kh_0)a_0 e^{-i(kx-\omega t)}\}.$$

The velocity components become

$$u = \Re\{i\omega a_0 e^{-i(kx-\omega t)}\}$$

and

$$v = \Re\{-f a_0 e^{-i(kx-\omega t)}\},$$

with the velocity vector found to satisfy trajectory

$$\frac{u^2}{\omega^2 a_0^2} + \frac{v^2}{f^2 a_0^2} = 1.$$

Since $\omega > f$, this is an ellipse elongated in the x-direction.
(ii) For $\omega < f$, we define

$$\gamma^2 = \frac{f^2 - \omega^2}{gh_0} > 0$$

whereupon the solution is obtained as

$$\hat{\eta} = Ae^{\gamma x} + Be^{-\gamma x}.$$

Since a growing solution with $+x$ can not be accepted, it is required that $A = 0$. The wave-maker boundary condition at $x = 0$ implies

$$B = a_0 \gamma h_0,$$

which yields

$$\eta = \Re\{\gamma h_0 a_0 e^{-(\gamma x - i\omega t)}\} = \gamma h_0 a_0 e^{-\gamma x} \cos \omega t.$$

The velocity components are

$$u = \Re\{i\omega a_0 e^{-(\gamma x - i\omega t)}\} = -\omega a_0 e^{-\gamma x} \sin \omega t$$

and

$$v = \Re\{-f a_0 e^{-(\gamma x - i\omega t)}\} = -f a_0 e^{-\gamma x} \cos \omega t,$$

with the trajectory of the velocity vector found as

$$\frac{u^2}{\omega^2 a_0^2} + \frac{v^2}{f^2 a_0^2} = e^{-2\gamma x}.$$

Since $\omega < f$, this is an ellipse elongated in the y-direction. The ellipse size gets smaller with distance from the wave-maker.

(iii) The waves are a mixture of longitudinal and transversal oscillations. In the frequency range $\omega > f$, longitudinal waves dominate and travel out without change in amplitude, while in the $\omega < f$ range transversal oscillations are dominant and decay with distance from the wave-maker, trapping energy near the wall.

For $\omega = f$ the wave motion becomes uniform everywhere with circular orbits and constant amplitude of sea surface and velocity.

As $\omega \to 0$

$$\gamma^2 = \frac{f^2}{gh_0}$$

and

$$\eta = f\sqrt{\frac{h_0}{g}} e^{\frac{-fx}{\sqrt{gh_0}}}, \qquad u = 0 \text{ and } v = -f a_0 e^{\frac{-fx}{\sqrt{gh_0}}}$$

where the length scale $\sqrt{gh_0}/f$ is defined as the Rossby radius of deformation.

It can be noted that as $\omega \to 0$ the motion becomes completely transversal and converts into a mean current in the $-y$ direction along the coast, decaying away from the coast, with an accompanying surface slope.

Chapter 6, Exercise 3

With the linear shallow water equations

$$\frac{\partial}{\partial t}\left\{\left(\frac{\partial^2}{\partial t^2} + f^2\right)\eta - gh\nabla^2\eta\right\} = 0$$

applied respectively in regions $x < 0$ and $x > 0$, where the depth h is replaced by the constants h_1 and h_2. The motion is only a function of x, independent of y.

(i) For an incident plane wave solution

$$\eta^I = Ae^{i(k_1 x+\ell y-\omega t)} \quad \text{in} \quad x < 0$$

to satisfy the above equation of motion with $h = h_1$ in $x < 0$, the following dispersion relation is found

$$(\omega^2 - f^2) = gh_1(k_1^2 + \ell^2)$$

so that for k_1 to be real-valued, the following relation needs to be satisfied

$$k_1^2 = \frac{(\omega^2 - f^2)}{gh_1} - \ell^2 > 0,$$

requiring

$$\omega^2 > f^2 + gh_1\, \ell^2.$$

In region $x < 0$, there is a possibility that a reflected wave

$$\eta^R = \hat{\eta}^R e^{i(\ell y-\omega t)} = Be^{i(-k_1 x+\ell y-\omega t)} \quad \text{in} \quad x < 0$$

with the negative value of the same wave-number $-k_1$ can exist, since it satisfies the same dispersion relation. The complex constants A and B are amplitudes yet to be determined. The total solution in this half-space is therefore

$$\eta_1 = \eta^I + \eta^R = (\hat{\eta}^I + \hat{\eta}^R)e^{(\ell y-\omega t)}.$$

In the other half region with $h = h_2$ in $x > 0$, the outgoing wave solution is

$$\eta^T = \hat{\eta}^T e^{i(\ell y-\omega t)} = Ce^{i(k_2 x+\ell y-\omega t)} \quad \text{in} \quad x > 0$$

and considering a similar dispersion relation for k_2 to be real-valued, leads to so that for k_1 to be real-valued, the following relation needs to be satisfied

$$k_2^2 = \frac{(\omega^2 - f^2)}{gh_2} - \ell^2 > 0,$$

and therefore

$$\omega^2 > f^2 + gh_2\, \ell^2.$$

Comparing with the dispersion relation on the incident region $x < 0$,

$$gh_1(k_1^2 + \ell^2) > gh_2\, \ell^2.$$

so that a criterion is found for reflected and transmitted waves to exist:

$$\left(\frac{k_1}{\ell}\right)^2 > \frac{h_2 - h_1}{h_1} \equiv \frac{\Delta h}{h_1}.$$

Note that this criterion is always satisfied for a negative value $\Delta h < 0$, i.e. when the depth is decreasing from region 1 to region 2 (up-step), since in this case $(k_1/\ell)^2 > 0$ is always true.

In the case that $\Delta h > 0$, i.e. when the depth is increasing from region 1 to region 2, above criterion is applied, and remembering that the ratio of the wave-numbers is related to the incidence angle θ^I, the tangent of which is the ratio of the wavelengths in x and y directions:

$$\tan\theta^I = \frac{2\pi/k_1}{2\pi/\ell} = \frac{\ell}{k_1} < \sqrt{\frac{h_1}{\Delta h}}.$$

The criteria for transmission of the wave to occur for any $\frac{\Delta h}{h_1} > 0$ (down-step) is therefore

$$\theta^I < \theta^B$$

where

$$\theta^I = tan^{-1}\frac{\ell}{k_1} \quad \text{and} \quad \theta^B \equiv tan^{-1}\sqrt{\frac{h_1}{\Delta h}}.$$

The critical incidence angle θ^B is called *Brewster's angle*. When

$$\theta^I > \theta^B$$

total reflection occurs.

(ii) The jump conditions that need to be satisfied at the abrupt change in depth at $x = 0$ are

$$\eta_1 = \eta_2 \quad \text{and} \quad h_1 u_1 = h_2 u_2 \quad \text{at} \quad x = 0,$$

where indices 1, 2 denote the regions $x < 0$ and $x > 0$ respectively. Since the y dependence of solutions in both half-spaces are the same, $expi(\ell y - \omega t)$, these are equivalent to equating the amplitudes. For sea surface elevation the first jump condition gives

$$\hat{\eta}^I + \hat{\eta}^R = \hat{\eta}^I \quad \text{at} \quad x = 0,$$

and the second condition is

$$h_1(\hat{u}^I + \hat{u}^R) = h_2\hat{u}^T \quad \text{at} \quad x = 0,$$

where $\hat{u}$ are the amplitudes of the incident, reflected and the transmitted waves.

The first condition is equivalent to

$$A + B = C$$

and for the velocity amplitudes of each component we can use the velocity equation (see last exercise) to obtain

$$(f^2 - \omega^2)\hat{u} = -g\left(-i\omega\frac{\partial\hat{\eta}}{\partial x} + if\ell\hat{\eta}\right)$$

for each of the components, from which the jump condition is reduced to

$$h_1\{f\ell(A+B) - ik_1\omega(A-B)\} = h_2\{f\ell - ik_2\omega\}C$$

Eliminating C between the two equations yield the reflection coefficient

$$R = \frac{B}{A} = -\frac{(h_1 - h_2)f\ell - i\omega(k_1h_1 - k_2h_2)}{(h_1 - h_2)f\ell + i\omega(k_1h_1 + k_2h_2)}$$

while the transmission coefficient can be obtained from

$$T = \frac{c}{A} = 1 + \frac{B}{A} = 1 + R = \frac{2i\omega k_1 h_1}{(h_1 - h_2)f\ell + i\omega(k_1h_1 + k_2h_2)}.$$

Note that both the reflection coefficient R and the transmission coefficient T are complex, carrying amplitude and phase information with them. Furthermore, the functions most likely need to be evaluated as functions of ω and incidence angle θ^I, from which the wave-number components have to be calculated and inserted above to evaluate R and T.

As for the angles of incidence, reflection and transmission we have

$$\theta^I = \frac{\ell}{k_1}, \qquad \theta^R = -\frac{\ell}{k_1}, \theta^I = \frac{\ell}{k_2}.$$

Using the dispersion relations for the two regions combined, i.e.

$$(\omega^2 - f^2) = gh_1(k_1^2 + \ell^2) = gh_2(k_2^2 + \ell^2)$$

and turning that around a little bit,

$$\frac{gh_2}{gh_1} = \frac{k_1^2 + \ell^2}{k_2^2 + \ell^2} = \frac{\left(\frac{k_1}{\ell}\right)^2 + 1}{\left(\frac{k_2}{\ell}\right)^2 + 1} = \frac{\frac{1}{\tan^2 \theta^I} + 1}{\frac{1}{\tan^2 \theta^T} + 1} = \frac{\sin^2 \theta^T}{\sin^2 \theta^I}$$

results in the refraction relation

$$\frac{\sin^2 \theta^T}{\sin^2 \theta^I} = \frac{gh_2}{gh_1} = \frac{c_2}{c_1} = \frac{h_2}{h_1}.$$

(iii) Proposing a propagating wave solution in the y-direction

$$\eta = \hat{\eta} e^{i(\ell y - \omega t)}$$

would require the amplitude $\hat{\eta}$ satisfy

$$\frac{\partial^2 \hat{\eta}}{\partial x^2} - \left(\frac{f^2 - \omega^2}{gh_i}\right)\hat{\eta} = 0$$

for the domains $i = 1, 2$ respectively the two half-domains. Evanescent modes are only possible in either of the domains if we set

$$-\gamma_i^2 = \frac{\omega^2 - f^2}{gh_i} - \ell^2 < 0$$

or if

$$\gamma_i^2 = \frac{f^2 - \omega^2}{gh_i} + \ell^2 > 0,$$

and that is typically possible in the regime with $f > \omega$ or as the inequality exactly describes.

Bounded solutions as $x \to \pm\infty$ are

$$\hat{\eta}_1 = Ae^{+\gamma x} \quad \text{and} \quad \hat{\eta}_2 = Be^{-\gamma x}.$$

Using the same matching conditions as in part `(ii)` gives

$$\hat{\eta}_1 = \hat{\eta}_2 \quad \rightarrow \quad A = B$$

$$h_1\hat{u}_1 = h_2\hat{u}_2 \quad \rightarrow \quad h_1(f\ell - \gamma_1\omega)A = h_2(f\ell - \gamma_2\omega)B.$$

From these equations A and B are eliminated, giving

$$\frac{\omega}{f} = \left(\frac{h_1 - h_2}{\gamma_1 h_1 - \gamma_2 h_2}\right)\ell.$$

From this equation, which becomes seriously complicated with substitutions of γ_1 and γ_2 from dispersion relations, the wave frequency ω can be determined as a function of wavelength.

Since $\gamma_1^2 > 0$ and $\gamma_2^2 > 0$, we note that

$$\omega^2 < f^2 + gh_1\ell^2 < f^2$$

and

$$\omega^2 < f^2 + gh_2\ell^2 < f^2.$$

Because solving the complex algebraic equation for frequency is complicated, approximation will simplify the result. Making the *rigid lid* or *non-divergence* assumption

$$\frac{f^2\lambda^2}{gh_i} << 1$$

where $\lambda = 2\pi/\ell$ is the wavelength, and since $\omega < f$ in general, γ_i are approximated as

$$\gamma_1 = \gamma_2 \simeq |\ell|,$$

so that the relationship for frequency becomes

$$\frac{\omega}{f} = \left(\frac{h_1 - h_2}{h_1 + h_2}\right) \mathrm{sgn}(\ell).$$

Using either the exact or the approximate formula for frequency ω as a function of ℓ, it can be observed that for

$$\text{for} \quad \ell > 0, \quad \frac{\omega}{f} < 0 \ \text{ if } \ h_1 < h_2$$

or

$$\text{for} \quad \ell > 0, \quad \frac{\omega}{f} > 0 \ \text{ if } \ h_1 > h_2$$

which shows that the wave propagation always takes the shallow domain to its right (in the northern hemisphere).

Chapter 6, **Exercise 4**

The equations of motion in the channel connecting the lagoon to the ocean can be reduced by taking x-derivative of the first equation multiplied by the depth h and subtracting it from the time derivative of the second, producing

$$\frac{\partial^2 \eta}{\partial t^2} + \nu \frac{\partial \eta}{\partial t} - gh \frac{\partial^2 \eta}{\partial x^2} = 0.$$

This equation, often called *telegraphers equation*, has a periodic solution of the form

$$\eta = \hat{\eta}(x) e^{i\omega t}$$

where ω is the oscillation frequency, reducing the equation to

$$(-\omega^2 + i\omega\nu)\hat{\eta} - gh \frac{\partial^2 \hat{\eta}}{\partial x^2} = 0.$$

The characteristic equation

$$(-\omega^2 + i\omega\nu) - ghk^2 = 0$$

has

$$k_{1,2} = \pm \left(\frac{\omega^2 - i\omega}{gh}\right)^{\frac{1}{2}} = \pm(\mu + im)$$

where μ and m are appropriate constants. The solution is then constructed as

$$\eta = (Ae^{+(\mu+im)x} + Be^{-(\mu+im)x})e^{i\omega t}.$$

`(i) case 1–large basin` The boundary conditions at the ocean end of the channel is

$$\hat{\eta}(0) = A + B = a_0$$

and at the basin side η_b is so small that we can take it as zero

$$\hat{\eta}(\ell) = (Ae^{+(\mu+im)\ell} + Be^{-(\mu+im)\ell}) = 0.$$

From these equations the two unknowns A and B can be determined:

$$A = -\frac{a_0}{2}\frac{e^{-(\mu+im)\ell}}{\sinh[(\mu+im)\ell]}$$

$$B = -\frac{a_0}{2}\frac{e^{+(\mu+im)\ell}}{\sinh[(\mu+im)\ell]}.$$

`(i) case 2–small basin`

The basin is assumed to have a uniform sea level response, $\eta_b = a_b e^{i\omega t}$ where a_b is a complex amplitude. At the junction of channel with basin, continuity equation has to be satisfied

$$uhw = A_b\frac{\partial \eta_b}{\partial t} = a_b A_b(i\omega)e^{i\omega t} \; at \; x = \ell.$$

The velocity in the channel can be obtained from the continuity equation

$$u(x,t) = \hat{u}(x)e^{i\omega t} = -\frac{i\omega}{\mu+im}(Ae^{+(\mu+im)x} - Be^{-(\mu+im)x})e^{i\omega t}$$

and therefore the boundary condition at $x = \ell$ becomes

$$(-Ae^{+(\mu+im)\ell} + Be^{-(\mu+im)\ell}) = \frac{a_b A_b}{hw}(\mu+im)$$

In addition, at the basin end of the channel, sea level has to satisfy

$$\hat{\eta}(\ell) = (Ae^{+(\mu+im)\ell} + Be^{-(\mu+im)\ell}) = a_b$$

and we also have at the ocean end

$$\hat{\eta}(\ell) = A + B = a_0,$$

so that we have three equations to solve for the the three unknowns A, B and a_b. Combining three equations, the elevation at ocean side is related to the lagoon level as

$$a_0 = \left[\cosh(\mu+im)\ell + \frac{A_b}{hw}\sinh(\mu+im)\ell\right]a_b.$$

Chapter 6, Exercise 5

Starting with linear shallow water equations and following the steps in Sect. 6.6.3, we obtain

$$\frac{\partial}{\partial t}\left\{\left(\frac{\partial^2}{\partial t^2}+f^2\right)\eta - g\nabla\cdot h\nabla\eta\right\} - fgJ(h,\eta) = -\frac{1}{\rho}\left\{\frac{\partial}{\partial t}\nabla\cdot\boldsymbol{\tau} + f\mathbf{k}\cdot\nabla\times\boldsymbol{\tau}\right\}.$$

Once the solution is obtained for surface displacement η, velocity is calculated from the re-organized form of the momentum equation

$$\left(\frac{\partial^2}{\partial t^2}+f^2\right)\mathbf{u} = -g\left(\frac{\partial}{\partial t}\nabla\eta - f\mathbf{k}\times\nabla\eta\right) + \frac{1}{\rho h}\left(\frac{\partial\boldsymbol{\tau}}{\partial t} - f\mathbf{k}\times\boldsymbol{\tau}\right).$$

(i) Considering time periodic solutions and forcing,

$$\eta(x,y,t) = N(x,y)e^{i\omega t}, \qquad \mathbf{u}(x,y,t) = \mathbf{U}(x,y)e^{i\omega t}, \qquad \boldsymbol{\tau}(x,y,t) = \mathbf{T}(x,y)e^{i\omega t}$$

the equations are reduced to those involving the amplitudes $\mathbf{T}$, N and $\mathbf{U}$:

$$i\omega\{(f^2-\omega^2)N - \nabla\cdot gh\nabla N\} - fgJ(h,N) = -\frac{1}{\rho}\{i\omega\nabla\cdot\mathbf{T} - \mathbf{k}\cdot\nabla\times\mathbf{T}\}$$

$$(f^2-\omega^2)\mathbf{U} = -g(i\omega\nabla N - f\mathbf{k}\times\nabla N) + \frac{1}{\rho h}(i\omega\mathbf{T} - f\mathbf{k}\times\mathbf{T})$$

(ii) With the given simplifications and geometry, the governing equations become

$$\frac{\partial^2 N}{\partial y^2} + \frac{\omega^2-f^2}{gh_0}N = 0$$

$$U = 0 \qquad (\text{no}x-\text{dependence})$$

and

$$(f^2-\omega^2)V = -g(i\omega)\frac{\partial N}{\partial y} + \frac{1}{\rho h_0}(i\omega T^y - fT^x).$$

where $V = \mathbf{U}\cdot\boldsymbol{\jmath}$ is the only velocity amplitude component. The no-flux boundary condition is applied at the coast as

$$V = \mathbf{U}\cdot\mathbf{n} = \mathbf{U}\cdot\boldsymbol{\jmath} = 0 \quad \text{at} \quad y = 0$$

or

$$\frac{\partial N}{\partial y} = \frac{1}{\rho g h_0}\left(T^y + i\left(\frac{f}{\omega}\right)T^x\right) \quad \text{at} \quad y = 0,$$

where $\mathbf{T} = (T^x, T^y)$.

The other boundary condition is that the solution should either remain bounded or have outgoing waves at most as $y \to \infty$.

We note that the spatially constant wind-stress appears only in the coastal boundary condition and not in the governing equation. This is a result of the vertical averaging performed earlier when the shallow water equations were developed, of including the top and bottom Ekman layers in the averaging, which in effect has distributed the wind-stress across total depth of the fluid. In fact, only the divergence and curl of the wind-stress appears in the equation for η, which disappear for constant wind-stress amplitude. Wind-stress only directly influences current velocity $\mathbf{u}$ including Ekman layer contributions in this configuration, and when the flux is intercepted by the coast, it produces forcing for the whole domain.

(iii) The solution is obtained with respect to the frequency range.

For $\omega > f$ we let

$$k^2 = \frac{\omega^2 - f^2}{gh_0} > 0$$

and the solution satisfying no-flux boundary condition at the coast and outward propagation at $y \to \infty$ is

$$\eta = -\frac{1}{\rho k g h_0}\left(iT^y + \frac{f}{\omega}T^x\right)e^{-i(ky-\omega t)}.$$

The solution is an outgoing wave in the y direction. It can be observed that the motion has elliptic trajectories that depend on the direction of wind-stress and the frequency.

For $\omega < f$ we let

$$\gamma^2 = \frac{f^2 - \omega^2}{gh_0} > 0$$

and the solution satisfying no-flux boundary condition at the coast and decaying out as $y \to \infty$ is

$$\eta = -\frac{1}{\rho \gamma g h_0}\left(T^y + i\frac{f}{\omega}T^x\right)e^{-\gamma y}e^{+\omega t}.$$

The solution in this case does not have wave propagation. Instead it has elliptic particle trajectories that depend on the direction of wind-stress and frequency, but the magnitude of the oscillations decays out in the y direction.

When $\omega \to f$ either $k \to 0$ or $\gamma \to 0$ in the above solutions and the solutions become infinite ! This is because forcing at the inertial frequency $\omega = f$, which is the natural frequency of the ocean, produces resonance. Unbounded solutions in nature are often damped by other effects not included in the above analyses, such as frictional and non-linear dynamics.

When $\omega \to 0$, direct substitution in above misleads one to infinite solutions. Instead, setting $\omega = 0$, the governing equations become

$$\frac{\partial^2 N}{\partial y^2} - \frac{f^2}{gh_0}N = 0$$

$$U = 0, \qquad V = -\frac{f}{\rho h_0} T^x,$$

with the bounded solution

$$N = Ae^{-fy/gh_0}, \qquad U = 0, \qquad V = 0$$

where we set also $V = 0$ for the solution to be bounded.

By also requiring that $\omega = 0$, the boundary condition at the coast becomes

$$\frac{\partial N}{\partial y} = \frac{1}{\rho g h_0} T^y \quad \text{at} \quad y = 0$$

which is then used to determine the constant A, with which the solution becomes

$$N = -\frac{T^y}{\rho f} e^{-fy/gh_0}, \qquad U = 0, \qquad V = 0.$$

This is a case with no currents, but the free surface piled up against the coast if the wind is towards the coast or depressed near the coast if the wind blows away from the coast.

`(iv)` to represent the effects of a periodic, propagating wind-stress in the alongshore direction x, we make the substitutions

$$\mathbf{T}(x) = \mathbf{T}' e^{-ikx}$$

$$N(x, y) = N'(y) e^{-ikx}$$

$$\mathbf{U}(x, y) = \mathbf{U}'(y) e^{-ikx},$$

where $\mathbf{T}'$ is constant, and the governing equations become

$$i\omega \left\{ (f^2 - \omega^2) N' - g h_0 \frac{\partial^2 N'}{\partial y^2} + g h_0 k^2 N' \right\} = -\frac{1}{\rho} i\omega(-ik)(T^x)'$$

$$(f^2 - \omega^2)\mathbf{U}' = -g \left(i\omega \frac{\partial N'}{\partial y} \jmath + f \frac{\partial N'}{\partial y} \imath \right) + \frac{1}{\rho h} (i\omega \mathbf{T}' - f \mathbf{k} \times \mathbf{T}')$$

or letting

$$\ell^2 = \frac{\omega^2 - f^2}{g h_0} - k^2 > 0$$

and dropping primes

$$\frac{\partial^2 N}{\partial y^2} + \ell^2 N = \frac{1}{\rho} \{-ikT^x\}$$

$$(f^2 - \omega^2) U = -fg \frac{\partial N}{\partial y} + \frac{1}{\rho h} (i\omega T^x + f T^y)$$

$$(f^2 - \omega^2)V = -i\omega g \frac{\partial N}{\partial y} + \frac{1}{\rho h}(i\omega T^y - fT^x)$$

The solution, matching the boundary condition at the coast and bounded as $y \to \infty$ is

$$\eta = Ae^{-i\ell y} - \frac{ik}{\rho \ell^2} T^x$$

substituting in the boundary condition $V = 0$ at $y = 0$ evaluates the constant A so that

$$\eta = \left\{ \frac{1}{\rho h g \ell} \left(iT^y - \frac{f}{\omega} T^x \right) e^{-i\ell y} - \frac{ik}{\rho \ell^2} T^x \right\} e^{-i(kx - \omega t)}$$

Note the above solution is valid for $\omega^2 > f^2 + gh_0(k^2 + \ell^2)$. Wave motion is generated that travels in the same direction of propagation of the wind-stress disturbance.

On the other hand, letting

$$\gamma^2 = \frac{f^2 - \omega^2}{gh_0} + k^2 > 0$$

in the regime $\omega^2 > f^2 + gh_0(k^2 - \gamma^2)$, the solutions become

$$\eta = Ae^{-\gamma y} - \frac{ik}{\rho \gamma^2} T^x$$

substituting in the boundary condition $V = 0$ at $y = 0$ and evaluating the constant A, the solution in this case becomes

$$\eta = \left\{ -\frac{1}{\rho h g \gamma} \left(T^y + \frac{f}{\omega} T^x \right) e^{-\gamma y} - \frac{ik}{\rho \gamma^2} T^x \right\} e^{-i(kx - \omega t)}$$

The solution depends on the ratio ω/f and the direction of the wind-stress vector as before. However, now there are two components, one forced wave component that is induced by both components of wind-stress that propagates or decays in the y-direction and and another one that is directly driven by the wind-stress x-component and travels with it in fine tune.

As $\omega -> f$ again the solution approaches the resonant inertial oscillations. As $\omega -> 0$ there are decaying modes in the y direction but with a decay scale that is slightly modified by the wave number of the wind forcing:

$$\gamma = -\sqrt{\frac{f^2}{gh_0} + k^2}.$$

Chapter 7, **Exercise 1**

The quasi-geostrophic vorticity equation developed in this section is principally a non-linear conservation of fluid, planetary and topographic vorticity components, and includes the divergence term as well.

The plane-wave solutions demonstrated here are divergent, combined planetary-topographic Rossby waves. The divergence term is related to the time rate of change of the free surface, i.e. stretching terms by the divergence of volume flux, and the topographic terms represent vorticity imparted by changes in topography. These are additions to the fluid and planetary vorticity terms.

Substituting the plane-wave solution

$$\psi = Ae^{i(\omega t - kx - ly)}$$

to the quasi-geostrophic vorticity equation, one gets the divergent Rossby wave dispersion relation

$$\omega = -\frac{\beta k}{k^2 + l^2 - \delta}.$$

A plane-wave solution can exist for the nonlinear equation, because the non-linear term

$$J[\psi, (\nabla^2\psi - \delta\psi + \eta_b)]$$

is identically equal to zero when the two arguments of the Jacobian are functionally dependent, as we have shown in (6.204).

The phase velocity components in the x and y-directions are obtained as

$$C_x = \frac{\omega}{k} = -\frac{\beta}{k^2 + l^2 - \delta}$$

$$C_y = \frac{\omega}{l} = -\frac{\beta}{k^2 + l^2 - \delta}\frac{k}{l}$$

and the group velocity components are

$$C_{gx} = \frac{\partial\omega}{\partial k} = \frac{\beta(l^2 - k^2 - \delta)}{k^2 + l^2 - \delta} = \frac{\beta[(k^4 - l^2 - \delta)^2]}{(k^2 + l^2 - \delta)^2}$$

$$C_{gy} = \frac{\partial\omega}{\partial l} = -\frac{2\beta kl}{(k^2 + l^2 - \delta)^2}$$

Both the phase and group velocity of the divergent Rossby wave is different from those formerly reviewed in Chap. 6, which excluded the divergence.

Chapter 7, **Exercise 2**

In Sect. 6.5.3 solutions were sought to the full nonlinear potential vorticity equations (6.173a,b) in the steady, free, inviscid case, addressing the initial value problem (6.186) for a finite step geometry.

Utilizing the quasi-geostrophic theory, we use Eq. (7.24)

$$\frac{\partial \zeta^{(0)}}{\partial t} + \mathbf{u}^{(0)} \cdot \nabla \zeta^{(0)} + \beta v^{(0)} - \delta \frac{\partial \eta^{(0)}}{\partial t} - \mathbf{u}^{(0)} \cdot \nabla(\delta \eta^{(0)} - \eta_b) = 0$$

and its equivalent form (7.26) in the text.

The evolution of velocity field in the region $x > 0$, past the small step topography η_b, is described by the above non-dimensional quasi-geostrophic equations, where $\zeta^{(0)} = \mathbf{k} \cdot \nabla^2 \mathbf{u}^{(0)}$ is the fluid vorticity and $\delta = f_0^2 L_0^2 / g H_0$ is the divergence parameter.

The quasi-geostrophic equation already represents the lowest order dynamics according to ordering in terms of the Rossby number. Because steady motion is investigated here, time dependent terms are dropped and because of the uniform incoming flow in the x-direction y-dependent terms can also be dropped. In addition, because of the abrupt change of depth, it is in order to keep the second derivative in x, which is much larger than the same in y, and therefore dropped in the vorticity term $\nabla^2 \mathbf{u}^{(0)}$. Furthermore since the disturbance would mainly occur in the velocity component $v^{(0)}$, the $u^{(0)}$ component is assumed to remain unchanged, substituting $u^{(0)} = 1$, scaled with U_0 in the dimensionless equations. These approximations in the right hand side domain $x > 0$ lead to

$$\frac{\partial^2 v^{(0)}}{\partial x^2} + \beta v^{(0)} = -\frac{\partial \eta_b}{\partial x}.$$

An important approximation in developing quasi-geostrophic theory is that the bottom topographic variations are assumed to be small, $h_b / H_0 = O(\epsilon)$, entering the equations at lowest order, where it is scaled as

$$\eta_b(x) = \begin{cases} 0 & \text{for} x < 0 \\ \Delta \equiv \frac{1}{\epsilon} \frac{\Delta h_b}{H_0} = \frac{f_0 L \Delta h_b}{U_0 H_0} & \text{for} x > 0. \end{cases}$$

The step-wise depth change implies a δ-function on the right hand side of the simplified vorticity equation

$$\frac{\partial^2 v^{(0)}}{\partial x^2} + \beta v^{(0)} = -\frac{\partial \eta_b}{\partial x} = -\Delta \, \delta(x).$$

Integrating from $x = 0_-$ to $x = 0_+$ (coordinates just before and after the step) yields

$$\frac{\partial v^{(0)}}{\partial x}|_{x=0_+} = -\Delta.$$

In addition, continuity of the velocity should be assured as a jump condition

$$v^{(0)}|_{x=0_+} = v^{(0)}|_{x=0_-} = 0$$

and

$$u^{(0)}|_{x=0_+} = u^{(0)}|_{x=0_-} = 1,$$

which are the initial conditions to obtain the solution in region $x > 0$ as

$$v^{(0)} = -\frac{\Delta}{\sqrt{\beta}} \sin(\sqrt{\beta} x)$$

and

$$u^{(0)} = 1.$$

In dimensional form, the solution is

$$v^{(0)} = -\frac{\Delta h_b}{H_0} \sqrt{\frac{U_0 f_0}{\beta}} \sin\left(\sqrt{\frac{f_0 \beta}{U_0}} x\right)$$

and

$$u^{(0)} = U_0.$$

By making use of the definitions (7.33) and (7.34), the stream-function is obtained as

$$\psi = -U_0 y + \frac{\Delta h_b}{H_0} \frac{U_0}{\beta} \cos\left(\sqrt{\frac{f_0 \beta}{U_0}} x\right).$$

The solution obtained from a linear, lowest order quasi-geostrophic approximation to the problem conserves the essential characteristics of the standing Rossby wave in a zonal current, although it is very different and less complicated, when compared to the nonlinear solution obtained in Sect. 6.5.3.

Chapter 8, Exercise 1

(i) The geostrophic interior is described by (8.10a), (8.10b)

$$\begin{aligned} \mathbf{u}_i &= \frac{1}{2} \mathbf{k} \times \nabla_h p_i, \\ \nabla_h \cdot \mathbf{u}_i &= 0. \end{aligned}$$

and the Ekman layers at the top and bottom are governed by Eqs. (8.19a), (8.19b)

$$\begin{aligned} 2\mathbf{k} \times \mathbf{u}_e &= \frac{\partial^2 \mathbf{u}_e}{\partial \zeta^2}, \\ \frac{\partial w_e}{\partial \zeta} &= -E^{-1/2} \nabla_h \cdot \mathbf{u}_e. \end{aligned}$$

where the distance between plates is 1 in non-dimensional variables and the thickness of the Ekman layers are $O(E^{\frac{1}{2}})$ with the expanded vertical coordinate $\zeta = E^{-\frac{1}{2}}z$ defined in (8.16) earlier.

The Ekman layer and interior velocity, pressure are additive to compose the total field by (8.12a), (8.12b) or simply satisfying (8.13a), (8.13b) separately near each boundary. Let the top and bottom boundaries S^t and S^b to be moving at with horizontal velocities $\mathbf{U}^t$ and $\mathbf{U}^b$ respectively, so that the corresponding boundary conditions would be

$$\mathbf{u} = \mathbf{u}_i + \mathbf{u}_e^t = \mathbf{U}^t \quad \text{on} \quad S^t$$
$$\mathbf{u} = \mathbf{u}_i + \mathbf{u}_e^b = \mathbf{U}^b \quad \text{on} \quad S^b$$
$$w = w_i + w_e^t \quad \text{on} \quad S^t$$
$$w = w_i + w_e^b \quad \text{on} \quad S^b$$
$$p = p_i + p_e^t \quad \text{on} \quad S^t$$
$$p = p_i + p_e^b \quad \text{on} \quad S^b$$

consistent with the earlier approximations. In Sect. 8.1, it was shown that Eqs. (8.19a), (8.19b) can be put into the form of (8.26), where by (8.27) we have

$$\mathbf{Q} \equiv \mathbf{k} \times \mathbf{u}_e + i\mathbf{u}_e$$

for either the top or bottom Ekman layers, which are then respectively designated with superscripts t and b.

It remains to apply the appropriate boundary conditions at the surface and bottom to obtain solutions. With no-slip boundary conditions, the Ekman boundary layer solutions can be obtained as

$$\mathbf{Q} = \{\mathbf{k} \times (\mathbf{U}^t - \mathbf{u}_i) + i(\mathbf{U}^t - \mathbf{u}_i)\}e^{(1+i)\zeta} = \mathbf{q}^t e^{(1+i)\zeta}$$

for the top and

$$\mathbf{Q} = \{\mathbf{k} \times (\mathbf{U}^b - \mathbf{u}_i) + i(\mathbf{U}^b - \mathbf{u}_i)\}e^{-(1+i)\zeta} = \mathbf{q}^b e^{-(1+i)\zeta}$$

for the bottom layers.

Following the development in text, the vertical velocity at the top and bottom boundaries of the fluid are integrated in the vertical by making use of (8.19b) and the vector identity (1.27c) to yield

$$\begin{aligned} w_e^t(0) - w_e^t(-\infty) &= \Im\left\{-E^{\frac{1}{2}}\nabla_h \cdot \mathbf{q}^t \int_{-\infty}^{0} e^{(1+i)\zeta} d\zeta\right\} \\ &= \Im\left\{E^{\frac{1}{2}}\left[-\mathbf{k} \cdot \nabla_h \times (\mathbf{U}^t - \mathbf{u}_i) + i\nabla_h \cdot (\mathbf{U}^t - \mathbf{u}_i)\right]\frac{1}{1+i}\right\} \\ &= -\frac{E^{\frac{1}{2}}}{2}\left[\mathbf{k} \cdot \nabla_h \times (\mathbf{U}^t - \mathbf{u}_i) + \nabla_h \cdot (\mathbf{U}^t - \mathbf{u}_i)\right] \end{aligned}$$

$$
\begin{aligned}
w_e^b(\infty) - w_e^b(0) &= \Im\left\{-E^{\frac{1}{2}}\nabla_h \cdot \mathbf{q}^b \int_0^\infty e^{-(1+i)\zeta} d\zeta\right\} \\
&= \Im\left\{E^{\frac{1}{2}}\left[-\mathbf{k}\cdot\nabla_h \times (\mathbf{U}^b - \mathbf{u}_i) + i\nabla_h\cdot(\mathbf{U}^b - \mathbf{u}_i)\right]\frac{1}{1+i}\right\} \\
&= -\frac{E^{\frac{1}{2}}}{2}\left[\mathbf{k}\cdot\nabla_h \times (\mathbf{U}^b - \mathbf{u}_i) + \nabla_h\cdot(\mathbf{U}^b - \mathbf{u}_i)\right]
\end{aligned}
$$

respectively for the top and bottom boundary layers, noting that the vertical velocity is the imaginary part of the expression, just as the horizontal velocity is the imaginary part of (8.27). Then, in analogy to (8.36) and (8.46), making use of the fact that $w_e^t(-\infty) = 0$ $w_e^b(\infty) = 0$, the following relations are obtained for the vertical velocity at the top and bottom surfaces:

$$
w_i^t = -w_e^t(0) = \frac{E^{\frac{1}{2}}}{2}\{\mathbf{k}\cdot\nabla_h \times (\mathbf{U}^t - \mathbf{u}_i) + \nabla_h\cdot(\mathbf{U}^t - \mathbf{u}_i)\},
$$

$$
w_i^b = -w_e^b(0) = -\frac{E^{\frac{1}{2}}}{2}\{\mathbf{k}\cdot\nabla_h \times (\mathbf{U}^b - \mathbf{u}_i) + \nabla_h\cdot(\mathbf{U}^b - \mathbf{u}_i)\}.
$$

Making use of the fact that vertical velocity should be constant throughout the water column in the f-plane approximation, by virtue of (8.9b) and (8.11), and making use of the geostrophic balance of the interior, we set $w_i^t = w_i^b$ to obtain

$$
\nabla_h^2 p_i = 2\mathbf{k}\cdot\nabla_h \times \mathbf{u}_i = \mathbf{k}\cdot\nabla_h \times (\mathbf{U}^b + \mathbf{U}^t) + \nabla_h\cdot(\mathbf{U}^b + \mathbf{U}^t),
$$

where it is noted that

$$
\mathbf{k}\cdot\nabla_h \times \mathbf{u}_i = \frac{1}{2}\nabla_h^2 p_i = -\nabla_h^2\psi
$$

is the fluid vorticity. The first term in the above vorticity equation is the torque applied by the boundaries on the interior through viscous forces, while the second term is the divergence imposed by the movement of top and bottom surfaces which could in reality only occur if fluid is vertically injected or sucked out from outside the boundaries. In effect, torque or injection applied at the boundary influence vertical velocity imposed by the boundary layers, which then imparts vorticity in the flow. It is noteworthy that if the vector sum of velocity on either the bottom or top would cancel out, or if either the curl or the divergence would vanish, then there is no effect on the interior, although a uniform vertical velocity is imposed to enter from one of the surfaces and leave from the other, without being sensed by the interior.

(ii) With the top plate stationary $\mathbf{U}^t = 0$ and the bottom plate rotating with dimensionless angular velocity $\omega_0 = \omega_0'/(f/2)$ relative to earth's rotation, where ω_0' is the dimensional angular velocity

$$
\mathbf{U}^b = U_r^b\mathbf{e}_r + U_\theta^b\mathbf{e}_\theta = \omega_0\mathbf{k}\times\mathbf{r} = \omega_0 r\mathbf{k}\times\mathbf{e}_r = \omega_0 r\mathbf{e}_\theta,
$$

or

$$U_r^b = 0, \qquad U_\theta^b = \omega_0 r.$$

From the vorticity equation we have

$$\begin{aligned} 2\mathbf{k} \cdot \nabla_h \times \mathbf{u}_i &= \mathbf{k} \cdot \nabla_h \times \mathbf{U}^b + \nabla_h \cdot \mathbf{U}^b \\ &= \frac{1}{r}\frac{\partial (rU_\theta^b)}{\partial r} = \frac{1}{r}\frac{\partial(\omega_0 r^2)}{\partial r} = 2\omega_0, \end{aligned}$$

calculated in polar coordinates, so that

$$\mathbf{u}_i = \frac{1}{2}\omega_0 r \mathbf{e}_\theta, \qquad \mathbf{u}_e(\zeta = 0) = \mathbf{U}^b - \mathbf{u}_i = \frac{1}{2}\omega_0 r \mathbf{e}_\theta$$

and

$$\begin{aligned} \mathbf{Q} &= \{\mathbf{k} \times (\mathbf{U}^b - \mathbf{u}_i) + i(\mathbf{U}^b - \mathbf{u}_i)\}e^{-(1+i)\zeta} \\ &= \{\mathbf{k} \times \mathbf{u}_e(0) + i\mathbf{u}_e(0)\}e^{-(1+i)\zeta} \\ &= \{\mathbf{k} \times \frac{1}{2}\omega_0 r \mathbf{e}_\theta + i\frac{1}{2}\omega_0 r \mathbf{e}_\theta\}e^{-(1+i)\zeta} \\ &= \left\{-\frac{1}{2}\omega_0 r \mathbf{e}_r + i\frac{1}{2}\omega_0 r \mathbf{e}_\theta\right\}e^{-(1+i)\zeta} \end{aligned}$$

yielding

$$\mathbf{u}_e = \Im\{\mathbf{Q}\} = \frac{1}{2}\omega_0 r\ e^{-\zeta}\{\cos\zeta\ \mathbf{e}_\theta + \sin\zeta\ \mathbf{e}_r\}$$

The total fluid horizontal velocity is given by

$$\mathbf{u} = \mathbf{u}_i + \mathbf{u}_e = \frac{1}{2}\omega_0 r\{(1 + \ e^{-\zeta}\ \cos\zeta)\ \mathbf{e}_\theta + \sin\zeta\ \mathbf{e}_r\}.$$

and

$$w_e(\zeta) = \Im\left\{E^{\frac{1}{2}}\nabla_h \cdot \mathbf{q}^b \int_\zeta^\infty e^{-(1+i)\zeta'}d\zeta'\right\} = E^{\frac{1}{2}}e^{-\zeta}(\cos\zeta + \sin\zeta)$$

so that the sum of interior and Ekman layers is

$$w(\zeta) = w_e(\zeta) + w_i = w_e(\zeta) - w_e(0) = E^{\frac{1}{2}}\{e^{-\zeta}(\cos\zeta + \sin\zeta) - 1\}.$$

The pressure $p = p_i$ in the interior, which is simply impressed on the boundary layer is given by

$$\nabla_h^2 p_i = 2\mathbf{k} \cdot \nabla_h \times \mathbf{u}_i = \frac{1}{r}\frac{\partial}{\partial r}r\frac{\partial}{\partial r}p_i = 2\omega_0,$$

which is integrated to give

$$p_i = \frac{1}{2}\omega_0 r^2.$$

In addition to the first order internal pressure distribution, there is of course a much smaller order Ekman layer pressure distribution according to the boundary layer vertical momentum equation

$$0 = -\frac{\partial p_e}{\partial \zeta} + E^{\frac{1}{2}}\frac{\partial^2 w_e}{\partial \zeta^2}$$

which by substituting for $w_e(\zeta)$ can be integrated to give

$$p_e = -2E\, e^{-\zeta}\sin\zeta,$$

which is seen to be of $O(E) < O(E^{\frac{1}{2}})$, so that total pressure is $p = p_i$ within that order.

The vertically integrated horizontal mass transport of the Ekman layer is

$$\begin{aligned}
\mathbf{M}_e &= E^{\frac{1}{2}}\int_0^\infty \mathbf{u}_e d\zeta = E^{\frac{1}{2}}\Im\left\{\int_0^\infty \mathbf{Q} d\zeta\right\} \\
&= E^{\frac{1}{2}}\Im\left\{\int_0^\infty \frac{1}{2}\omega_0 r(-\mathbf{e}_r + i\mathbf{e}_\theta)e^{-(1+i)\zeta}d\zeta\right\} \\
&= E^{\frac{1}{2}}\Im\left\{\omega_0 r(-\mathbf{e}_r + i\mathbf{e}_\theta)\frac{1}{1+i}\right\} \\
&= E^{\frac{1}{2}}\frac{1}{2}\omega_0 r(\mathbf{e}_r + \mathbf{e}_\theta).
\end{aligned}$$

This transport is clearly at $\pi/4$ (45°) angle to the right of the bottom surface velocity vector $\mathbf{U}^b = \omega_0 r\mathbf{e}_\theta$ with an outward component relative to the center of rotation.

The total mass transport, which is the sum of the interior and Ekman layer transports, with a total fluid depth of 1 is

$$\mathbf{M} = \mathbf{M}_i + \mathbf{M}_e = \omega_0 r\left\{\frac{E^{\frac{1}{2}}}{2}\mathbf{e}_r + \left(1 + \frac{E^{\frac{1}{2}}}{2}\right)\mathbf{e}_\theta\right\}$$

which makes a small angle

$$\alpha = \tan^{-1}\frac{E^{\frac{1}{2}}}{2 + E^{\frac{1}{2}}}$$

with the direction $\mathbf{e}_\theta$.

The similarity to atmospheric and oceanic vortices, *e.g.* weather systems can be established, first if the angular velocity of rotation is set to $\omega_0 = \omega_0'/(f/2) = 1$, so that the dimensional angular velocity is the local vertical component of the earth's rate of rotation $\omega_0' = f/2$.

Secondly, the forcing in the present problem is provided by the bottom surface, which is rotating at a rate $\omega r\ \mathbf{e}_\theta$. In order that we establish similarity with an atmospheric system we have to transfer the motion to a rotating frame of reference rotating at the opposite direction with a the same but negative rate of rotation, $\omega r\ \mathbf{e}_\theta = r\ \mathbf{e}_\theta$, also setting $\omega = 1$.

Calculating the horizontal velocity $\mathbf{u}_i^R$ in the rotating frame of reference, we have

$$\mathbf{u}_i^R = \mathbf{u}_i + \mathbf{u}_e - r\mathbf{e}_\theta = \frac{1}{2}r\{(-1 + e^{-\zeta}\ \cos\zeta)\ \mathbf{e}_\theta + \sin\zeta\ \mathbf{e}_r\},$$

and for pressure p^R in the rotating frame

$$p^R \simeq p_i^R = p_i - r^2 = \frac{1}{2}r^2 - r^2 = -\frac{1}{2}r^2,$$

while the vertical velocity remains unchanged.

Similarly the mass transport transferred to the rotating frame of reference will be

$$\mathbf{M}_e^R = E^{\frac{1}{2}}\frac{1}{2}r(\mathbf{e}_r - \mathbf{e}_\theta).$$

and

$$\mathbf{M}^R = \mathbf{M}_i + \mathbf{M}_e - r = \mathbf{M} - r = r\left\{\frac{E^{\frac{1}{2}}}{2}\mathbf{e}_r - \left(1 + \frac{E^{\frac{1}{2}}}{2}\right)\mathbf{e}_\theta\right\}.$$

The angle α will also be reversed so that it is the same with negative sign, relative to the direction $-\mathbf{e}_\theta$.

With the transfer to the rotating frame we see that the pressure distribution is reversed to become anticyclonic (greater at the center) and the flow velocity reversed to be anticyclonic, as displayed in Fig. 10.9.

(iii) With the angular velocity of the rotating plate reversed, the non-dimensional rate of rotation takes on a negative value $\omega = -\omega_0$,

$$\mathbf{U}^b = -\omega_0 r\mathbf{e}_\theta = r\mathbf{e}_\theta$$

and therefore the sign of everything changes. With these changes, we obtain

$$\mathbf{M}_e = -E^{\frac{1}{2}}\frac{1}{2}\omega_0 r(\mathbf{e}_r + \mathbf{e}_\theta),$$

and

$$\mathbf{M} = \mathbf{M}_i + \mathbf{M}_e = -\omega_0 r\left\{\frac{E^{\frac{1}{2}}}{2}\mathbf{e}_r + \left(1 + \frac{E^{\frac{1}{2}}}{2}\right)\mathbf{e}_\theta\right\}$$

If we are on earth rotating with $\omega_0 = 1$, by fixing the coordinate system in the rotating frame we have

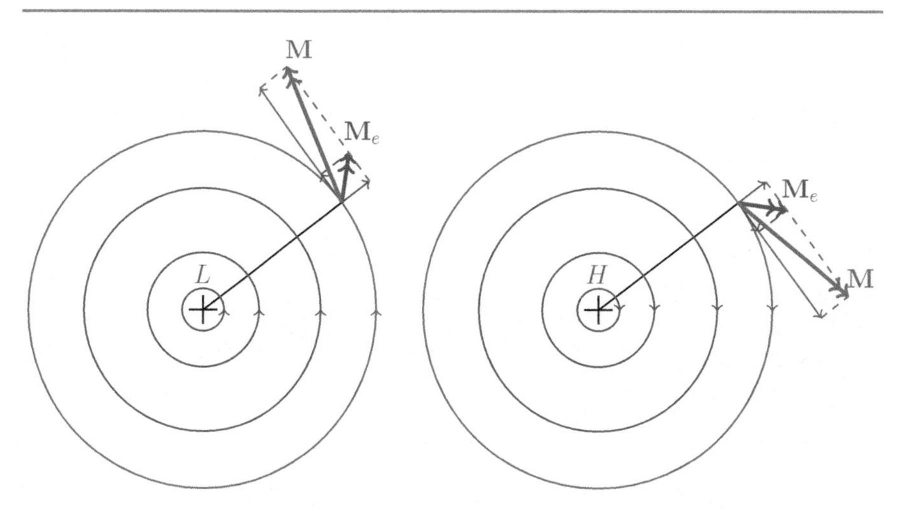

Fig. 10.9 (left) Motion on a plate with anti-clockwise rotation, (right) the same motion observed on a rotating frame of reference fixed to the plate, that simulates a vortex with anticyclonic circulation

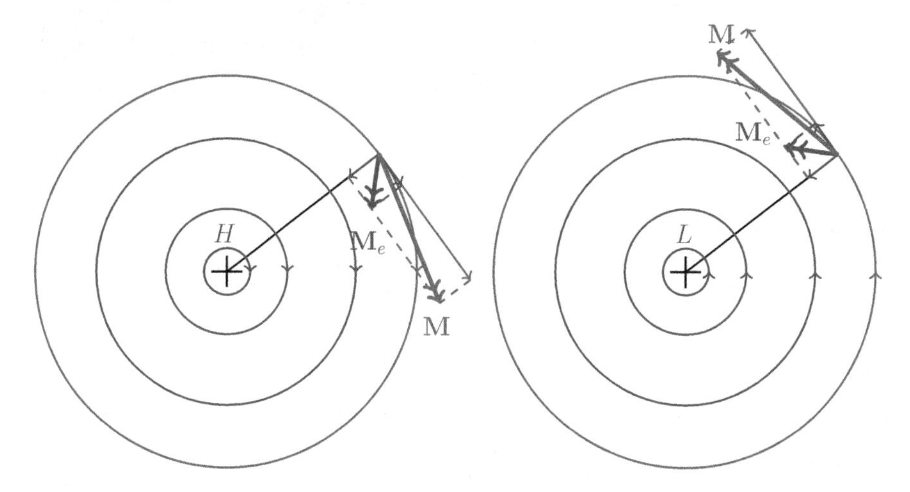

Fig. 10.10 (left) Motion on a plate with clockwise rotation, (right) the same motion observed on a rotating frame of reference fixed to the plate, that simulates a vortex with cyclonic circulation

$$\mathbf{M}_e^R = -E^{\frac{1}{2}}\frac{1}{2}r(\mathbf{e}_r - \mathbf{e}_\theta).$$

and

$$\mathbf{M}^R = \mathbf{M}_i + \mathbf{M}_e + r = \mathbf{M} + r = r\left\{-\frac{E^{\frac{1}{2}}}{2}\mathbf{e}_r + \left(1 + \frac{E^{\frac{1}{2}}}{2}\right)\mathbf{e}_\theta\right\}.$$

Fig. 10.11 An intense "medicane" (Mediterranean hurricane) approaching the eastern Mediterranean on 28 September 2018, 12 UTC, by EUMETSAT Monitoring of Weather and Climate from Space, made available to world-wide users (https://www.eumetsat.int/website/home/Data/index.html)

This case, in the rotating frame of the earth corresponds to the cyclonic vortex, a low pressure system as shown in Fig. 10.10.

In atmospheric low and high pressure systems, respectively with cyclonic and anticyclonic centers of circulation, it is quite often observed that surface winds within the boundary layer veer from the geostrophic flow pattern that would normally follow the isobars. Observations indicate wind components spiraling inward to the cyclonic center and outward from the anti-cyclonic center to be quite common. The above analysis confirms these observations.

In Fig. 10.11, a satellite image of medicane, a very intense meso-scale cyclone, in the eastern Mediterranean shows spiraling clouds towards center of the intense storm, in the same way as predicted by the analyses in this exercise.

Swirling flows with apparently spiral trajectories are clearly demonstrated in large scale vortices such as the cyclonic *great red spot of Jupiter* and adjacent anti-cyclonic centers observed by Voyager in 1979 and other NASA missions since then, as shown

Fig. 10.12 Image showing the "great red spot" on Jupiter with swirls transporting materials between cyclonic and anti-cyclonic vortices. Credits: The Bruce Murray Space Image Library supplying NASA Voyager 1 Mission images (https://www.planetary.org/multimedia/space-images/jupiter/voyager-1-view-of-the-great-red-spot.html), under creative commons license

in Fig. 10.12. The red spot, known to be churning since at least hundreds of years of astronomers' first observations, has a size multiple that of the earth and flow velocities of greater than 600 km/h are estimated within these vortices of immense size.

Recommended Text Books

Aris, R., 1962. Vectors, Tensors, and the Basic Equations of Fluid Mechanics, Dover.
Badin, G. and F. Crisciani, 2018. Variational Formulation of Fluid and Geophysical Fluid Dynamics: Mechanics, Symmetries and Conservation Laws, Springer.
Batchelor, G. K., 1970. An Introduction to Fluid Dynamics, Cambridge University Press.
Cavallini, F. and F. Crisciani, 2013. Quasi-Geostrophic Theory of Oceans and Atmosphere: Topics in the Dynamics and Thermodynamics of the Fluid Earth, Springer.
Chemin, J.-Y., Gallagher, I. and E. Grenier, 2006. Mathematical Geophysics: An introduction to rotating fluids and the Navier–Stokes equations, Clarendon Press, Oxford.
Cullen, M., 2006. A Mathematical Theory of Large-Scale Atmosphere/ocean Flow, Imperial College Press.
Cushman-Roisin, B., 1994. Introduction to Geophysical Fluid Dynamics, Prentice-Hall.
Cushman-Roisin, B. and J.-M. Beckers, 2009. Introduction to Geophysical Fluid Dynamics: Physical and Numerical Aspects, Academic Press.
Csanady, G. T., 1982. Circulation in the Coastal Ocean, Springer.
Daily, J. W. and D. R. F. Harleman, 1966. Fluid Dynamics, Addison-Wesley.
Debnath, L. 2008. Sir James Lighthill and Modern Fluid Mechanics, Imperial College Press.
Dijkstra, H. A., 2008. Nonlinear Physical Oceanography: A Dynamical Systems Approach to the Large Scale Ocean Circulation and El Niño, Springer.
Dijkstra, H. A., 2008. Dynamical Oceanography, Springer-Verlag.
Durran, D. 1999. Numerical Methods for Wave Equations in Geophysical Fluid Dynamics, Springer.
Egbers, C. and G. Pfister (Eds.), 2000. Physics of Rotating Fluids, Springer.
Friedunder, S., (ed.), 1980. An Introduction to the Mathematical Theory of Geophysical Fluid Dynamics, North-Holland.

© Springer Nature Switzerland AG 2020

E. Özsoy, *Geophysical Fluid Dynamics I*, Springer Textbooks in Earth Sciences, Geography and Environment, https://doi.org/10.1007/978-3-030-16973-2

Gertenbach, J. D., 2001. Workbook on Aspects of Dynamical Meteorology: A Self Discovery Mathematical Journey for Inquisitive Minds, First Edition, Author's copyright.
Ghil, M. and S. Childress, 1987. Topics in Geophysical Dluid Dynamics: Atmospheric Dynamics, Dynamo Theory, and Climate Dynamics, Springer.
Gill, A. E., 1982. Atmosphere-Ocean Dynamics, Academic Press.
Greenspan, H. P., 1980. The Theory of Rotating Fluids, Cambridge University Press.
Holton, J. R., 2004. An Introduction to Dynamic Meteorology, Academic Press.
Hoskins, B. J. and I. N. James, 2014. Fluid Dynamics of the Midlatitude Atmosphere, Wiley - Blackwell.
Huang, R. X., 2010. Ocean Circulation: Wind-Driven and Thermohaline Processes, Cambridge University Press.
Hutter, K. and Y. Wang, 2016. Fluid and Thermodynamics Volume 1: Basic Fluid Mechanics, Springer.
Hutter, K. and Y. Wang, 2016. Fluid and Thermodynamics Volume 2: Advanced Fluid Mechanics and Thermodynamic Fundamentals, Springer.
Ibragimov, N. K. and R. N. Ibragimov, 2011. Applications of Lie Group Analysis in Geophysical Fluid Dynamics, World Scientific.
Kantha, L. H., Clayson, C. A., 2000. Small Scale Processes in Geophysical Fluid Flows, Academic Press.
Kundu, P. K. and I. M.Cohen, 2002. Fluid Mechanics, Academic Press.
Landau, L. D. and E. M. Lifshitz, 1959. Fluid Mechanics, Pergamon Press.
LeBlond, P. H. and L. A. Mysak, 1981. Waves in the Ocean, Elsevier.
Majda, A. J. and X. Wang, 2006. Non-linear dynamics and statistical theories for basic geophysical flows, Cambridge University Press.
Malanotte-Rizzoli, P. and A. R. Robinson, 1994. Ocean Processes in Climate Dynamics: Global and Mediterranean Examples, Kluwer.
Marshall, J. and R. A. Plumb, 2008. Atmosphere, Ocean, and Climate Dynamics: An Introductory Text, Elsevier Academic Press.
McWilliams, J. C., 2011. Fundamentals of Geophysical Fluid Dynamics, Cambridge University Press.
Monin, A. S., 1990. Theoretical Geophysical Fluid Dynamics, Kluwer Academic Publishers.
Müller, P. and H. von Storch, 2004. Computer Modelling in Atmospheric and Oceanic Sciences: Building Knowledge, Springer.
Müller, P., 2006. The Equations of Oceanic Motions, Cambridge University Press.
Olbers, D., Willebrand, J. and C. Eden, 2012. Ocean Dynamics, Springer.
Paldor, N., 2015. Shallow Water Waves on the Rotating Earth, Springer.
Pedlosky, J., 1987. Geophysical Fluid Dynamics, Springer-Verlag.
Pedlosky, J., 1998. Ocean Circulation Theory, Springer.
Pedlosky, J., 2003. Waves in the Ocean and Atmosphere. Introduction to Wave Dynamics, Springer-Verlag.
Pratt, L. J. and J. A. Whitehead, 2007. Rotating Hydraulics: Nonlinear Topographic Effects in the Ocean and Atmosphere, Springer.

Reid, W. H., 1971. Mathematical Problems in the Geophysical Sciences: Geophysical fluid dynamics, American Mathematical Society.
Salmon, R., 1998. Lectures on Geophysical Fluid Dynamics, Oxford University Press.
Samelson, R. M. and S. Wiggins, 2006. Lagrangian Transport in Geophysical Jets and Waves: The Dynamical Systems Approach, Springer.
Sarkisyan, A. S. and J. E. Sündermann, 2009. Modelling Ocean Climate Variability, Springer.
Shevchuk, I. V., 2016. Modelling of Convective Heat and Mass Transfer in Rotating Flows, Springer.
Schlichting, H., 1968. Boundary Layer Theory, Mc Graw-Hill.
Shames, I. H., 1962. Mechanics of Fluids, Mc Graw-Hill.
Stern, M. E., 1975. Ocean Circulation Physics, Academic Press.
Stocker, T. and K. Hutter, 1987. Topographic Waves in Channels and Lakes on the f-Plane, Springer-Verlag.
Vallis, G. K., 2017. Atmospheric and Oceanic Fluid Dynamics: Fundamentals of Large-Scale Circulation, Cambridge University Press.
Vanyo, J. P., 1993. Rotating Fluids in Engineering and Science, Butterworth-Heinemann.
Velasco Fuentes, O. U., Sheinbaum, J. and J. Ochoa, (eds), 2003. Nonlinear Processes in Geophysical Fluid Dynamics: A tribute to the scientific work of Pedro Ripa, Kluwer Academic.
Von Schwind, J. J., 1981. Geophysical Fluid Dynamics for Oceanographers, Prentice-Hall.
White, F. M., 1974. Viscous Fluid Flow, Mc Graw-Hill.
Whiteman, C. D., 2000. Mountain Meteorology Fundamentals and Applications, Oxford University Press.
Yang, H. 1991. Wave Packets and Their Bifurcations in Geophysical Fluid Dynamics, Springer.
Yih., C.-S., 1969. Fluid Mechanics, Mc Graw-Hill.
Zeitlin, V., 2018. Geophysical Fluid Dynamics: Understanding (Almost) Everything with Rotating Shallow Water Models, Oxford University Press.

Printed in the United States
By Bookmasters